Yates
Garden Guide

FOR AUSTRALIAN GARDENERS

FONTANA/COLLINS

ACKNOWLEDGEMENTS

We wish to thank the following for providing photographs for this book: Gordon Butler, Stirling Macoboy, Malcolm O'Reilly, Adrian Plowman, Brian Hart, Ed Ramsay, Murray Fagg, Mutual Pools, the *Australian Women's Weekly* and *Your Garden* magazine.

Line drawings by Peter Nettlefold

First published 1895
Thirty-sixth edition published 1984
Reprinted 1985
by William Collins Publishers Pty Ltd, Sydney
Type set by Asco Trade Typesetting Ltd, Hong Kong
Printed by Dai Nippon Printing Co. (Hong Kong) Ltd

Yates garden guide for Australian gardeners

Rev. ed.
Includes index.
ISBN 0 00 636721 6.

1. Gardening–Australia–Handbooks, manuals, etc.

635'.0994

CONTENTS

FOREWORD

When my grandfather, in the closing years of Queen Victoria's reign, used the convalescence from an illness to write a guide for home gardeners in Australia he could not have realized that by 1984 almost six million copies would have been bought by Australian gardeners and that his book would have become one of this country's all-time best sellers. This, however, is what happened.

This thirty-sixth edition has been revised and enlarged. It brings a mass of information and practical advice to Australia's gardeners from Darwin to Hobart, and from Perth to Sydney. It is right up-to-date with the latest plant varieties and modern methods of cultivation.

The company acknowledges with thanks its sincere appreciation to the many contributors for their assistance in the preparation and writing of this edition. The result of their labours is a concise, easy-to-read, practical guide to successful gardening in Australia.

I hope it will help you to enjoy many productive hours in your garden.

Peter B. Yates
Chairman of Directors
Yates Seeds Limited

METRIC CONVERSION

To help the home gardener convert from metric units to imperial units, simple, direct conversion charts for length, area, weight, weight per area, volume for liquids and temperature appear overleaf. In the text of the book, metric measurements have been chosen which correspond approximately to imperial units with which the average reader is familiar. For example, in sowing seeds depths of 6, 12 and 25 millimetres correspond to $\frac{1}{4}$, $\frac{1}{2}$ and 1 inch. For plant spacing, or describing the height or spread of plants, 15, 30, 50 and 90 centimetres correspond to 6, 12, 24 and 36 inches, or 3, 6 and 12 metres correspond to 10, 20 and 40 feet. Similarly with weight units, 30, 60 and 250 grams correspond to 1, 2 and 8 ounces, and 2, 4 and 8 kilograms correspond to 4, 9 and 18 pounds.

For most practical purposes, capacity equivalents will suffice. With Gro-Plus Complete, for example, 30 g corresponds to $1\frac{1}{2}$ metric tablespoons, 60 g to 3 metric tablespoons, 90 g to $\frac{1}{4}$ metric cup, 100 g to $\frac{1}{3}$ metric cup, and 300 g to 1 cup.

These equivalent values are sufficiently accurate for practical use in the home garden but it is a good idea to have a set of kitchen scales (as well as metric measuring cups and spoons) for weighing quantities of powdered chemicals, and a measuring jug or cylinder for liquids. It is important to apply the correct amounts of insecticides, fungicides and weedicides so they must be weighed or measured accurately. A set of scales to weigh 5 kilograms by 25 grams is suitable. Several makes are available; most models have a graduated scale in imperial units (pounds and ounces) as well. For liquid measures, a kitchen jug to hold 1000 millilitres (almost 2 pints) is suitable; most are graduated at 100-millilitre intervals and also in fluid ounces.

For very small quantities of liquid, a measuring cylinder to hold 100 millilitres (about $3\frac{1}{2}$ fluid ounces) is necessary. These are usually graduated at intervals of 5 or 10 millilitres. With both kitchen jugs and measuring cylinders, glass containers are better than plastic ones. The plastic containers tend to discolour with use and it becomes difficult to measure accurately.

METRIC-IMPERIAL EQUIVALENT UNITS

LENGTH

millimetres (mm)	3	6	12	20	25	40	50	75	100
inches (in)	$\frac{1}{8}$	$\frac{1}{4}$	$\frac{1}{2}$	$\frac{3}{4}$	1	$1\frac{1}{2}$	2	3	4

centimetres (cm)	2.5	5	7	10	12	15	20	25	30
inches (in)	1	2	3	4	5	6	8	10	12

centimetres (cm)	40	50	60	75	90	120	150
inches (in)	16	20	24	30	36	48	60

LENGTH

metres (m)	1	2	3	4	5	6	9	12	15	18	21	24	30
feet (ft)	3¼	6½	10	13	16	20	30	40	50	60	70	80	100

AREA

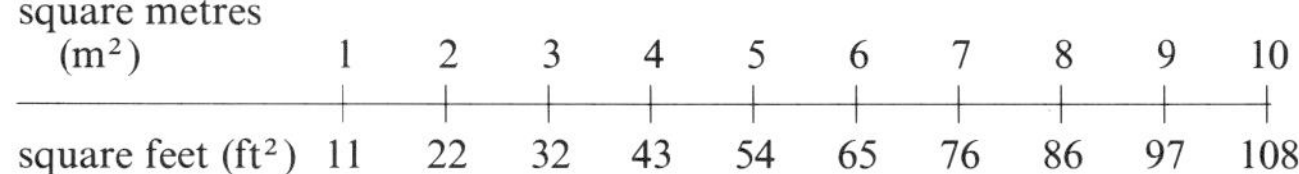

square metres (m^2)	1	2	3	4	5	6	7	8	9	10
square feet (ft^2)	11	22	32	43	54	65	76	86	97	108

Note: Larger areas are usually expressed as hundreds of square metres (100 m^2). Add two noughts to each of the figures above and below the line to give 100 m^2 expressed in thousands of square feet (1000 sq. ft.), e.g. 500 m^2 equals 5400 sq. ft.

WEIGHT (MASS)

grams (g)	10	15	20	30	60	90	125	150	185	220	250	375	500
ounces (oz)	⅓	½	⅔	1	2	3	4	5	6	7	8	12	18

kilograms (kg)	0.5	1	2	3	4	5	6	7	8	9	10
pounds (lb)	1	2	4	7	9	11	13	15	18	20	22

WEIGHT/AREA

kg/100 m^2	1	2	3	4	5	6	7	8	9	10	15	20	25
lb./1000 sq. ft.	2	4	7	9	11	13	15	18	20	22	33	44	55
g/m^2	10	20	30	40	50	60	70	80	90	100	150	200	250
oz./sq. yd.	⅓	½	1	1⅓	1¾	2	2½	2¾	3	3½	5	7	8

VOLUME

millilitres (ml)	10	20	30	40	50	100	250	500	750	1000
fluid ounces (fl. oz)	⅓	⅔	1	1⅓	1¾	3½	9	18	26	35

litres	1	2	3	4	5	6	7	8	9	10
pints	1¾	3½	5¼	7	9	10½	12	14	16	18

litres	10	20	30	40	50
gallons	2¼	4½	6¾	9	11¼

TEMPERATURE

°C	–5	0	5	10	15	20	25	30	35	40	45	50
°F	23	32	41	50	59	68	77	86	95	104	113	122

Chapter 1

STARTING A GARDEN

A garden is for plants—and for people too. It is a place for enjoyment and relaxation—but it must be be functional and cater for family needs and activities. As children grow up their interests change, sand pits, swings and seesaws quickly giving way to barbecues, swimming pools or space to park another car. Your garden may have to change with them.

Your degree of enthusiasm for gardening is an important point. Some people are never happier than when digging, planting, mowing or just pottering. Others are interested in golf, fishing or other pursuits and want a garden which is pleasant but maintained without too much effort. Whatever your attitude, a garden should never become a burden or its upkeep a demanding chore.

These days, few home owners favour a formal garden design with straight-edged square or rectangular beds and close-clipped hedges. The trend is towards flowing, natural lines and a better blending of house and garden. Home design is changing too. A patio, terrace or deck for outdoor living and entertaining is often the focal point of modern houses. Many new homes are built in bushland settings with the native plants and natural features retained wherever possible. New landscape materials—paving, pebbles and pine bark—have added new interest to planning a garden. Many of these are easy do-it-yourself materials, so you need not be an expert to use them.

But no matter what sort of garden you choose, you do need a garden plan. The purpose of this chapter is to outline some of the principles in planning a pleasant and functional garden.

First steps for a successful garden—study the site, appraise the possibilities, plan for your needs

When confronted with a situation like this, start with a plan

PATTERNS OF CLIMATE

Climate determines what plants you can grow. This applies to all plants—trees, shrubs, lawns, flowers and vegetables.

Climate is simply defined as a combination of temperature and rainfall. The highest and lowest temperatures are more critical than the average temperature, because it is the extremes which can cause problems with plant growth. This is important with more permanent plants which may be damaged or destroyed by heat or cold. As a general rule, it is best to select plants for their particular temperature range. Rain and when it falls is important too. Although we cannot protect plants from excessive rain, we can supplement low rainfall in dry districts or in dry seasons by watering. In most parts of Australia, gardens need extra water.

Australia has three broad climate zones—tropical/sub-tropical; temperate and cold. In all zones, rainfall decreases from the coast to the dry inland. These zones are illustrated in the map on the next page; cities and main towns are also shown.

TROPICAL/SUBTROPICAL ZONE

The *tropical zone* covers North Queensland, the Northern Territory and parts of Western Australia. It includes Rockhampton, Mackay, Townsville, Cairns, Darwin, Broome and Carnarvon.

The *subtropical* zone covers the east coast from Rockhampton to Coffs Harbour in New South Wales. It includes Bundaberg, Maryborough, Brisbane, Lismore and Grafton.

In the tropical zone, heavy summer rains are usual from December to March. Temperatures are high throughout the year. The subtropical zone also has a summer-wet, winter-dry pattern, but rainfall is not so heavy or concentrated and winter months are cooler.

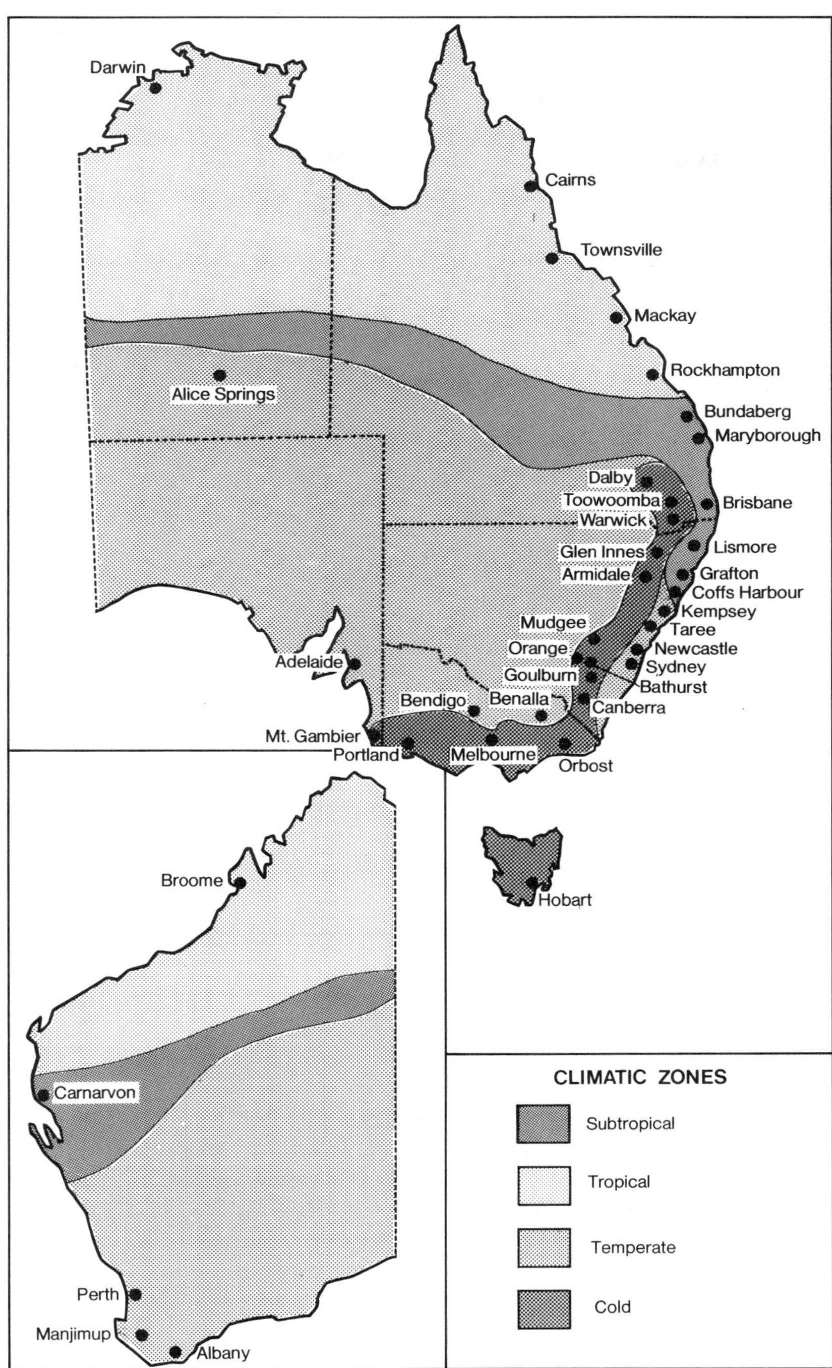
Darwin
Cairns
Townsville
Mackay
Rockhampton
Alice Springs
Bundaberg
Maryborough
Dalby
Toowoomba
Brisbane
Warwick
Glen Innes
Lismore
Armidale
Grafton
Coffs Harbour
Kempsey
Mudgee
Taree
Orange
Newcastle
Adelaide
Sydney
Goulburn
Bathurst
Bendigo
Benalla
Canberra
Mt. Gambier
Portland
Melbourne
Orbost
Hobart
Broome
Carnarvon
Perth
Manjimup
Albany
CLIMATIC ZONES
Subtropical
Tropical
Temperate
Cold

TEMPERATE ZONE

The *temperate* or *mild zone* covers the coast of New South Wales from Coffs Harbour to the Victorian border. It includes Kempsey, Taree, Newcastle, Sydney and the south coast. Adelaide, Alice Springs and Perth also have a temperate climate, as do many inland towns in New South Wales, Victoria, South Australia and Western Australia. Summer temperatures are higher and frosts more likely in inland districts than on the coast. Rainfall in the northern part of the temperate zone is fairly well distributed throughout the year but changes to a winter-wet, summer-dry pattern in the south.

COLD ZONE

In Queensland and New South Wales the *cold zone* follows the Great Dividing Range to the Victorian border. It includes Dalby, Toowoomba, Warwick, Glen Innes, Armidale, Mudgee, Orange, Bathurst, the Blue Mountains, Goulburn and Canberra. In Victoria the cold zone extends from Orbost through Melbourne to Portland, and as far as Bendigo and Benalla in the north. Small pockets of cold include Mount Gambier in South Australia, and Manjimup and Albany in Western Australia. All of Tasmania has a cold climate. All districts in the cold zone have low temperatures and frosts (and sometimes snow) in winter, and a short summer growing season. Rainfall in northern parts of the cold zone is well distributed throughout the year, but again changes to the winter-wet, summer-dry pattern in southern areas.

MICROCLIMATE

This description of climate zones in Australia is very general. Any district, city or town includes some locations where the climate is different from the general pattern. This difference may be due to one of many factors—including elevation, slope, aspect, proximity to the coast or lakes, prevailing winds.

For example, northern aspects are warmer than southern aspects. Hollows or flat sites at the bottom of slopes are cold and more liable to frost because cold air flows down hill. These small local climates are called *microclimates*. You can often create a microclimate in your own garden by providing extra warmth and shelter from a brick, stone or concrete building or wall. You can build artificial microclimates too—bush or shadehouses, glasshouses and glass frames. These are discussed in Chapter 2.

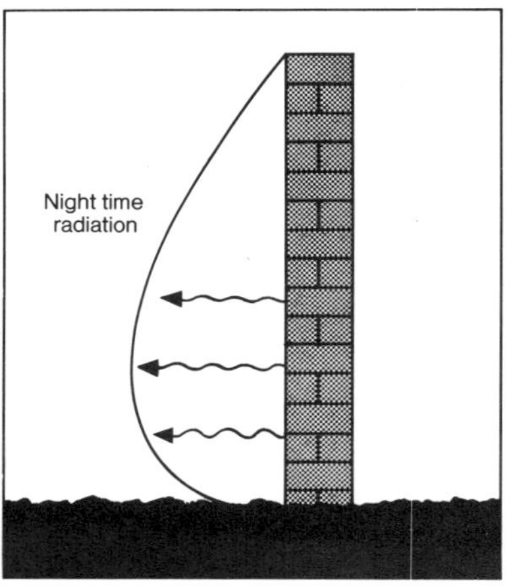

Walls of brick or stone absorb heat during the day and radiate this heat back at night to create a warm microclimate for plants

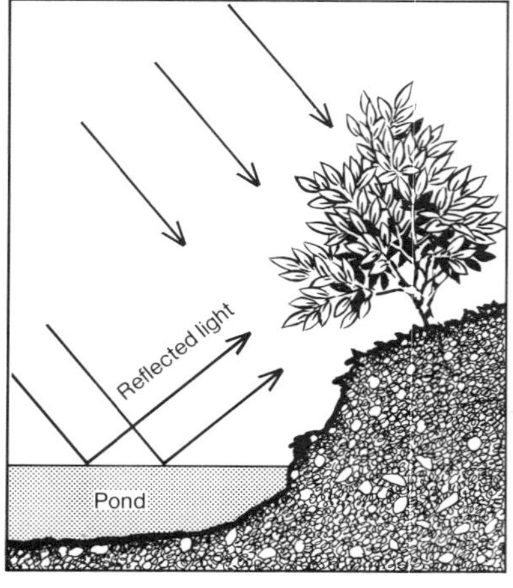

A pond reflects sunlight from the surface of the water and provides extra light for plants growing near it

ASPECT, SUN AND SHADE

Plants need light to carry out the process called photosynthesis. This allows them to make sugars (the starting point for more complex plant foods) from carbon dioxide in the air and water from the soil.

Most plants require full sunlight, but others tolerate shade in varying degrees. The preferred aspect for most Australian houses and gardens is north-east, which receives sun in the early morning and for most of the day. A north-east slope is even better, because it is warmer and protected from cold southerly and south-west winds. It is possible to build a house on almost any block of land in such a position that the garden surrounding it gets sufficient sunlight. In fact, many people prefer the front of the house to face south so that they can design the main garden and outdoor living areas at the back of the house for more privacy.

When planning a garden it is important to know the sunny spots and the areas shaded by buildings, trees and shrubs. The diagrams show the position and length of shadows at 8 a.m., noon and 4 p.m. in mid-summer and in mid-winter. Shadows in spring and autumn will be somewhere in-between in shape and size.

A deciduous tree such as liquidambar or prunus planted on the north of the house will give summer shade but will let through winter sun. Alternatively, you could cover a north-facing terrace with a pergola and grow a deciduous climber like wisteria over it. The south and west sides of the house are suitable for shade-loving plants such as hydrangeas, azaleas and camellias. Individual trees, shrubs and smaller flowering perennials also compete for light, one with the other, so it is important to know their shape, spread and the height to which they will grow.

WIND PROTECTION

Plants grow better and people enjoy outdoor living more if they are sheltered from strong winds. In the southern part of Australia, where most of us live, the worst winds are the cold southerlies of winter and early spring and the hot westerly winds of summer. North-east winds may be a problem in exposed areas on the eastern coast.

Windbreaks of trees, shrubs, hedges or fences on the south or west sides of a garden will give good protection for distances up to ten times the height of the barrier, and some protection at twenty times the height. Fortunately, trees and shrubs planted on the south or west will not create a shade problem and will

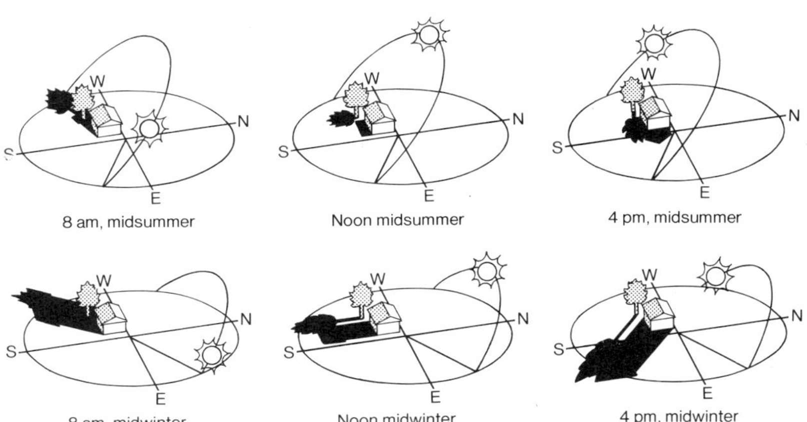

The position and length of shadows changes with the different seasons of the year. This is important in selecting plants. Some prefer full sunlight, others prefer shade

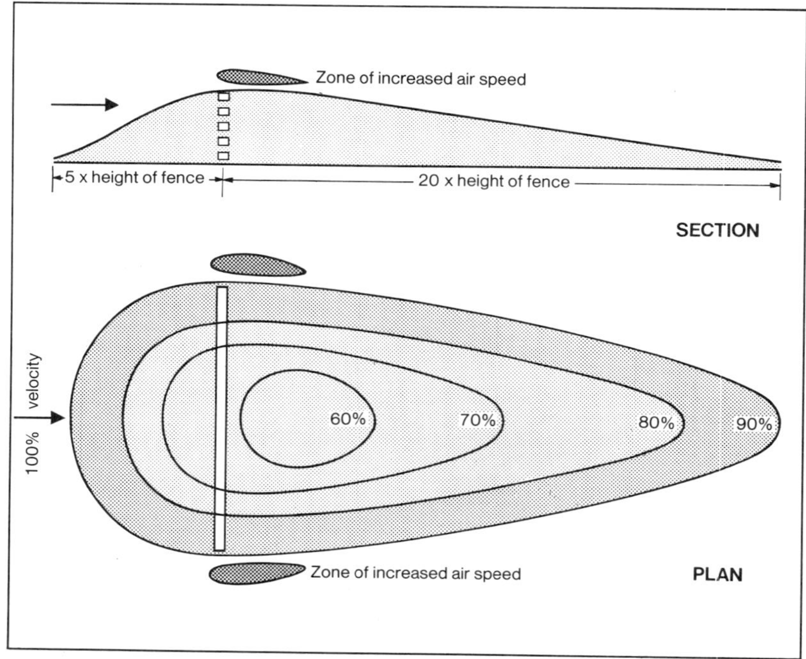

A windbreak gives protection to plants for a distance of up to twenty times its height on the leeward side. But wind speed increases just above and at the sides of the windbreak, as shown above

provide an effective shelter zone when planted well back from garden or vegetable beds.

The best wind-barriers are those which filter the wind. Slatted fences, lattice or perforated concrete blocks (with about 50 per cent opening) are more effective than solid paling fences or brick walls. Solid barriers create wind turbulence on both sides. In the same way a tree or shrub with dense foliage will create turbulence but lightly foliaged plants will filter wind and reduce its speed. The correct placement of suitable windbreaks will prevent plant damage and create a better microclimate for the garden. Some of the effects of different windbreaks are shown in the diagrams.

CONSIDER WHAT YOU HAVE

Bushland sites may include native trees or shrubs which you want to keep. Disturb the ground as little as possible until you are quite ready to make improvements—otherwise the weeds move in as soon as the natural cover is removed. Rocky outcrops may also be worked into the garden design. Even a low-lying wet area may be useful—for an ornamental pool, perhaps. Subdivisions in older settled areas usually contain worthwhile specimens of trees and shrubs, which should be retained if possible to give your garden a flying start. Subdivisions in areas which were previously farming land usually have little in the way of ornamental plants or existing features, so you have to start your design from scratch.

Another point to consider is the existing outlook from the house or from the garden. Pleasant outlooks can often be improved by making suitable plantings as a 'frame', while the reverse situation

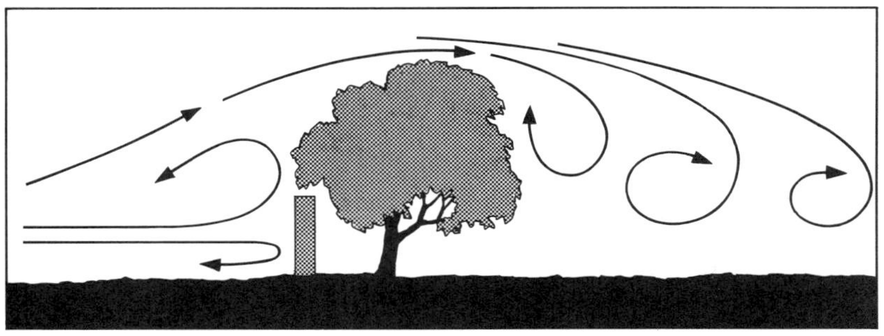

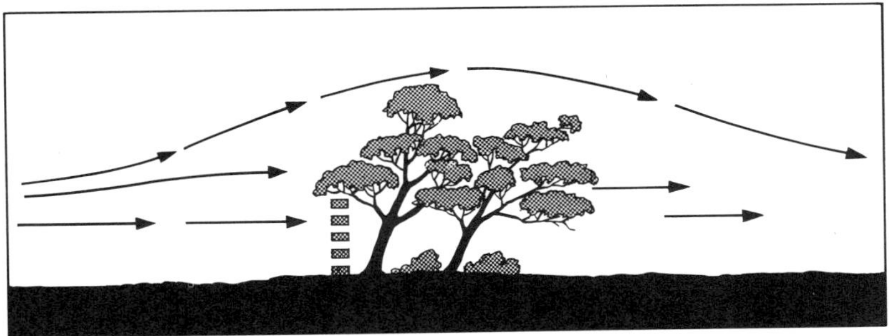

Solid barriers such as a paling fence, a brick wall or trees with dense foliage create turbulence on both sides of the wind barrier. Slatted fences or trees and shrubs with light foliage will filter the wind and reduce its speed

may apply to an unsightly outlook. This should be screened from view with trees, shrubs, a hedge or a garden wall. Sometimes more privacy is required from adjacent buildings or from public view. This may be the case if the block of land is overlooked by neighbour's windows or is located on the low side of a roadway where traffic noise and perhaps headlight glare will be a further disadvantage. These unsightly or disturbing aspects must be noted for future screening.

SOIL TYPE, DRAINAGE

This topic is described fully in Chapter 3 but a few points here may help in the planning stage.

Heavy clay soils are more fertile than light sandy soils, but may present drainage problems. Sandy soils will often have a clay subsoil, or a rock layer close to the surface which traps water—even on sloping ground. Usually a deep rubble drain on the high side of the block will divert run-off water or seepage from above. On very steep slopes, surface run-off water may be excessive. This is best overcome by terracing the slope with retaining walls or rockery treatment.

Whatever type of soil you have—clay, sand or loam—remember the topsoil is the most valuable. It is a good idea to remove about 5 cm of topsoil from the actual house-site, if you are building a

The sympathetic mixing of native and exotic species on this sloping site gives an air of permanence to the garden

new house. Heap this good soil in an out-of-the-way corner for later use. Excess clay from excavations or trenches should be kept in the building area or heaped separately. It can be used for the base of rockeries or built-up beds, but in most cases it is best removed from the site entirely.

Any part of the garden which has been excavated to clay subsoil will need to be covered with a generous layer of imported garden soil. It is best to buy light-coloured sandy soils from a reliable supplier. These sandy soils are less likely to contain weed seeds or the underground parts of hard-to-get-rid-of weeds like couch grass, paspalum, nut grass and onion weed. Sandy soils are poor in plant nutrients but are easy to work and can be improved quickly by adding manure or compost and fertilizer.

MAKING YOUR PLAN

After you have considered all the features of the block—both good and bad—the next step is to make a plan to scale. Use stiff white paper or graph paper for the base plan and you'll need several sheets of tracing paper for transparent overlays.

House and site plans are now drawn on a metric scale of 1:100 (a line one centimetre long on the plan will represent one metre on the ground). Pads of graph paper are available from stationers, measuring 18 cm by 30 cm, a suitable size for an average block with a frontage of 15–18 m and a depth of 30 m. Use two sheets joined together for larger blocks.

Make the base plan very clear so it can be seen through the tracing paper on which your garden designs will be drawn. On the base plan show the boundaries of the land, the position of the house and existing features such as trees and shrubs. It is also a good idea to indicate the general contours or slope and note the position of steep areas, low-lying spots and shallow soil over rocky sections, because these may influence future planting schemes. Mark in the positions of all services, including electricity, gas, telephone, water (with tap positions),

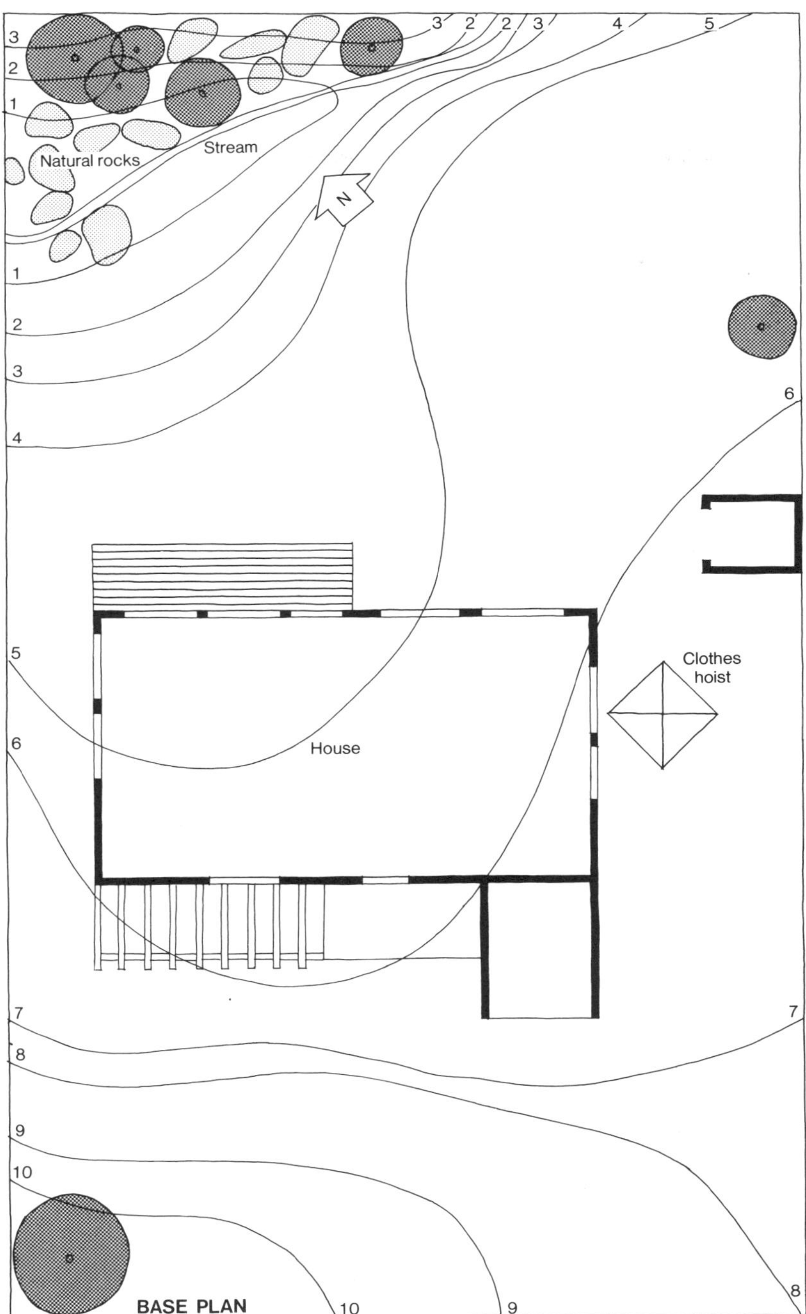

Base plan for a new garden

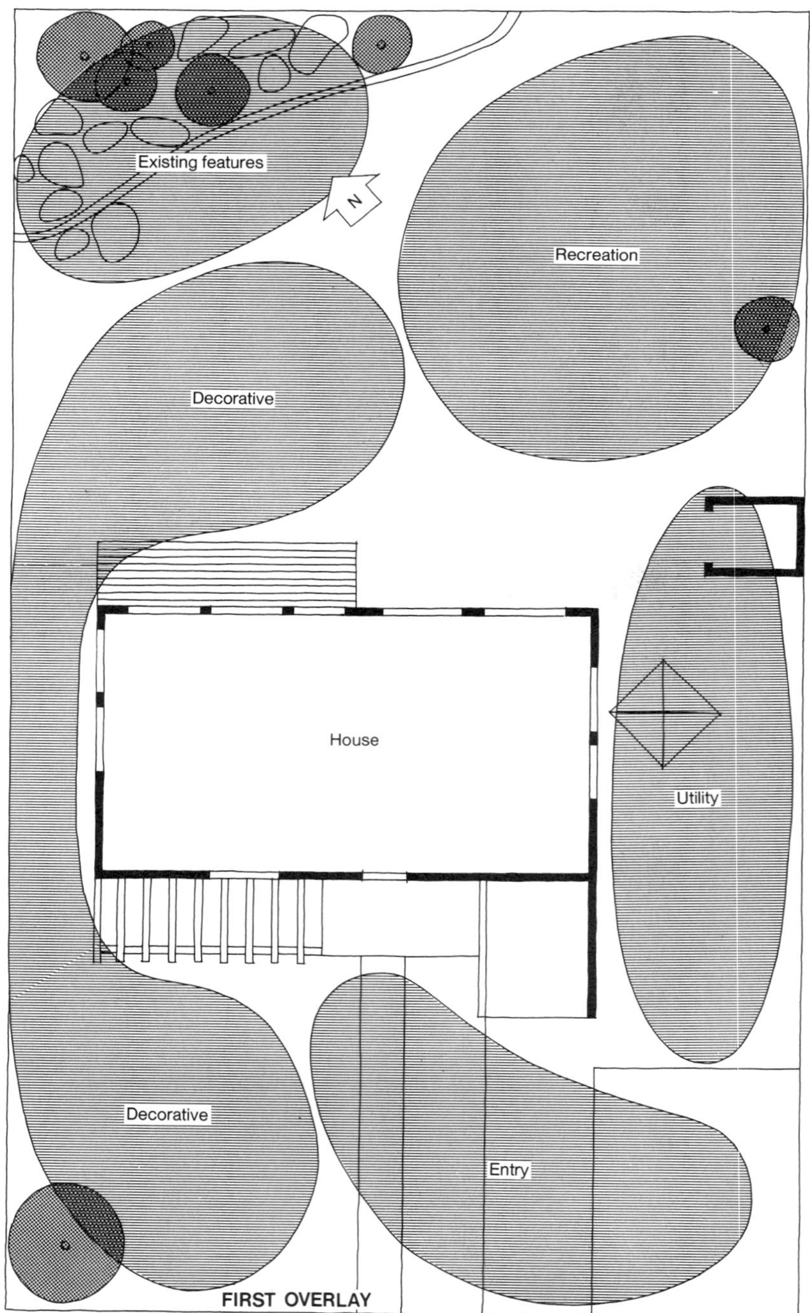

First garden plan overlay showing main purpose and activity areas

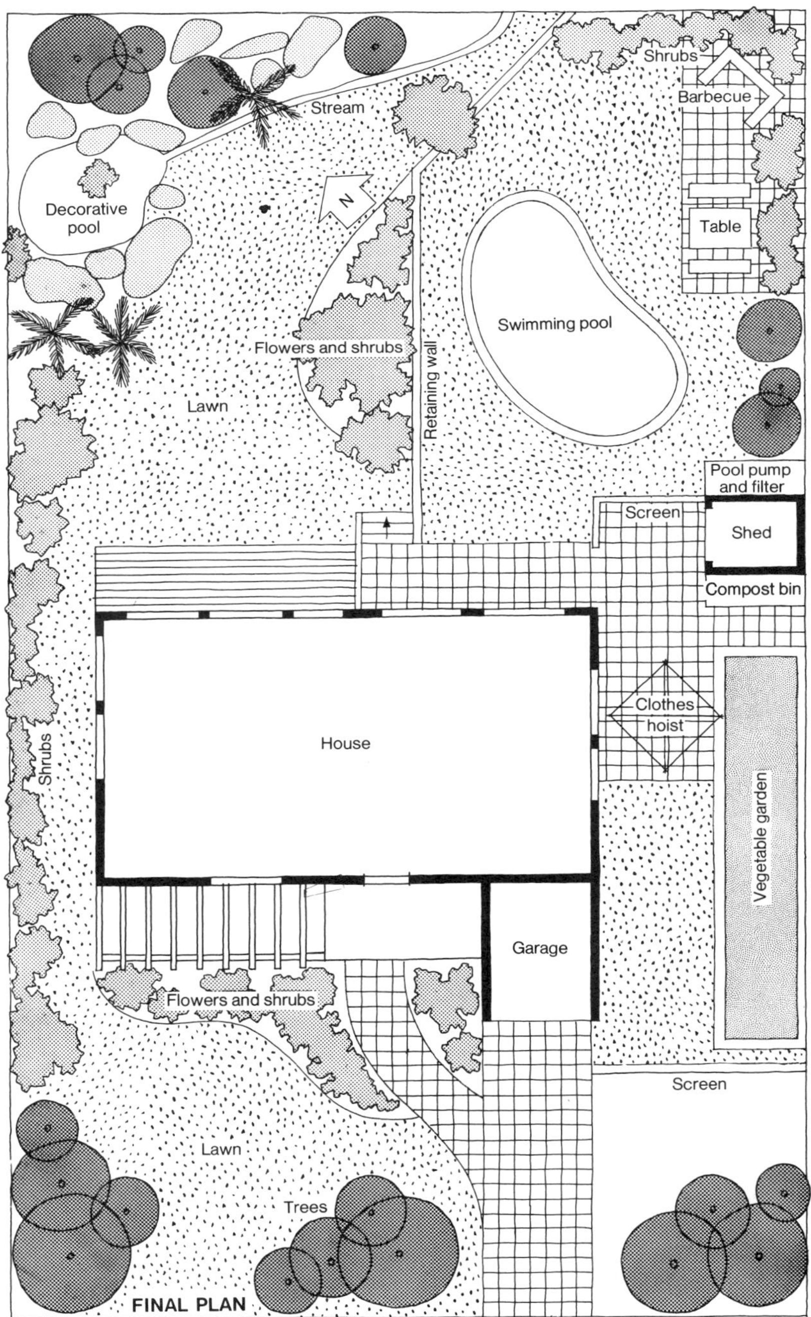

Second garden plan overlay showing final garden layout

sewerage or septic systems.

Outside the boundary show compass points to give you an idea of sun and shade areas, and mark in the prevailing wind direction. Also indicate any unsightly aspects to be screened or pleasant views to be retained.

Use the first overlay of tracing paper to broadly define suitable areas for different purposes or activities. Decorative areas include trees, shrubs, lawns and flower beds used to frame the house and provide a pleasing view from the entrance drive or frontage. Recreation and children's play areas should be close to the kitchen and preferably in view from kitchen or living room windows.

Utility areas include garage or carport, clothes-line or hoist (close to laundry), garden shed, incinerator and compost heap. A sunny position for the vegetable garden handy to the kitchen, a water tap, and a few fruit trees complete the utility area.

Use further sheets of tracing paper to plan the positions of trees, shrubs, lawns, garden beds, paths, steps, walls and screens. Plan the positions of the trees first, because they are large and permanent. Shrubs are the 'second storey' of plants, but again you must know their shape, spread and height. A brief description of the more popular trees and shrubs is given in Chapter 12, but this is by no means a complete coverage. You can get a good idea of the size and shape reached by advanced specimens in the gardens of your friends and neighbours. Nursery catalogues and your local nursery will also provide valuable information.

You will certainly draw many plans on the tracing paper (and scrap them) before you decide on the final layout. Always remember that a garden is a place for you and your family to relax and enjoy. Your own planned garden will help you achieve this pleasure.

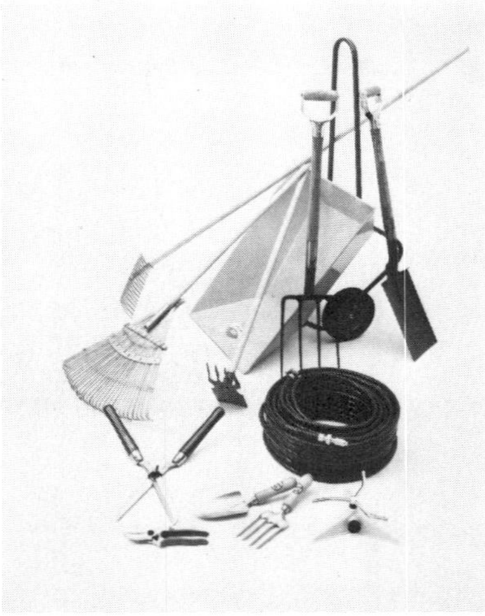

Basic tools and equipment for small gardens

Use of height variation to enhance display.

Chapter 2

FEATURES FOR YOUR GARDEN

Someone with imagination and ingenuity can introduce many special features to make a garden more interesting and attractive. There are lots of opportunities for creative thinking. Anyone at all 'handy' can build many of these features unaided—and there is a wealth of landscaping materials from which to choose. A do-it-yourself job may take a little longer, but it will give much more satisfaction than calling in a landscape consultant or contractor. This chapter aims to give you some first principles for the design and construction of garden features.

PATIOS AND TERRACES

Patio is a Spanish word meaning the inner court of a house. A *terrace* is a promenade or place for leisurely walking. In the Australian garden scene, both words have come to mean an uncovered area which is an extension of the house with access to living room, dining room or kitchen.

A patio or terrace usually faces east, north-east or north to trap sun from early morning to mid-afternoon. Both are ideal for outdoor meals—breakfast, lunch or dinner. On flat sites it is usual to build them flush with garden beds or lawn, or with one or two shallow steps to the garden. A railing or low wall is needed on sloping sites, especially if young children are around.

The patio or terrace is ideal for outdoor pot and tub plants which need open sunlight and extra warmth. Container-grown vegetables and herbs are also popular.

The paving surface of patio or terrace should be smooth (for walking and for placing outdoor tables and chairs) and easy to sweep and clean. Brick paving, which lends itself to many patterns, is popular but may absorb stains. A less expensive alternative is pre-cast concrete paving blocks or slabs. Quarry tiles, which come in a wide variety of colours, shapes and textures, give a more formal appearance but must be laid on a concrete base. The tiles do not absorb stains and are easy to clean. Large split sandstone slabs need careful laying for an even surface, but dressed (sawn) slabs make one of the best paving materials. These have an attractive, non-slip, easy-to-clean surface which dries quickly after rain.

COURTYARDS

A paved courtyard is enclosed by a building, or walls which may be 1.5 to 2.5 m high for privacy. They also act as a barrier against wind and outside noise.

Courtyards are usually open to the sky and are really an outside room. Some people prefer a covered courtyard—a vine covering a pergola, or a roof of opaque fibreglass as protection against direct sunlight. You can build a courtyard in a very small space—between house and garage or between the house and a boundary fence or wall.

A courtyard usually contains more plants than a patio or terrace—in many cases it becomes a sort of bush-house. Select plants carefully because often there is limited sunlight. Shade or semi-shade plants often do best, for example, ferns, begonias, fuchsias and balsam. Large areas of paving (the same materials for patio or terrace) may be broken by plant beds with tubs or pots of shrubs and flowers for accent. Small trees or shrubs—camellias, azaleas and citrus—are good for large tubs. Small tubs and pots of annual flowers can be moved into position in season. Trailing plants in hanging baskets add to the 'mini-garden' effect, and creepers can be grown over the walls or on a trellis.

DECKS

A deck or timber platform offers another alternative for outdoor living. It is a natural for steep or rocky sites where flat land is scarce. It is less expensive to build a deck than to level the ground and erect a massive retaining wall.

You can use low-level decks on flat sites as an alternative to paving. This way you can extend the house at floor level, provide a non-slip surface adjacent to a swimming pool or make a shady dining area around the trunk of a large tree. High-level decks need a safety rail. These can be of a ranch-style design, with the opportunity of incorporating screens, overhead frames, tables and seats as well.

Decks of stained timber are simple to maintain. Small gaps between decking boards make for quick drainage after rain and easy sweeping. Most do-it-yourself enthusiasts can build simple low-level decks, but a high-level deck is usually a job for an experienced carpenter. The foundation area must be well drained to carry water away from the timber or metal posts or the brick or concrete piers supporting the deck. Container-grown plants are most suitable for the deck itself but space below decks can often be utilized for ferns, bush-house plants and creepers.

PEBBLE GARDENS

The great advantage of a pebble garden is low maintenance. Pebble gardens have many uses—for dry spots (underneath eaves), very shady areas where grass or groundcover plants do poorly, or just as a garden feature. The strip between car tracks or a narrow garden between side path and boundary fence are good for pebble treatment.

Pebbles come in different sizes. Large grades of 4–5 cm are rather difficult to walk on and too large to rake easily if covered with fallen leaves, so smaller grades are often preferred. Pea gravel, red ridge gravel, crushed brick or crushed tile are attractive alternatives. Stepping stones, river boulders, large rocks and well-placed low growing shrubs add interest and charm to a pebble garden.

Where persistent weeds (couch, kikuyu, nut grass, onion weed, oxalis) are not a problem, a 5 cm layer of gravel will discourage most other weeds. Spot-spraying with weedicides is very effective against persistent weeds. (See Chapter 17.) Alternatively, you can lay down black polythene sheeting before you place the pebbles

A pebble garden turns an awkward dry corner into an attractive feature

or gravel in position. Set plants through slits in the sheeting with a small depression in the gravel around them so that water seeps to the roots. If you are using large rocks or stepping stones in your design, place them on sisalcraft or tarred paper so they won't puncture the plastic.

PINE BARK

Pine bark, pine flakes, wood chips and leaf litter are now widely used as a mulch for gardens or for spreading on outdoor living areas. These materials conserve soil moisture, deter weeds and reduce maintenance, and add a rustic touch to the garden.

Pine bark, a by-product of timber and pineboard milling, comes in four or five grades. The large grades (2.5 cm and 5 cm) are very attractive when combined with plantings of Australian native trees and shrubs. The wood mulches do not last as long as pebbles or gravel. Even the large grades of pine bark will decompose with time.

DRIVEWAYS, PATHS, STEPPING STONES

Driveways are a strictly utilitarian part of the garden but ideally should blend with the general design. There are practical considerations too. For the sake of economy, keep the driveway as short as possible. For convenience and safety, avoid sharp curves. The angle of approach to or from the street should not be too sharp, nor should the entrance be obscured by large trees or shrubs. If possible, provide a turning area in the garden so you don't have to back your car into the street. (A turning area can also be used for car-wash space, children's play area or just as parking space for a visitor's car.) Sloping sites present special problems for driveways. The construction of driveway and garage may call for excavating, filling and wall-building. On very steep slopes falling away from the street, a suspended car ramp may be the answer.

Driveways need solid foundations to prevent cracking and sinking. Split sandstone slabs or bricks must be bedded on a 10 cm sand base. Gravel in various colours is quite a good surface, but is prone to weed invasion and to erosion on slopes. A bitumen driveway is best constructed by a contractor; you need an area of about 60 square metres to make it an economical proposition.

Concrete is cheap and needs minimal aftercare. Its appearance can be greatly improved by pressing coloured gravel into the surface before it sets. Concrete tracks are the least expensive solution. Fill the centre strip with pebbles or gravel or plant groundcovers or dwarf perennials like gazania or mesembryanthemum (pigface).

A path need not be the shortest distance between two points except between house and utility areas—clothes-hoist, incinerator, garage or carport. A curved path is often preferable, provided the garden scene is attractive and interesting. Planning and building some paths can often be deferred until the garden is well established, when the best positions can be decided.

There are as many materials for paths as there are for driveways. The surface should be easy to walk on, but foundations need not be as solid as for driveways and large paved areas. Gravel of many grades and shades is popular and may be contained by a brick or concrete border adjacent to lawn or flower beds. Other paths are less formal, with groundcovers spilling on to them from either side.

Stepping-stones may form a sufficient path across a lawn. The dotted-line effect is often more pleasing than a solid one, and does not divide the lawn into separate sections. Set flagstones or concrete paving slabs flush with the lawn on a sand base 4 cm deep. An occasional trim around the edges will keep them neat and tidy. Flat rocks or discs of sawn hardwood make good stepping-stones for informal paths or rustic areas laid with pebbles, gravel or pine bark.

STEPS AND RETAINING WALLS

Both steps and retaining walls are almost inevitable on sloping land. Steps can be an attractive garden feature as well as giving access to a change in level. They should be wide and shallow for ease of walking up and down, especially for older people and very young children.

The riser (vertical part) should not exceed 15 cm in height, and the tread (horizontal part) should not be less than

Colourful mesembryanthemums spill over a retaining wall in gay profusion

30 cm wide. For brick steps, two courses for the riser and a tread as wide as one brick length plus one brick width is a comfortable combination.

Deep steps are often necessary on steep slopes. For long flights, incorporate a landing every ten to twelve steps. The landing can also be used as a place to change direction. A handrail is an asset where steps are steep. Steps are awkward to negotiate with lawnmowers and wheelbarrows, so a ramp may be a practical alternative. Ramps, however, take much more space to give an easy grade.

Split or dressed stone steps are probably the most popular. Bricks, concrete blocks or slabs are used in more formal situations. Horizontal logs, sawn timber decking, railway sleepers and large hardwood discs blend with informal or bush settings.

Retaining walls resist the downhill thrust of earth behind them. Again, brick, stone, concrete and timber may be used. Dry walls—any form of packed stone walling, whether cement is used or not—

Retaining wall and rockery blend admirably on a sloping site

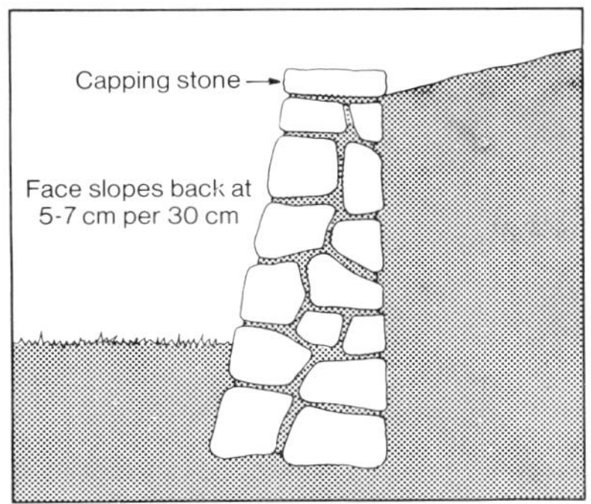

Dry walls are attractive and more stable if less than 1.5 m high. The face of the wall should slope slightly backward

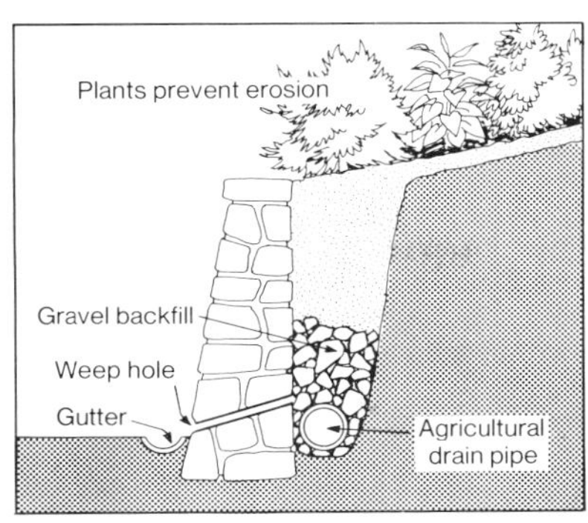

On steep slopes where high walls are necessary, it is important to provide drainage if the wall is to be completely stable

are more attractive and quite stable if they are less than 1.5 m high. The face can be sloped slightly backward for greater stability. Drainage is important, too—if the wall is not to become a dam. Provide weep holes every 2.5–3 m along the base of the wall.

Lay the largest rocks or stones at the bottom, bedding them in the soil. Between each row of stones, distribute a 4 cm layer of soil and bring the filling soil up behind them. Then lay another row of stones and repeat the process. Each row of stones should be slightly behind the row below to make the batter or backward slope—about 5 cm for every 30 cm of height.

If trailing or rockery plants are to be grown in the 'wall garden', plant them as the wall is built, spreading their roots into the soil in the crevices. The best soil is a loam which can be easily spread and worked into crevices. Soil should be damp enough to hold together, but not wet. Sandy soils are difficult to use for this job and tend to flow out of the cracks when dry.

ROCK GARDENS

You can build a rock garden or rockery anywhere, if you are prepared to get the rocks together. A rock garden offers hundreds of possibilities for creative planning. What is more, building it is a do-it-yourself job for any average homeowner.

On level sites, a rock garden to add relief to the flat landscape can be constructed by mounding soil in a corner or at the base of a large tree. Most rockeries are built on sloping ground, on a rock face or on an earth embankment where they serve the same purpose as a retaining well. Rockeries are also useful in odd-shaped corners, places which may be difficult to deal with otherwise.

Rockeries built above ground level or on slopes seldom present a drainage problem. On flat sites and on clay soils drainage may be needed, or you may need to take water away from the foundations of a building or a wall. In these situations, grade the base soil on which the rockery is constructed in the direction you wish the water to flow, and then cover it with a 7–10 cm layer of gravel or coarse sand. Mound the soil to form the filling for the rockery on top of this drainage layer. You may have suitable soil for this on the property, but if not buy some weed-free garden loam from a reliable supplier.

Place rocks on the mound almost flat but with a slight slope to lead moisture to, rather than away from, the plants you will grow. For low maintenance, surround the rockery with a concrete mower strip (or a pebble strip) if adjacent to a lawn, to prevent grass and weeds invading it.

On sloping ground, rocks are used to

form pockets of terraced beds. Bury their large ends (to about one-third of their depth) in soil. Always use the largest rocks at the bottom of the mound or slope. Rocks will need to touch across the slope to retain soil. They need not all be buried deeply, but they should be firm, not wobbly. You can create a random effect by varying the distance between rocks up and down the slope.

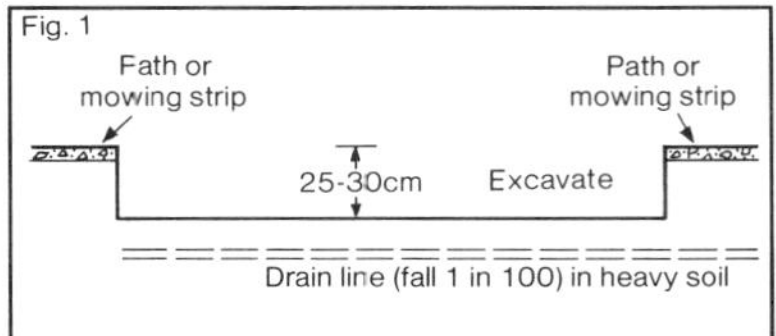

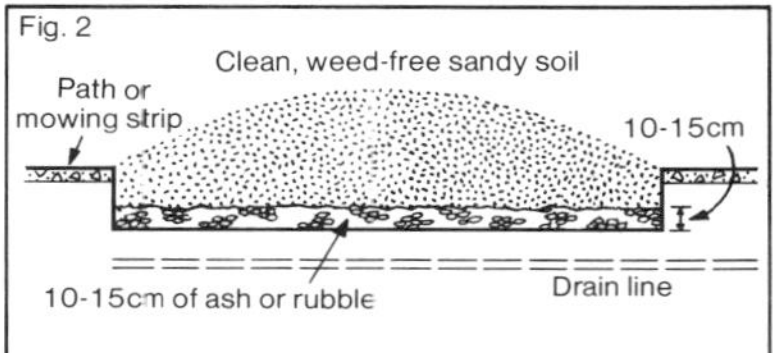

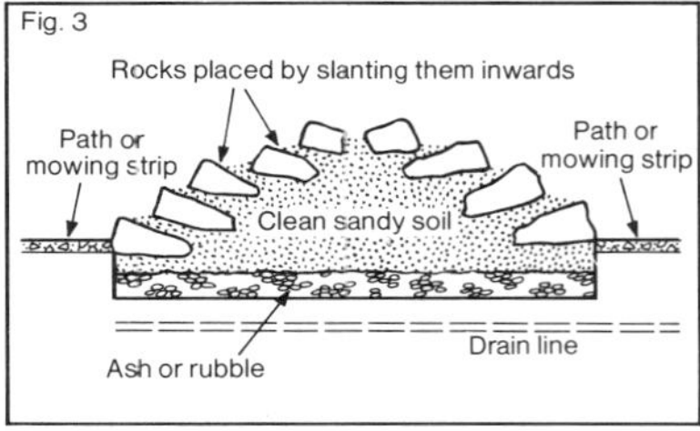

Fig. 1, 2, 3. Building a rockery on a flat site (see text)

Many rockeries are over-planted in a desire to have an instant garden, but they will finish up as a shapeless mass of greenery. A happier approach is to plant alternate pockets with suitable perennials and keep the remaining ones for colourful annuals to give a more interesting and varied effect throughout the year.

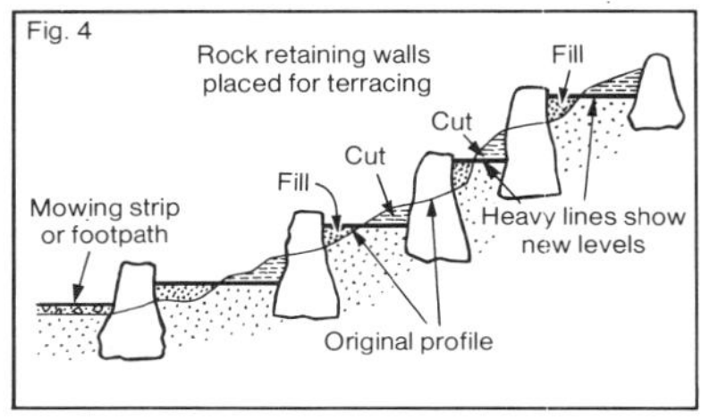

Fig. 4. Rockery pockets on a sloping site (see text)

There are hundreds of plants suitable for rockery treatment. These are often specially listed in nursery catalogues. A selection of the more popular rockery plants is given in Chapter 13.

DECORATIVE POOLS

Decorative pools or ponds introduce movement and sparkling light to a garden and give a sense of coolness in summer heat. Informal pools are best located in shady, secluded areas. Ready-made fibreglass liners are available in various shapes and sizes. Excavate a hole for them and—instant pool. More popular are the do-it-yourself pools which are constructed by lining the excavation of your own design with black polythene sheeting. A suitable grade of polythene can be bought at nurseries and hardware stores. Also available are kits for poolmaking which include instructions and diagrams.

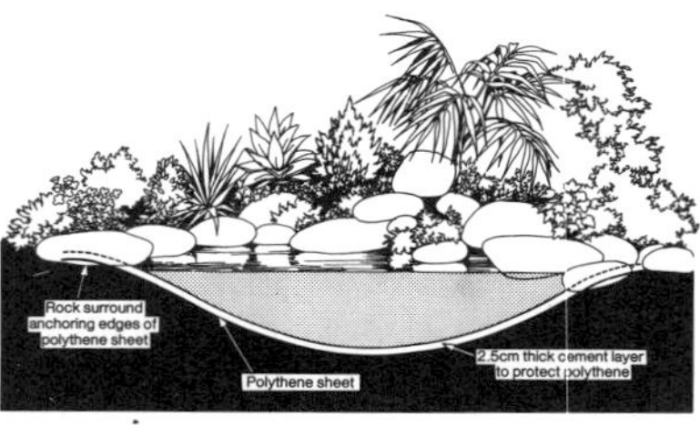

Cross-section of a decorative pool lined with polythene

A skilful blending of rocks, annuals and conifers

A pool adds movement and sparkle to any garden

Having decided on the shape and size of your pool, you should allow enough polythene for the sloping sides, plus about 20 cm overlap all around so you can anchor it down. If you wish to keep and breed fish (which keep the pool clean and eat mosquito wrigglers) you will need to excavate part of the pool to a depth of about 45 cm, and ensure that this deep part is overhung with rocks or vegetation.

Some of the excavated soil can be used to form a rocky mound behind the pool or to build up the edges on the low sides. In the latter case, tread the soil down firmly. Lay the polythene loosely, anchoring the edges with small stones, and fill with water. This gives firm contact between the sheet and the base, and allows you to check water levels accurately at all points. Build up low spots and shave off high spots as required.

Now lap the edges with stones to cover and anchor the polythene. Use thinner stones at the front and wider slabs overhanging the water at the back. One or two large stones can project well over the water, but make sure they are counterbalanced because some adventurous child is sure to stand on them at some time or other. All stones and edges should be firm and well settled at this stage. When you are happy with the effect, siphon out the water and apply a 2.5 cm layer of cement (one part cement to three parts sand) as a permanent protection to the polythene. Keep the cement mix fairly dry. Starting at the lowest point, gradually spread the cement upward and tuck under the stones. For easier working place a corn sack, an old mat or piece of carpet over the cemented bottom so you can stand on this without causing damage. When almost dry, smooth the cement by rubbing with a piece of hessian.

When completed blend the pool into the surroundings with scattered rocks, boulders, pebbles or gravel. Ferns, Nile grass (*Cyperus papyrus*), New Zealand flax, iris and dwarf conifers are good subjects for pool surrounds. Do not plant large trees close by because the roots may damage the pool.

Formal pools, which may be circular, oval, square or rectangular, are often more appropriate in small gardens, especially in paved areas. The surround can be built in brickwork on a concrete slab, and can be topped with a coping slab. The brickwork should be cement-rendered (one part cement to three parts sand) to which a waterproofing compound has been added. If the edge of the pool is built up 40–50 cm above ground level the coping slab can be used as a seat. A high surround is an effective barrier for small children.

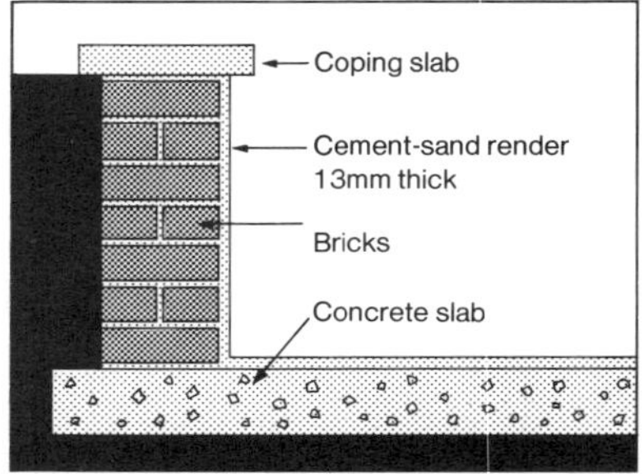

Construction details of a formal pool with cement-rendered brickwork

In both kinds of pool, informal and formal, pockets of brick or stone can be positioned on the bottom to hold aquatic plants, or plants can be grown in sunken pots or wire baskets. Water lilies (*Nymphea* spp.) are the most beautiful, but there are many other attractive and useful oxygenating plants which will keep the water clean and fresh. Among them are Egyptian lotus (*Nelumbium nuciferum*), arrowhead (*Sagittaria sagittifolia*), water thyme (*Anachris canadensis*), pickerel weed (*Pontederia cordata*), duckweeds (*Lemna spp.*) and *Salvinia braziliensis*. The last two are floating plants without true roots. If water plants become too rampant, as some species may do, thin them out before they take over the pool completely.

PERGOLAS

A pergola was originally an arbour or covered walk at the bottom of the garden. Today, a pergola is commonly an extension of the house over a patio or terrace. In many respects it has taken over from the verandah of colonial days. A well-designed pergola is not only an

A simple garden pergola adds a touch of old-world charm

architectural addition to your house but a structure to support vines or creepers and hang pots or baskets.

The structure must be strong if it is to carry a heavy creeper like wisteria or a grape vine. Wooden posts 100 mm × 100 mm square, with load-bearing rails of 100 mm × 50 mm timber on edge and 75 mm × 50 mm cross-rails on edge, are standard construction. You can use pine posts and rails for a more rustic look. Posts can also be brick piers or moulded concrete. Roofing is not usual but some people use timber slats, Sarlon shade cloth or opaque fibreglass sheets.

SCREENS

Building sites are likely to get smaller rather than bigger in future. The average home-site now has 15–18 m frontage and 30–45 m depth. This means that privacy from the street or from your neighbours is often hard to come by—unless you create it yourself. Trees, shrubs and hedges come to mind immediately, but in some situations there is not sufficient space for them to grow, and screens are a useful substitute..

There are many kinds of screen—brick, concrete blocks, stone, slatted timber, lattice, translucent plastic sheets and toughened glass panels. The handyman can cope with most of them. Screens must look good from both sides—your own and your neighbours'! Check with them first, then check local council regulations to see how high and how near the boundary your screen can go. Many screens will support attractive, useful creepers and vines. (See Chapter 11.)

BUSH-HOUSE

A bush-house reduces direct sunlight, provides a cool, moist atmosphere and protects plants from hot, dry winds. When you build one you create a microclimate for plants which may not survive in the open. Select a sunny aspect for it as it should not be dark and dismal. The size of the structure will depend on your enthusiasm for bush-house plants and ferns. The frame must be solid—timber or galvanized iron—and the uprights are best set in concrete. The floor can be concrete too but ashes, gravel, metal dust or pine bark are less expensive alternatives.

Until recently, lattice, wooden slats or 10 cm wire netting laced with teatree pieces were used for walls and roof, but Sarlon shadecloth is an effective and long-lasting modern substitute. Sarlon is made in many grades giving different percentages of shade ranging from 32 per cent to 92 per cent. Choose a grade which allows more rather than less light—32 per cent and 50 per cent are both good. A standpipe for watering should be located near the door. Overhead mist sprinklers are more professional, but also more costly.

Older layouts for bush-houses included beds with paths in between. Today most plants are grown in containers so benches figure prominently in the modern layout. Staggered benches give a much better display. Use hanging pots, baskets or troughs as well. This way you use the space between benches and roof to better advantage.

The most popular bush-house plants are orchids (see Chapter 14) and ferns,

This lean-to glasshouse is attached to a blank wall of the house. It is made of standard glazing bars fixed to a timber framework. A useful structure for raising cuttings and potting plants

but there are hundreds of other species with brilliant flowers and coloured or variegated leaves which are suitable. Large nurseries and those which specialize in shade plants usually have a wide range from which to choose.

GLASSHOUSE

In Australia, glasshouses are popular in cold climate zones, especially Victoria and Tasmania. If your hobby is growing delicate tropical and warmth-loving plants you will need one too.

The construction of a glasshouse is not difficult and the use of standard glasshouse fittings simplifies the job considerably. The frame is made from 75 mm × 50 mm timber, or an all-steel structure can be made using galvanized pipe and pressed metal fittings. Width, length and height should be multiples of standard glass sheet plus an allowance for glazing bars. A growing trend is to build a small lean-to glasshouse on a blank wall of the house. In this case, make sure the glasshouse is in a position to receive plenty of sun.

A number of firms in capital cities cater for the needs of intending builders. They will be pleased to forward drawings and specifications on request. You can also buy complete do-it-yourself glasshouse kits. These come in various sizes with Galvabond or aluminium frames. They are easy to assemble. Kits for glass cold frames are also available. Other off-the-shelf models include the Multix Greenhouse and the Multix Growing Dome. Both are plastic-covered.

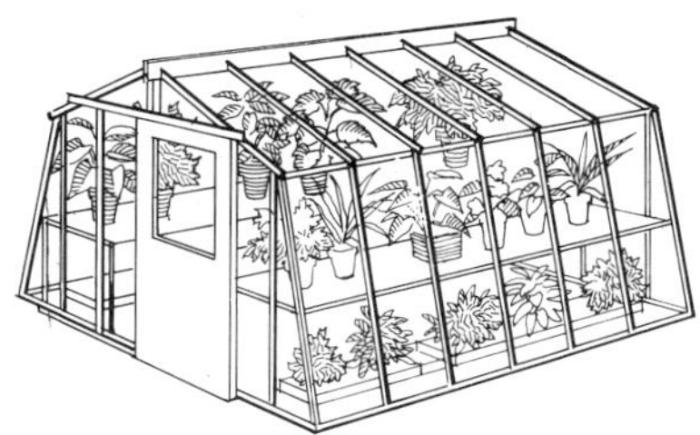

'Derby built' glasshouses with 'Galvabond' or aluminium frames are available in various sizes. The complete kits are easily assembled

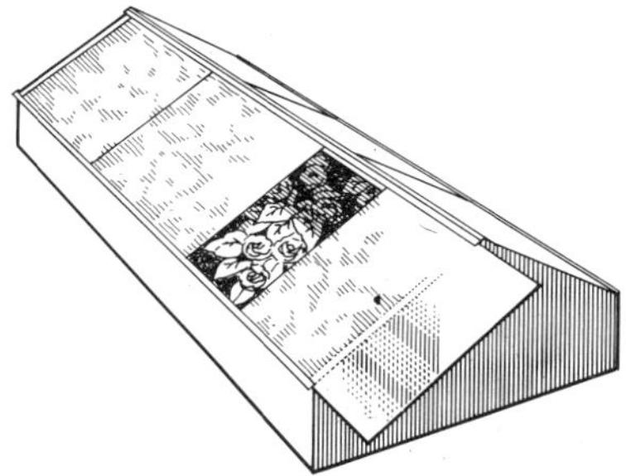

A glass cold frame for raising seedlings or striking cuttings

BARBECUES

Very often a barbecue is an afterthought—the decision to buy or build one is taken well after the garden is established. There are four things to consider when choosing the best site for it.

1. It should be close to kitchen or dining room—that is, within easy reach of crockery, cutlery, food and drinks.
2. A sheltered area is a must. Steak and chops sizzled quietly and slowly are better than charred pieces of meat which might have been through a bushfire! So consider prevailing winds, especially those from south and west. A screen may be the answer.
3. A combination of shade and sun is needed because the barbecue will be used in both summer and winter. Deciduous trees nearby will let through winter sun when they lose their leaves.
4. The barbecue area must be large enough to accommodate your family and guests without crowding. You will need space for outdoor furniture too.

SWIMMING POOLS

Most home owners consider a pool as an addition—few are fortunate enough to include a pool in the house plans. So you will need to select a suitable site in much the same way as selecting a site for a barbecue. You will need proximity to the house, preferably with access to an outside bathroom and toilet. Plenty of sunlight, shelter from the wind, and privacy are other considerations.

The surround to the pool should be at least 1 m wide with more space at the deep end for jumping and diving. Concrete probably makes the best surround and its appearance can be improved with embedded pebbles or gravel. Quarry tiles or dressed sandstone are non-slip and attractive but more expensive. The surround should slope slightly outwards so that rainwater does not carry dirt and leaves into the pool. Outside the surround, a lawn 3–4 m wide on at least one side is a good place for children to run and scuffle. A lawn also softens the glare from reflected sunlight from the pool and its masonry surround. Rockeries close to the pool can be a hazard to children's horseplay, as can prickly or spiky plants in them. Avoid planting large trees near a pool, as their roots may damage the foundations. Eucalyptus trees should never be planted close by, as they shed leaves (and bark) continuously. Small deciduous trees like Japanese maple and flowering plums (*Prunus* spp.) which shed all their leaves at once are far less trouble.

Most local government authorities require a pool to be fenced. Your pool contractor will be able to advise you or you can make enquiries from your local council. The fence you choose should blend with the pool surroundings.

PLAY AREAS

A playground is a great asset. It encourages youngsters to play out of doors and gives them a feeling of ownership to have their own small part of the garden. For toddlers, a sandpit is a must. Locate it where spilled sand does not matter and construct it on a base of unmortared bricks or concrete slabs for good drainage after rain. A surround of smooth boulders, concrete blocks or dressed timber (no splinters) can be used. Seesaws and swings suit older children, but make sure they are soundly built and unlikely to cause mishaps. Piles or stacks of smooth timber logs are quite attractive and kids will invent their own games climbing over or around them. For children who are old enough to use it safely, a rope ladder on the limb of a large tree is a great attraction, especially if it leads to a cubby-house. Play areas should be carpeted with rough lawn. It is soft enough for jumping and falling on and good for rough-and-tumble games. If it is too shady for grass use pine bark or pine flakes to soften falls.

Wisteria and flowering shrubs add charm to a gracious home

This interesting Adelaide suburban garden blends well with the surrounding area

Attractive use of plants and flowers in a confined space

Wisteria epitomizes the glory of spring

An inviting and attractive side entrance

Chapter 3

KNOW YOUR SOIL

To know your soil you should first consider how plants grow and manufacture food. Plants—from the smallest seedlings to the largest trees—are factories which take raw materials from the air, water and soil to build carbohydrates, proteins and fats. To do this they need a constant supply of the raw materials and a source of energy—sunlight—to form roots, leaves, stems, flowers, fruits and seeds. Each part of the plant has a special job to do but its performance depends on the co-operation of every other part.

Leaves and young stems are the real manufacturing departments. They contain a green pigment called chlorophyll which, in the presence of light, allows them to build sugars from carbon dioxide in the air and water in the soil. This sugar building is called photosynthesis and is the starting point for more complex substances such as starches and cellulose. The plant proteins and amino acids which are so important for both human and animal nutrition are nitrogenous compounds synthesised in the plant tissues from nitrogen absorbed from the soil. Other plant nutrients are absorbed from the soil too.

HOW ROOTS DO THEIR WORK

Roots anchor a plant in the soil. Some plants have a long, strong tap-root with smaller lateral roots. Others have a branched fibrous root system. Whatever form it takes, the root system absorbs water and dissolved nutrients from the soil through the root hairs (elongated cells just behind the root tip) and passes them to conducting tissue in the root, then on to the stem and other parts of the plant. Because roots must respire or breath to perform their task of absorbing and conducting efficiently, a soil must be able to provide air as well as water and nutrients.

SOIL TYPES

Some soils are better for growing plants than others. The terms rich and poor, good and bad, fertile and infertile are commonly used to describe these differences. The soil in your garden largely depends on the type of parent rock it came from but it also depends on the climate during the hundreds or even thousands of years in which it was formed. It is remarkable how poor soils can be improved if you know something about them and learn to manage them.

Soil is made up of mineral particles which vary in shape, size and chemical composition. Sand particles are quite large because they break down slowly. Other minerals break down more quickly into clay particles. These are extremely small—many thousands of times smaller than coarse sand—and they have an important effect on the physical and chemical properties of the soil. The size of the particles—coarse sand, fine sand, silt and clay—and the proportions in which they occur determine soil texture.

SANDY SOILS

Sandy soils have large particles with large spaces—called pore spaces—between them. They hold water and nutrients very badly and drain readily, but they do have good aeration and are easy to cultivate. For this reason they are often called 'light' soils.

CLAY SOILS

Clay soils have small particles and little pore space. They store water well—often too well for good drainage and aeration—and retain plant nutrients. Clay particles act as soil 'colloids'—the word means 'glue-like'—and attract and hold nutrients on their surface. Clay soils are difficult to cultivate and are often called 'heavy' soils.

AND THOSE IN BETWEEN

There are many kinds of soil texture between the extremes shown by sandy soil and clay soil. Loam is a soil with intermediate texture, that is, a mixture of coarse and fine particles. Others are described as sandy loam (more sand than clay) and clay loam (more clay than sand). You can identify the soil in your garden by its feel in your hand when the soil is slightly moist.

Sandy soil —does not stick together, coarse and gritty

Sandy loam—sticks together, friable (easily crumbled), slightly gritty

Loam —sticks together, friable, not gritty

Clay loam —sticks together, slightly friable but plastic

Clay soil —sticks together, not friable, plastic and sticky

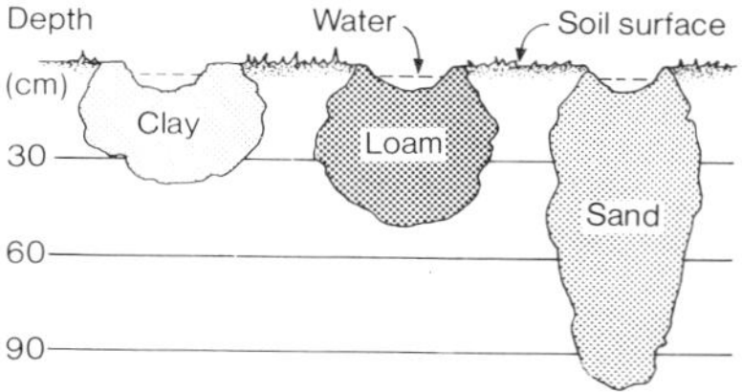

Penetration of equal amounts of water in furrows on different soils: (a) clay soil, (b) loam, (c) sandy soil. At the same depth, clay holds the most water, sand holds the least

ORGANIC MATTER AND SOIL STRUCTURE

Organic matter is plant or animal remains which are broken down by bacteria or other micro-organisms. It also includes the vast range of micro-organisms, insects and earthworms which inhabit the soil.

Carbohydrates and proteins decompose quickly into simple chemicals which can be absorbed by both plants and bacteria, but the more resistant parts of organic matter remain as small fragments which form a dark brown material called humus. Particles of humus attract and hold nutrients on their surface in the same way as clay colloids. Humus particles also help mineral particles (especially fine sand, silt and clay) to bind together to form crumbs which give the soil a 'crumb structure'. The crumbs have relatively large pore spaces between them which helps drainage and aeration. The addition of humus to heavy soils makes them easier to cultivate.

HOW TO IMPROVE SOILS

Sandy soils which hold water and nutrients poorly can be improved by adding organic matter. Animal manure, spent mushroom compost, leaf mould, garden compost and green manure crops, which are dug into the soil, are all excellent additives. Animal manures are probably the best because they contain useful quantities of nutrients as well. Animal manures and mushroom compost are not always easy to buy and are relatively expensive, but garden compost is cheap, easy to make and can be enriched by adding fertilizer. Green manure crops are also a good source of organic matter but few home gardens have space to grow them nowadays. For details of making compost and green manure crops, see Chapter 4.

Peat moss and vermiculite (also called soil improver) are good moisture-holding materials but contain negligible quantities of plant nutrients. Dry peat moss should be moistened before adding it to soil. Vermiculite will hold many times its own weight in water and has a more lasting effect on soil because of its mineral origin.

All organic materials will eventually decompose in soil so must be renewed from time to time, especially in annual flower and vegetable beds which are continually cultivated.

Clay soils benefit from organic matter too, because it improves structure and binds clay particles into crumbs. By adding coarse sand to heavy soils, you make a permanent improvement in their texture. Spread the sand to a depth of 5–8 cm, then dig it into the topsoil to a depth of 15–20 cm. Take small 'bites' with the spade so that sand and soil become well mixed.

The most important point in maintaining a good crumb structure on heavy soils is to dig them only when they are just damp—'dark damp' is a good way to describe it. Test this by taking a handful of soil and squeezing it gently. It should be damp enough to mould into a ball but dry enough to separate again on impact. If the soil remains in a ball, let the garden dry out for a day or two before digging. When cultivating any soil, only dig as deep as the topsoil. Do not dig so deeply as to bring subsoil (especially clay) to the surface.

SOIL WATER AND SOIL AIR

If a soil is saturated, the pore spaces are full of water and there is no room for air. As the soil drains, excess water—called 'gravitational water'—moves down wards, leaving a film of water around each soil particle. The soil is now at 'field capacity', that is, it is holding as much water as it can against gravity. Sandy soils have a very low field capacity. Clay soils may have a field capacity ten times as great.

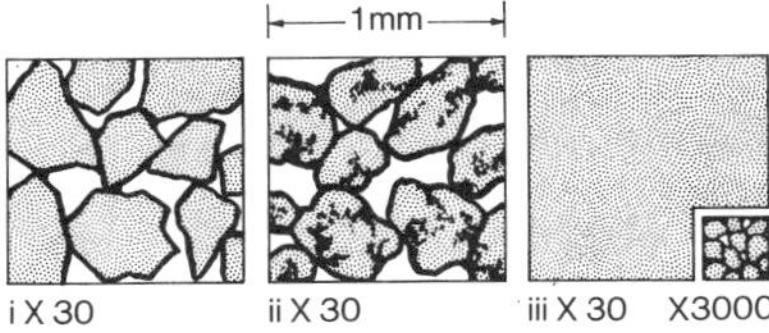

Three soils at field capacity compared. The diagrams show the proportions of air (white), water (black) and soil particles (grey). The soils are (i) sandy soil, (ii) clay soil with good crumb structure, (iii) clay soil with poor structure. The inset in (iii) is magnified by 100 to show that most of the pore spaces are filled with water and there are only a few which are large enough to contain air as well

Plants draw on the water film around the particles, the film becomes thinner and the force needed to remove water increases. When root hairs cannot extract any more water from the film, the soil is at 'wilting point'. Although some water remains in the soil, plants cannot use it. This is called 'hygroscopic water'. Unless the soil is again watered, plants growing in it begin to wilt and eventually die. Some plants (especially vegetables with large leaves, like cabbage, cauliflower, lettuce and vine crops) wilt readily in very hot weather. This is because they are losing water faster than the roots can take it up. They recover in the cool of the evening and the leaves are back to normal the following morning. The way plants use soil water is shown diagrammatically in the 'soil water barrel'.

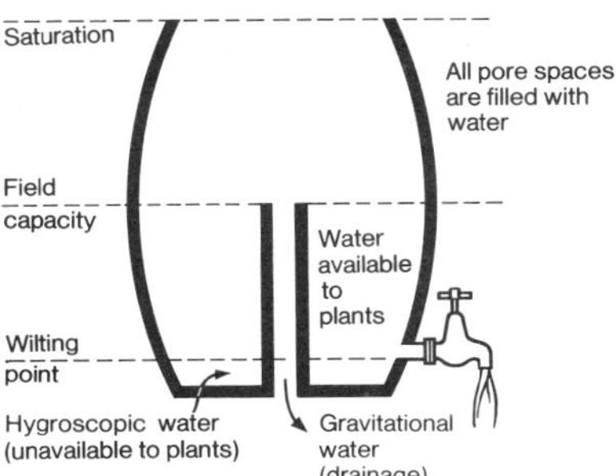

The 'soil water barrel'

DRAINAGE

Too much water is also harmful. In saturated soil, root respiration slows down due to a lack of air. Soil bacteria and other soil organisms also need air to function, so organic decay is restricted in wet soil. Another harmful aspect of wet soil is a decrease in soil temperature. Well drained, friable soils are warmer than wet soils. A warm soil increases root respiration and plants grow faster. Bacterial activity also increases.

Fortunately, many garden soils have good natural drainage. Soils with a high proportion of sand seldom present a problem unless there is a clay subsoil or rock layer close to the surface. On heavy soils, drainage of some kind is usually needed. A few test holes, dug to a depth of 40–50 cm and inspected after heavy watering or rain, will give the answer. If water remains in the holes for 24 hours, some artificial drainage is required.

On sloping sites, a rubble drain running diagonally at the top of the slope to divert run-off water or seepage away from the garden may be sufficient. Dig the trench 40–50 cm deep and fill with rough stones to a depth of 15–20 cm. Cover with a layer of aggregate or gravel and replace the top soil.

Another simple drainage method is to raise garden beds 15–20 cm above the surrounding surface. The sides of the beds should slope at 45–60 degrees. If gardens are surrounded by stone or brickwork, make provision for weep holes at the bottom of the surround.

If drainage is a major problem, a permanent herring-bone system made of terracotta, porous cement or slotted polythene pipes may be needed. Set the drains 6–9 m apart and dig trenches about 60 cm deep with a fall of not more than 1 in 200 so that the pipes are self-cleaning. Pipes must be bedded in porous material such as aggregate, gravel or coarse sand. In very wet situations, it may be best to consult a drainage expert.

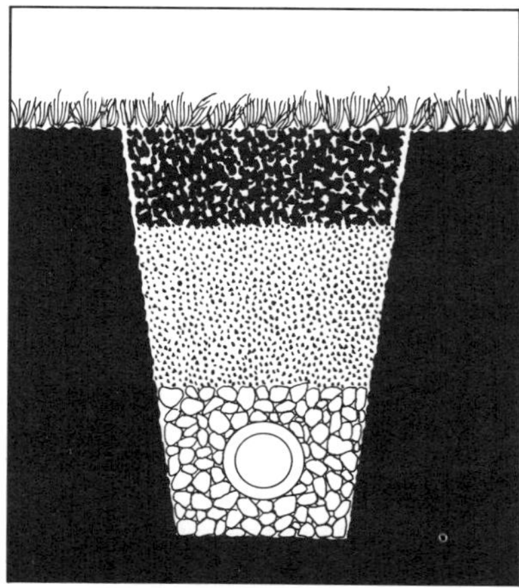

Drainage pipes—terracotta, cement or polythene—must be bedded in porous material such as aggregate, gravel or coarse sand before the subsoil and topsoil are replaced

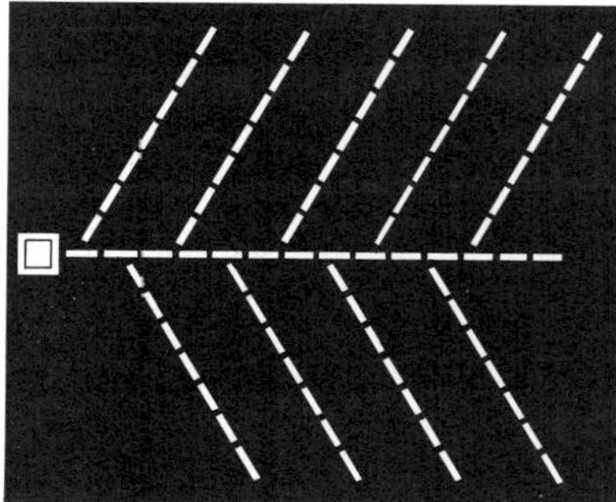

Herring-bone drainage pattern for wet situations. The lateral drains join the main drain (centre) which carries water to a stormwater outlet or soakage pit. A grade of 1 in 200 should be sufficient for the pipes to be self-cleaning

WATERING AND MULCHING

How much and how often you water plants depends on the soil type of your garden, the kind of plants you grow, your general climate and the time of the year. Microclimatic factors, such as slope, aspect or exposure to wind, will also affect loss of water from plants and soil. It is difficult to lay down rules for watering, but there are a few general principles worth knowing.

Sandy soils need more water than clay soils, so they must be watered more often. Whatever the soil type, do not let it reach wilting point before you water again. Plants do not thrive on such 'on and off' treatment. Remember, a good soaking encourages deep rooting and soil stays moist for a long time. Light sprinkling, no matter how often you do it, encourages roots to stay at the surface. The subsoil gets drier and drier and becomes difficult to wet again without a lot of heavy soaking. Early morning or evening is considered the best time for watering, thus avoiding high evaporation rates likely in the middle of the day. It is usually less windy, too, so the spray goes where you want it. In very dry weather, morning watering may help pollination and seed setting in some vegetables, particularly sweet corn and beans.

Mulching is another way of conserving soil moisture, especially in summer. A mulch is any soil covering which protects surface roots and prevents evaporation. It also helps maintain an even soil temperature and discourages many weeds from growing. On sandy soil, a mulch prevents heavy rain from washing out nutrients and humus particles. On clay soil, it maintains the soil in the dark damp condition so necessary for crumb structure.

The best mulches are loose materials which do not interfere with normal watering or light showers of rain. They must also allow free access of air to soil and roots. For this reason do not apply a mulch more than 3–5 cm in thickness. Garden compost, leaf mould, dry lawn clippings, well rotted animal manure, straw, tan-bark, pine bark, spent hops and sawdust are common mulching materials. Do not dig mulches into the soil (especially bark materials and sawdust). All will eventually decompose naturally and become integrated with the topsoil. Mineral substances—stones, pebbles, gravel and sand—are used as permanent mulches in gardens planted with trees and shrubs. More recently, black polythene has been used with success in mulching strawberry beds.

Over-watering is wasteful because it washes out plant nutrients. A common mistake is to over-water in winter when plant growth is slow. (In the case of deciduous shrubs and trees, the plants are dormant.) Over-watering often happens with summer-growing lawns like couch grass. If a plant is not growing actively, little or no water is needed.

Wet soil may also favour the spread of some fungus diseases, so it is best to keep a mulch some distance from the stems of plants or the trunks of shrubs and trees. Such a moist area may encourage root rots to attack the plant at or near soil level.

WATERING METHODS

There are dozens of kinds of sprays and sprinklers for watering gardens. Those which give a fine spray are generally preferred because large drops of water tend to pack the surface of heavy soils. Fine sprays also have a low watering rate which give better penetration and less run-off on sloping sites.

The Soakit hose, with small holes about 30 cm apart, is useful for slow watering but must be handled carefully in a crowded garden of shrubs or annuals. The small holes may become blocked and this needs watching. 'Trickle irrigation', in which holes are replaced by microtubes about 1 mm in diameter, and Via-Flo porous plastic tubing (water is delivered slowly throughout its length) are refinements of the soaking hose system.

Ordinary garden hose has stood the test of time and is as popular as ever. Nozzles adjustable for spray width and droplet size can be hand-held or attached to a stand. The Dram water-breaker nozzle is excellent for watering seed beds, small seedlings and plants in pots. It delivers a full volume of water in a soft and even spray.

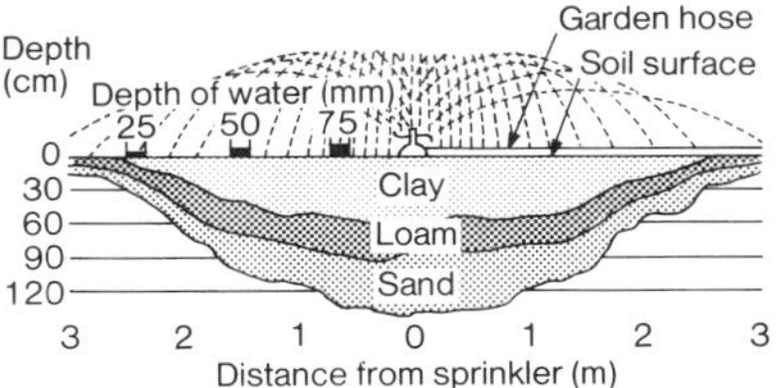

When sprinklers are used for watering, the sprays should overlap, otherwise the soil at the outer edge of the spray will not receive enough water

Careful attention to your soil will pay dividends in good crops

Chapter 4

PLANT NUTRIENTS AND FERTILIZERS

About 250 years ago, farmers, gardeners and men of learning in England and Europe started to ask the question 'What makes plants grow?' It was widely believed that soil humus was the source of carbon which makes up the sugar and starch in plants and that substances like saltpetre, lime and phosphates helped the humus to be more useful. Saltpetre (potassium nitrate) is the main ingredient in gunpowder, which had been known in England since the fourteenth century. Saltpetre was also used to preserve ham and corned beef and no doubt it had been noticed that leftover brine thrown on plants apparently caused them to grow more luxuriantly. It was not until 1840 that the German chemist Justus von Liebig proved that the carbon came from carbon dioxide in the air and not from humus in the soil. He proved too that other nutrients were absorbed by plant roots as simple chemicals dissolved in soil water.

Great strides in the study of plant nutrition have been made in the last one hundred years. We now know that apart from carbon, hydrogen and oxygen, which plants get from air and water, about a dozen nutrients or elements are essential for plant growth. These nutrients are contained in organic matter in the soil—but the organic matter must be decomposed so that plant roots can absorb the nutrients as simple chemicals or parts of molecules called 'ions'. Alternatively we can add fertilizers to the soil, in which case the nutrients are readily available. It really makes no difference whether plants obtain nutrients from decomposed organic matter or from fertilizers. Plants grow best when a combination of organic matter and fertilizers is available to them.

ESSENTIAL ELEMENTS

Plants require elements from three groups. The symbol for each element must be shown on fertilizer labels so it is useful to know their meaning. In the table below the symbols are shown in brackets.

Group 1—Major Elements
Nitrogen (N), phosphorus (P), potassium (K)

Group 2—Secondary Elements
Calcium (Ca), magnesium (Mg), sulphur (S)

Group 3—Minor or Trace Elements
Iron (Fe), manganese (Mn), copper (Cu), zinc (Zn), boron (B), molybdenum (Mo)

Recently three other trace elements—chlorine (Cl), cobalt (Co) and sodium (Na)—have been added to the list, but none of these is likely to be deficient in home garden soils. Many other elements—aluminium, arsenic, fluorine, iodine, mercury and silicon, to name but a few—occur in soils and are absorbed by plant roots, but none of them is an essential element.

LIME AND pH

Even if all essential elements are present in a soil, it cannot be taken for granted that they are all available to plants. Availability depends very much on the acidity or alkalinity (amount of lime) in the soil. Just as we can measure temperature with a thermometer, so we can measure whether a soil is sour (acid) or sweet (alkaline). Acidity or alkalinity (sometimes called 'soil reaction') is measured on a scale of pH units. This scale ranges from pH 0.0 (the most acid) to pH 14.0 (the most alkaline). The half-way mark, pH 7.0, is neither acid nor alkaline. Pure distilled water has pH 7.0.

In soils, the pH ranges from pH 4.0 (strongly acid) to pH 10.0 (strongly alkaline) so soils can be rated accordingly.

pH Value	Soil Reaction
4.0—5.5	Strongly acid
5.5—6.0	Medium acid
6.0—6.5	Slightly acid
6.5—7.0	Very slightly acid
7.0—7.5	Very slightly alkaline
7.5—8.0	Slightly alkaline
8.0—8.5	Medium alkaline
8.5—10.0	Strongly alkaline

If soil pH is too high or too low, some elements may not be available. This is shown in the diagram in which the width of the horizontal bars gives the relative availability of each element at different pH levels. You will see that on strongly acid soils (pH 4.0–5.0) all the important elements—nitrogen, phosphorus, potassium, calcium, magnesium and sulphur—and the trace element molybdenum are in short supply. On soils which are slightly to medium alkaline (pH 7.5–8.5), phosphorus again becomes unavailable, and so do the other five trace elements—iron, manganese, boron, copper and zinc. All plant nutrients are available between pH 6.0 and pH 7.0, with the best availability at pH 6.5—a soil which is very slightly acid.

HOW PLANTS REACT TO pH

Most plants grow happily if the soil pH is between 6.0 and 7.0, but there are exceptions. The lime-hating plants—azalea, camellia, erica, gardenia and rhododendron—prefer strongly acid soil (pH 5.0–5.5). Some others prefer a medium acid soil—pH 5.5–6.0—including amaryllis, cineraria, clematis, cyclamen, dianthus, ferns, fir trees, junipers, lupin, magnolia, orchids, veronica, weeping willow and most bulbs. We could add many Australian native shrubs and trees to this group. Most vegetables and herbs thrive on soils with a pH between 6.0 and 7.0. Exceptions are potato, sweet potato and

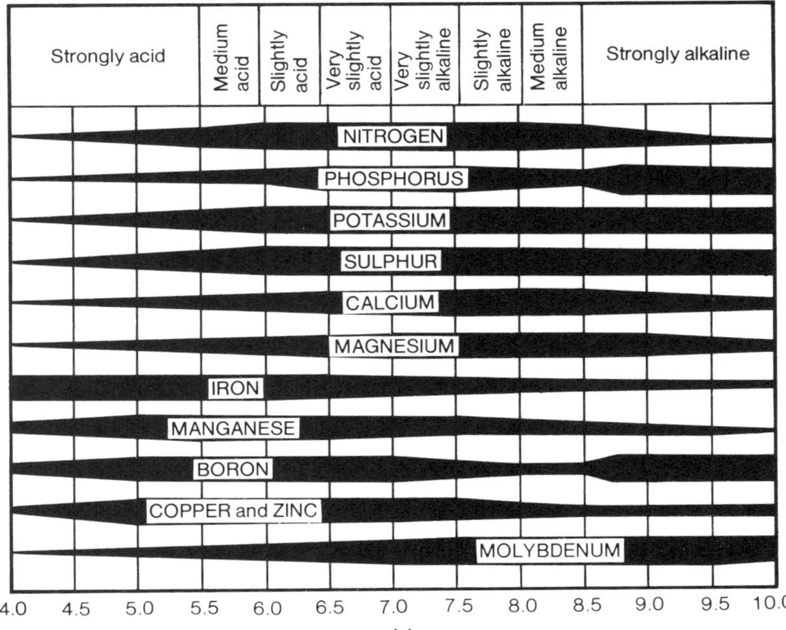

watermelon (pH 5.0–5.5). Pineapple is the only fruit which prefers a medium acid soil (pH 5.5–6.0). This is by no means a complete list. Many garden plants will tolerate a medium acid soil (pH 5.5–6.0).

HOW TO CHECK pH

In Australia, soils in high rainfall districts on the coast and eastern highlands are naturally acid. Most soils in these districts would have a pH between 5.0 and 6.5, with some as low as pH 4.0. In drier inland areas, the pH of soils approaches the neutral point and some are slightly to medium alkaline (pH 7.5–8.5).

This means that most soils in or near Australian capital cities will be medium to slightly acid, as they will be in many large towns in the eastern states. The exceptions are Adelaide, and towns in low rainfall districts where soils are neutral or alkaline. The Department of Agriculture in your own state or its local offices will have a good idea of the general level of soil pH in your district. The colour of hydrangeas growing in your neighbourhood is a good guide too—blue flowers in acid soils, pink flowers in alkaline soils.

There are several colour-chart pH testing kits available from nurseries and garden stores. The best of these is the CSIRO soil pH test kit, in which liquid indicator is mixed with a small soil sample to form a paste. A second chemical (a white powder) is puffed onto the wet soil which then changes colour and can be compared with the colours on the chart supplied. The colours range from orange (pH 2.0) through yellow, lemon, green, to violet (pH 10.0) at 0.5 unit intervals. Full instructions are included with the kit, which also contains a useful list of recommended pH range for flowers, ornamental shrubs and trees, vegetables and fruit.

HOW TO RAISE pH

Lime contains calcium, one of the base or 'alkali' elements, and is used to raise the pH of acid soils.

Agricultural lime (calcium carbonate) is the best to use and does not react with fertilizers if applied at the same time. It is finely ground, but takes some time to break down in the soil. Some ground limestones also contain small but variable amounts of magnesium carbonate—these limestones are called dolomite.

Hydrated lime or slaked lime (calcium hydroxide) can also be used. It is quick-acting in raising soil pH but reacts badly with some fertilizers—such as sulphate of ammonia, which loses the ammonia. Because it contains more calcium, 75 units of hydrated lime are equivalent to 100 units of agricultural lime.

Gypsum (calcium sulphate) is often used as a 'soil improver' on soils containing high levels of sodium. It contains 23 per cent calcium and 15 per cent sulphur but it is not a 'liming material'—in fact it has a neutral or very slightly acid effect on soil. However, by displacing sodium, gypsum improves the structure of soil.

Agricultural lime (or dolomite) can be applied at any time of the year and watered into the topsoil. The quantity required depends on soil type—sand, loam or clay—and the amount of organic matter in the soil. It is best to raise pH slowly rather than apply massive doses. The following quantities of agricultural lime are a guide to raising pH by one unit, say from pH 5.5 to pH 6.5.

Soil Type	**Quantity of Lime** (grams per square metre)
Sandy soil	150–200
Loam	200–280
Clay soil	280–450

In gardens where the soil is known to be acid, lime is often applied at the above rates every year or two. This maintains a desirable pH, especially in annual flower beds and vegetable plots which are cropped continuously throughout the year.

There is no practical evidence that lime binds sand or clay particles into crumbs in a direct way. However, it does have an effect on soil chemistry which favours bacterial activity and the decomposition of organic matter. The formation of humus particles, in turn, promotes a good crumb structure.

HOW TO LOWER SOIL pH

The best acidifying agent to lower pH on alkaline soils is sulphur (available as Flowers of Sulphur). Sulphates such as aluminium sulphate (alum) and iron sulphate are sometimes recommended. Sulphate of ammonia also has an acidifying effect if used over a long period. Materials such as leaf mould, peat moss, pine needles and well rotted sawdust will also lower pH, but the results are less predictable. The following quantities of sulphur are a guide for lowering the pH by one unit, say from pH 7.5 to pH 6.5.

Soil Type	**Quantity of Sulphur** (grams per square metre)
Sandy soil	30–60
Loam	60–90
Clay soil	90–120

When making soil tests, always record the area from which samples were taken, and the date of the test. If lime or sulphur is applied to correct pH levels, make a note of how much and when it was applied.

MAJOR ELEMENTS

Nitrogen, phosphorus and potassium, often called 'the big three', are most important. Each is required in large amounts and the presence or absence of any one has a dramatic effect on plant growth, as shown below in the photograph of cabbages grown in sand.

NITROGEN (N)

Nitrogen is an essential part of the proteins in plant (and animal) cells. It is also a necessary part of the green pigment chlorophyll in plants, and is extremely important to leaf growth. Plants deficient in nitrogen are stunted, with pale green or yellow leaves, often with reddish tints.

Soil bacteria break down protein nitrogen in organic matter to ammonia which is released as ammonium ions. Some ammonium ions are absorbed by plant roots, but most of them become attached to soil colloids. Another group of bacteria then converts ammonia nitrogen to nitrites and finally to nitrates, which are quickly absorbed. Because nitrates are very soluble they are easily leached (lost in drainage water) after heavy rain or over-watering. The breakdown of proteins to nitrates is called 'nitrification'. This process, shown in the diagram, is hastened by favourable conditions of moisture, aeration and soil temperature (12–25°C).

Another group of bacteria lives in nodules or swellings on the roots of

A dramatic demonstration of the need for fertilizers. These cabbages are growing in sand. The cabbage on the right has had no fertilizer added to it, while the one at the far left has benefited from the provision of nitrogen, phosphorus and potassium

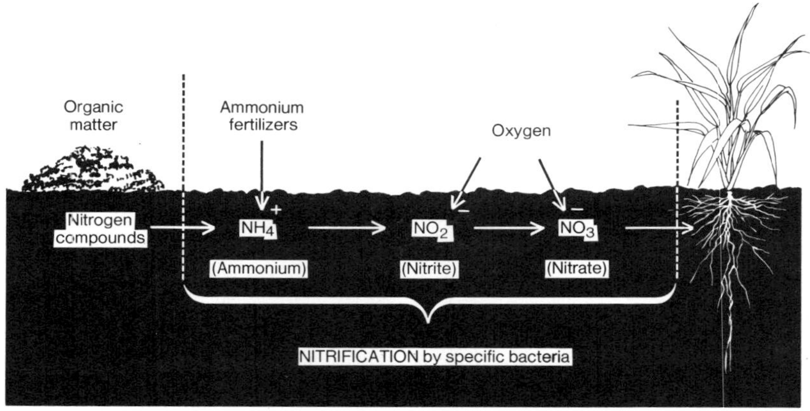

legumes (peas, beans, lupins, clovers, lucerne, acacias, cassia and many others). These bacteria are able to 'fix' nitrogen in the air between the soil particles. The 'fixed' nitrogen is used by the host plant but when the roots die, the nodules disintegrate and release nitrogen which can then be used by following crops. (See the section on Green Manure Crops, page 46.)

Nitrogen fertilizers vary in the amount of nitrogen they contain and its availability to plants. Sulphate of ammonia (21% N) is the best known. The ammonia nitrogen is converted to nitrates. This may take only a few days in warm soil but up to or four weeks in winter. Sulphate of ammonia also contains a useful amount (24%) of sulphur. Most powdered and granular fertilizer mixtures contain sulphate of ammonia as the source of nitrogen.

Calcium ammonium nitrate (26% N) is also known as Calnitro, Nitrogreen or C.A.N. The particles are granular, extremely soluble and give a quick response even in winter.

Ammonium nitrate (34% N), also known as Nitram, is similar in action to Calnitro but has a higher nitrogen content.

Urea (46% N) is a soluble, quick-acting form of nitrogen which is widely used in water-soluble mixtures such as Thrive, Aquasol and Zest. When applied as a spray some urea is absorbed through the leaves but most is washed into the soil and converted to nitrate nitrogen.

Potassium nitrate (13% N) is a soluble, quick-acting source of nitrogen and potassium (36% K). It is commonly used in water-soluble fertilizer mixtures.

Urea formaldehyde (38.4% N) is also known as Ureaform, Nitroform or U.F. It releases some nitrogen within a few weeks but the remainder is broken down by soil bacteria over a period of three or four months. The rate of release varies with soil temperature, and is slower in winter than summer.

PHOSPHORUS (P)

Phosphorus forms part of the nucleoproteins in plant cells, so it is important in growing tissue where cells are actively dividing. It promotes the development of seedlings, root growth, flowering and the formation of fruits and seeds. A deficiency of phosphorus leads to poor root development, stunted growth and often a purplish discoloration of the leaves. Most soils in Australia are deficient in phosphorus, especially those in higher rainfall areas.

Superphosphate (9.6% P) is the most common phosphorus fertilizer and is used in all powdered and granular mixtures. It is not completely soluble. It is a mixture of calcium phosphate and calcium sulphate, containing useful quantities of calcium (22% Ca) and sulphur (11% S) as well as phosphorus.

Mono-ammonium phosphate (22% P) contains both phosphorus and nitrogen (12% N). It is completely soluble, quick-acting and is a common ingredient in water-soluble mixtures such as Thrive, Aquasol and Zest.

POTASSIUM (K)

Potassium promotes chlorophyll formation and plays an important part in the strength of cells and the movement of water in plants. It also helps plants resist disease and improves the quality of flowers, fruits and seeds. Plants deficient in potassium have weak stems; their leaves—especially older leaves—may be floppy with yellow or brown tips or scorched margins. Sandy soils in high rainfall areas are most likely to be deficient.

Potassium chloride (49.8% K), also known as Muriate of Potash, is the most widely used potassium fertilizer. Used alone it can cause 'fertilizer burn', but is safe in powdered and granular mixtures because it is less concentrated.

Potassium sulphate (40% K) also contains sulphur (16% S) but is more expensive than potassium chloride. It is preferred for some crops, for example, strawberries and potatoes. It is widely used in the water-soluble mixtures.

SECONDARY ELEMENTS

CALCIUM (Ca)

Calcium is important in the construction and strength of cell walls in plants—similar to its role in the bones of animals. It also promotes proper functioning of growing tissue, especially in root tips. Calcium neutralizes acids in the cell sap and when applied to soil as lime plays an important part in reducing soil acidity. (See previous section, Lime and pH) Calcium as a nutrient is rarely deficient unless the soil is extremely acid.

MAGNESIUM (Mg)

Magnesium forms a small part of the chlorophyll molecule so is important in photosynthesis. Lack of magnesium causes leaf yellowing, especially on older leaves, because magnesium, like potassium, is mobile in the plant and young leaves have first call on these two nutrients. In most soils, magnesium is present in adequate quantities and deficiencies rarely occur. Many powdered and all water-soluble fertilizer mixtures contain adequate quantities of magnesium sulphate or Epsom Salts (10% Mg). Dolomite (3–8% Mg) is a poor source of magnesium. It must be applied in large quantities and is slowly released over two or three years.

SULPHUR (S)

Sulphur forms part of many plant proteins. It does not occur in chlorophyll but it is involved in its formation. Sulphur-deficient plants are stunted and yellow with symptoms similar to those of nitrogen deficiency. Australia has some sulphur-deficient soils in parts of the highlands of the Great Dividing Range. In garden soils a deficiency is extremely unlikely because superphosphate (11% S), sulphate of ammonia (24% S) and all powdered and granular mixtures contain sulphur in adequate quantities. Organic matter also contains sulphur, but like nitrogen it is only available following bacterial breakdown to the sulphate form. In districts where sulphur deficiencies are known to occur, Gypsum or calcium sulphate (15% S) can be used. It may be best to apply SF 45 (sulphur-fortified) superphosphate (45% S), as this fertilizer also contains phosphorus (5.9% P).

TRACE ELEMENTS

The importance of minor or trace elements is not questioned but we must understand that they are needed only in minute quantities. Their excessive use may do more harm than good. In a garden situation, trace element deficiencies are not common.

All organic materials—animal manures, blood and bone, bone dust, compost and leaf mould—contain trace elements. Even superphosphate, which is made from rock phosphate, contains useful amounts. Trace elements are added to some proprietary fertilizer mixtures, and also to water-soluble mixtures (Thrive, Aquasol and Zest) in small but adequate quantities. It is much easier and safer to use these products than to mix your own. Trace elements act as growth regulators or enzymes (starters) in building chemical compounds inside plant cells.

IRON (Fe)

Iron is not a part of the chlorophyll molecule but small quantities must be present for its formation. Symptoms of iron deficiency include yellowing of younger leaves, because iron is relatively immobile in the plant. A deficiency is more likely on alkaline soils. For this reason, many trace element mixes contain iron chelate, which is a more stable form than iron sulphate.

MANGANESE (Mn)

Manganese plays a similar role to iron but is needed to form proteins. Deficiency symptoms—again more likely on alkaline soils—are yellowing of younger leaves, especially between the veins. Plants may also be stunted. An excess of manganese may produce toxic effects on very acid soils. These are overcome by liming to raise pH above pH 5.5.

COPPER (Cu) AND ZINC (Zn)

Both elements are enzyme activators. Lack of either element leads to leaf mottling, and yellowing in younger leaves. In citrus trees zinc deficiency causes an abnormality called 'little leaf'. Copper and zinc deficiencies are more likely on very acid, sandy coastal soils but may also occur on fertile 'black earth' soils in dry inland districts.

BORON (B)

Boron is important to growing tissue in young shoots, roots, flower buds and fruits. A deficiency leads to breakdown of internal tissue and corkiness, especially in apples, beetroot and turnips—often called 'brown heart'. Tissue breakdown may also occur in the stems of celery and silver beet and in the flower buds of cauliflower and broccoli. Boron deficiency is more likely on alkaline soils or on those which have been limed heavily.

MOLYBDENUM (Mo)

Molybdenum is a growth regulator for building protein from nitrates. Deficiencies are most prevalent in high rainfall areas on acid soils. Very often an application of lime will release sufficient soil molybdenum to correct a deficiency. As little as 25 grams per hectare of molybdenum will correct deficiencies. Molybdenum deficiencies are common in cauliflowers, brussels sprouts and other members of the cabbage family, causing the disease known as 'whiptail'. In home gardens, the use of one of the water-soluble fertilizers—all of which contain molybdenum in adequate quantities—will correct any deficiency.

ORGANIC FERTILIZERS

These fertilizers include animal manures and animal or vegetable by-products. Animal manures contain small quantities of nitrogen, phosphorus and potassium which vary with the kind of animal, its diet and the amount of straw or litter mixed with it. They are first class materials for improving the texture and structure of soils, but they must be added in large quantities to benefit the soil in this way. Spread them to a depth of 5–7 cm over the surface and dig them into the topsoil. Animal manure containing a lot of straw may cause a temporary nitrogen deficiency because bacteria decomposing the straw have first call and plants may suffer. So add extra nitrogen or, better still, a complete fertilizer which contains at least 10% nitrogen plus phosphorus and potassium. Liquid animal manure watered on to leafy vegetables every week or two was a popular fertilizer many years ago. The liquid was made by suspending a hessian bag filled with fresh manure in a large cask or drum of water. Nutrients released by bacterial fermentation dissolved in the water. After a week the liquid was diluted with water (1 part to 3 parts) for use. The drum was again filled and a week later the liquid was diluted with water (1:1). This was repeated a third time and the liquid used without dilution. This messy (and rather smelly) practice has been superseded by the use of water-soluble fertilizer mixtures—except, perhaps, in country areas where fresh manure is available.

Organic by-products of animal origin are bone dust, bone meal and blood and bone. They contain higher quantities of nitrogen and phosphorus than animal manure, but no potassium. Nutrients are released slowly. They are not as bulky as animal manures and contain no fibrous material, so have little effect on soil texture or structure. Spread them at 125–250 grams per square metre and dig into the topsoil. Blood and bone is usually an ingredient of mixed fertilizers where a slow release of nutrients is desirable. For this reason it is included in Gro-Plus for Camellias and Azaleas and also in Gro-Plus for Roses.

Analysis of Organic Manures and Organic By-products

Manure or Fertilizer	Approximate Nutrient Content %		
	Nitrogen (N)	Phosphorus (P)	Potassium (K)
Animal Manure			
Cow	1.0	0.4	0.5
Fowl	2.1	1.6	1.0
Fowl (pelleted Slow Release)	5.0	3.3	1.5
Horse	0.7	0.4	0.5
Pig	1.1	0.7	0.1
Sheep	1.8	0.4	0.5
Organic By-products			
Bone dust	3.0	10.9	—
Blood and bone	6.0	7.0	—
Castor meal	5.5	0.8	0.9
Cotton-seed meal	6.0	1.3	0.9
Linseed meal	3.0	0.4	1.7

Organic by-products of plant origin are the seed meals—castor, cotton and linseed—but unfortunately they are not always available because there is a large demand for them as stock foods. They have more nitrogen than animal manures but contain about the same amount of phosphorus and potassium. Apply them at the same rate as bone dust or blood and bone.

The analysis of fertilizer materials is expressed as the percentage of nitrogen, phosphorus and potassium they contain, often called the N.P.K. ratio. An average analysis of animal manures and organic by-products discussed in this section is shown in the table above.

GARDEN COMPOST

Compost is the cheapest source of organic matter in the home garden. It also provides a convenient and hygienic way of recycling waste plant material and kitchen refuse. Almost any plant material can be composted—lawn clippings, spent plants from vegetable or flower gardens, soft shrub or hedge prunings, fruit peelings, egg shells and non-fatty kitchen scraps. Manure, if available, can be added too. Fibrous or woody materials do not make good compost. Do not use paper, straw, rice hulls, sawdust, wood shavings or tough or oily leaves such as those from eucalypts and conifers.

Diseased plants must not be composted as they may carry over the disease to cause future trouble. Weeds which have gone to seed or weeds with tough underground rhizomes or bulbs should be avoided, too. Although considerable heat is developed in well made compost, disease organisms, weed seeds, etc. may survive, so it is not worth taking the risk.

Composting is simply a way of providing good conditions for organic matter to break down. This means adequate moisture (but not too wet), good aeration, a pH close to neutral, plus a ration of complete fertilizer. These are ideal conditions for bacteria, fungi and other micro-organisms.

Choose a shady site well protected from hot winds for the compost heap. A heap about one metre square is a good size. It is best to surround the heap on three sides with cement blocks, timber planks or chain wire and galvanised pipe. Such bins keep the heap tidy and compact and make the composting process more efficient. Three bins are preferred, one for new material, one for turning and one for finished compost.

Start the heap by spreading the material to a depth of 15–20 cm, then give a generous dusting of agricultural lime and sprinkle a handful of a complete fertilizer containing about 10% nitrogen. Cover with a thin layer of garden soil (this is not absolutely necessary). Build successive layers in this way until the heap is about 1 m high. Turn the heap every two or three weeks to hasten decomposition. After

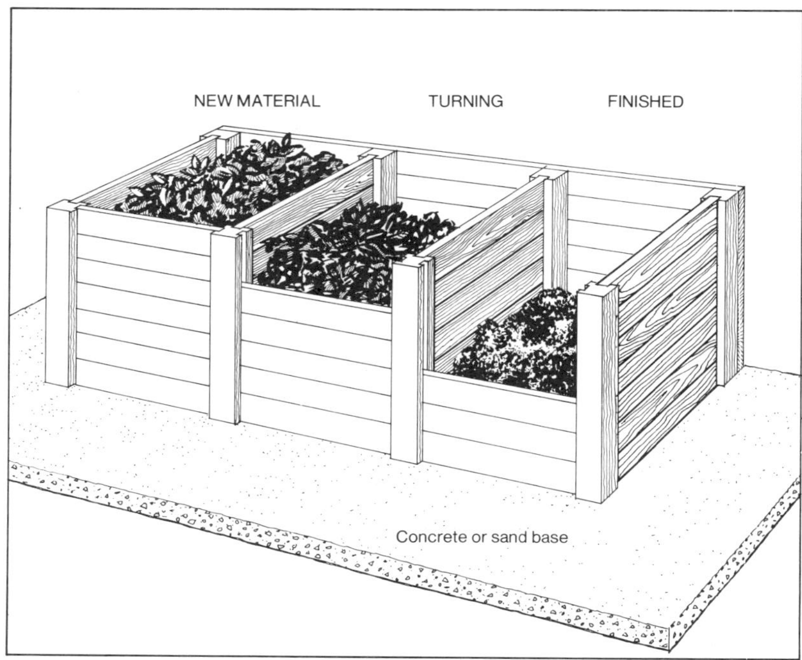

Three compost bins made from hardwood timber. These keep the compost in tidy, compact heaps. Each bin has a floor area of 1.2 m² and a height of 90 cm

turning, give the heap a light watering to settle it down again. Compost heaps will not work efficiently if too wet or too dry, so to keep out rain, it is a good idea to cover the material with sacks, an old piece of carpet or plastic sheet. Compost should be ready to use in 8–10 weeks in warm weather, but will take longer in winter. Spread compost over garden beds to a depth of 5–8 cm and spade it into the topsoil.

GREEN MANURE CROPS

Green manuring is another inexpensive way of adding organic matter to the soil, but the system is usually confined to vegetable gardens when empty beds are lying idle in winter. However, there are many vegetables which can be grown in winter so there is often little space left over for a green manure crop.

If you do have an empty bed there are a number of green manure crops to use. Seeds of wheat, barley or oats can be broadcast at 30–60 grams per square metre in autumn to provide a large bulk of material to dig into the soil in spring. Legume crops which add nitrogen through nodule bacteria in their roots are usually preferred. Suitable crops for autumn planting are field peas, New Zealand blue lupins and vetches. Recommended seed rate is 15–30 grams per square metre. Dig the crops in when they commence to flower in spring. Good summer-growing legume crops for warm northern climates include dolichos (lab lab bean) and cow peas. These are sown in spring and dug in mid-summer when they flower. A complete fertilizer must be broadcast at $\frac{1}{3}$ cup per square metre when sowing all green manure crops.

Water the crops a day or two before digging them in. If the crop is very tall, flatten it and chop up with a sharp spade. After digging keep the soil damp but not wet, then after three weeks dig the soil over again. It will take another three or four weeks for the organic matter to decompose. If there is any sign of yellowing in the following crop give side dressings of a nitrogen fertilizer.

INORGANIC FERTILIZERS

Single-element or 'straight' fertilizers have already been described under the headings nitrogen, phosphorus and potassium. Home gardeners will find it easier and more convenient to use complete N.P.K. mixtures than to spend time and effort in mixing their own.

N.P.K. mixtures come in powdered, granular and water-soluble form. There are dozens of brands and many formulations. Some powdered mixtures contain blood and bone as well as inorganic chemicals. Granular mixtures have the advantage that they are free-running and do not set hard in storage, so you can buy them in large quantities and have them on hand when you need them. Some of the more popular home garden packs contain a trace element mix.

POWDERED AND GRANULAR FERTILIZERS

There are four basic groups or types of these fertilizers.

Group 1 Many of these, including Gro-Plus Complete, have an approximate analysis N.P.K. 5:7:4. This analysis, which is high in phosphorus, is recommended as a general garden fertilizer, especially for pre-planting or pre-sowing. Gro-Plus Lawn Starter has a similar analysis and is designed as a pre-sowing mixture for lawn seed.

Group 2 Other mixtures contain more nitrogen and potassium and less phosphorus. For example, N.P.K. 10:4:6 or N.P.K. 10:4:10 are designed to promote leaf growth in vegetables and perennial plants such as shrubs and fruit trees.

Group 3 Lawn fertilizers contain even higher quantities of nitrogen with an approximate analysis N.P.K. 17:0.5:5. These, including Gro-Plus for Lawns, are designed to replace the large amounts of nitrogen removed in grass clippings. They are also useful to apply as side dressings (fertilizer scattered around or along rows of plants) for leafy vegetables such as cabbage, cauliflower, lettuce and silver beet.

Group 4 The last group of mixtures are those designed for shrubs with shallow fibrous roots, such as azalea, camellia, daphne and rhododendron. These fertilizers contain large quantities of slow-acting organic material like blood and bone. They can be used safely without the risk of root damage, which can often occur when more concentrated mixtures are used. An example of these organic-based mixtures is Gro-Plus for Camellias and Azaleas, with an approximate analysis N.P.K. 7:5:2.

WATER-SOLUBLE FERTILIZERS

Thrive, Aquasol and Zest are well known water-soluble fertilizers. The main ingredients are urea, mono-ammonium phosphate and potassium sulphate or potassium nitrate. Each one also contains a balanced trace element mix. All the ingredients dissolve completely in water, so it is easy to apply them in dilute solutions by watering-can or through a hose-spray attachment. They can be used safely as nutrient boosters for flowering annuals, vegetables, shrubs, indoor and outdoor pot plants—in fact, for every plant in the garden. The N.P.K. analysis of all three is higher, especially in nitrogen, than for powdered or granular fertilizers. The approximate analysis of Thrive is N.P.K. 31:5:9 and is excellent for promoting quick growth in flowering annuals and vegetables.

SLOW-RELEASE FERTILIZERS

There is nothing really new about slow-release fertilizers. We have been using them for years—bone dust, blood and bone, animal manures and compost. All these organic materials must be decomposed before the nutrients are available. It is interesting to note that superphosphate—the first quick-acting fertilizer—may also be regarded as 'slow-release' in some situations. On very acid or alkaline clay soils, the soluble phosphorus is 'fixed' but may be released to plants many months after it is applied. Most of the inorganic fertilizers (sulphate of ammonia, ammonium nitrate, urea and the potassium salts) and fertilizer mixtures which contain them are soluble and can be lost by leaching.

Here was a dilemma for horticulturists and chemists. They had replaced slow-release organics with quick-acting chemicals which, unless applied regularly, could be lost in drainage water. Hence they argued that the perfect fertilizer should release nutrients in a steady stream to be absorbed by plants as required.

Ureaform and Plant Pills Because nitrogen is easily lost by leaching, chemists concentrated on this element. Urea-formaldehyde (Ureaform) was the first slow-release nitrogen fertilizer. It has been used for fertilizing lawns because one application lasts several months. It has also been used in the many brands of slow-release plant pills or tablets such as Yates Slow Release, Rite-Gro Slow Release, Aquasol Slow Release and Gard-n-Tabs. Some brands have an approximate analysis N.P.K. 10:2:5 but there is some variation, particularly in the nitrogen component.

Plant pills have been most successful in feeding pot plants of all kinds but can be used for small shrubs and trees too. Use them according to directions—the number of pills increases with size of pot or height of shrub or tree. Do not exceed the dose or apply them more frequently than at three-monthly intervals. For delicate indoor plants the dose should be halved.

Mag-Amp and K-Mag Ammonium phosphate can combine with heavy-metal elements like magnesium to form magnesium ammonium phosphate (Mag-Amp). This chemical is slightly soluble and releases nutrients to the soil over many months. K-Mag is the same material with potassium sulphate added. The rate of release of nutrients depends on temperature and granule size. One brand (Mag-Amp with K) has granules almost the size of pea seed so nutrients are very slowly released.

The Mag-Amp slow-release fertilizers are good for shrubs, trees, fruit trees, citrus and pot plants (including orchids and indoor plants). They are also recommended for Australian native plants. It is better to rake or dig the granules into the top soil than to scatter them on the surface.

Coated Fertilizers and Osmocote® A new approach to slow-release fertilizers is to protect the soluble fertilizer particle with a coating or capsule which dissolves slowly. Paraffins, waxes, resins, polythene and sulphur have been used as coating materials.

The best known is Osmocote® Control Release Fertilizer, manufactured by the Sierra Chemical Company of California. The name Osmocote comes from the phenomenon called 'osmosis' in which dissolved substances diffuse through a membrane. This is the way nutrients are absorbed by the root hairs of plants and passed from cell to cell in plant tissue.

In Osmocote, the polymer resin coating acts as the membrane. Water penetrates the membrane to dissolve the nutrients and the coating swells into a small plastic bead. The dissolved nutrients then diffuse slowly into the soil solution. When all the nutrients are released the beads collapse into small husks which are decomposed by soil bacteria. The rate of release depends on soil temperature. This is an advantage because plants grow faster and take up more nutrients in warm soils. Release is not influenced by soil acidity or alkalinity or by bacterial activity in any way.

Osmocote is now available in Australia in bulk and in small home garden packs. It is available in two formulation for 100-day feeding and 280-day feeding. Both are complete N.P.K. mixtures, but do not contain trace elements. These can be added if necessary by spraying plants with water-soluble fertilizers. Osmocote is recommended for soil mixtures for seed beds, boxes and punnets and for a wide range of container-grown plants.

HYDROPONICS

Hydroponics is a system of growing plants without soil. If there is sufficient light, the right temperature, air to supply carbon dioxide to leaves and oxygen for the plant and its roots, we should be able to feed them water and nutrients for healthy growth. So why do we need soil at all?

In the search for answers to plant nutrition, many experiments were carried out in laboratories using water-culture methods. These gave precise information on the effect of each nutrient on plants. In true water-culture, roots dip into a nutrient solution through which air is bubbled regularly so roots can breathe. More often, plants are grown in a combination of free-draining sand or gravel and an inert moisture-holding material like vermiculite, and fed with a nutrient solution. Hydroponic gardens have been successful in America, England, South

Africa and India. In Australia, carnations and strawberries have been grown commercially in hydroponic beds.

There is new interest in this system of growing plants for vegetables, flowers and small shrubs. It has a special appeal to people living in flats or home units where there is a sunny balcony or patio but no outside garden. Hydroponics is much the same as growing plants in tubs or pots—but without soil. Containers must have free drainage and plants must be watered and fed regularly. Several kits for hydroponic beds and planter boxes are available.

What are the advantages of hydroponics? No digging, no weeds, and very little maintenance—apart from regular watering and feeding. If you want to try your hand at hydroponics, fill a Hanimex foam styrene trough with a mixture containing two parts coarse river sand, one part crushed charcoal and one part vermiculite, plus a tablespoon of Gro-Plus Complete per bucket of mix. Sow three rows of baby carrots and water them regularly. The nutrient mixtures for hydroponic gardening are made up to special formulas. Your local Department of Agriculture will be able to give you advice on these mixtures.

Correct plant nutrition is particularly important in the vegetable garden

Chapter 5

SOWING SEED AND RAISING SEEDLINGS

Most garden plants are grown from seed but many perennials are best multiplied by plant division, layering, cuttings, budding and grafting. These vegetative methods are described in Chapter 6.

All flowering annuals and most vegetables must be grown from seed. Many biennial flowers are best treated as annuals. Perennial vegetables, like potatoes and sweet potatoes, are grown from tubers (underground stems).

Seeds are formed in the flowers of plants by male pollen cells fertilizing the female ovules or egg cells. Most plants have male and female parts in the one flower. Some, like sweet corn and vine crops, have male and female flowers on different parts of the same plant. Asparagus, date palms and papaw trees are even less romantic—they have male flowers on one plant and female flowers on another.

Many annual plants are cross-pollinated—the pollen is transferred by wind or insects (mostly bees) to other plants. The other plants may be the same variety, a different variety and sometimes a different, but closely related, species. For this reason, plants grown from seed of cross-pollinated plants may not be true to type unless special care is taken by seedsmen to prevent 'crossing'. Plant breeders use controlled cross-pollination to develop new varieties and more recently Fl (first filial generation) hybrids. Hybrid flowers and vegetables have special qualities of uniformity, vigour and tolerance to unfavourable conditions. Many hybrids are disease-resistant too. The seed of hybrid plants should not be saved because these advantages will be lost in the second or F2 generation.

SEED STRUCTURE AND SEED LIFE

All seeds have two parts—an embryo, in which the shoot and root of the new plant are already formed, and storage tissue to feed the embryo when germination starts. The seeds you buy may look dry and lifeless but they do contain some moisture (8–10%) and respiration (breathing) is going on at a slow rate—very much like a motor idling.

Like all living things, seeds will eventually die. They die faster when stored in warm, humid conditions, so it is difficult to keep them alive in tropical climates. Even in a mild climate like Sydney, short-lived seeds (aster, carnation, gerbera, onion, parsley and parsnip) will begin to lose germination vigour in eighteen months or less. Seedsmen have overcome this storage problem by drying seeds to a low moisture content (4–7%) and sealing them in moisture-proof packets of aluminium foil. This way, respiration is slowed down further and seeds maintain germination vigour for many years. Long-lived seeds (lupins, sweet peas, zinnia, capsicum, tomato and vine crops) maintain germination for several years under favourable conditions. They are not sealed in foil packets unless they are to be sold in tropical areas. If you are storing seeds at home—whether in foil or paper packets—always keep them in a cool, dry place.

HOW SEEDS GERMINATE

WATER AND AIR

Seeds must absorb 40–60 per cent of their weight in water to trigger germination. When germination starts they respire (breathe) faster so they need air as well. When you sow seeds in soil, they take up moisture from the films of water surrounding the soil particles. The space between the particles (pore space) supplies the air. If the pore spaces are very small, as in silty or clay soils, there is too much water and not enough air. Sandy soils have large pore spaces and a good air supply but hold moisture badly. If you add moisture-holding materials (peat moss, vermiculite, compost or seed raising mixture) to sandy soils, then you have the ideal combination for seeds to germinate —sufficient water and sufficient air.

TEMPERATURE

Soil temperature is important too. Most garden seeds will germinate if soil temperature is 20°C. As soil temperature decreases, germination is slower. There are some exceptions, however. A few spring flowering annuals—alyssum, cornflower, gypsophila, larkspur, linaria, nemesia, polyanthus, poppy (Iceland) and primula —germinate well at 15°C. Spinach (not silver beet) also germinates in cool soil.

Many summer flowering annuals—amaranthus, celosia, coleus, gerbera, petunia, portulaca, salvia and zinnia—need a soil temperature of 25°C to germinate quickly. This also applies to warm-season vegetables such as beans, capsicum, sweet corn, tomato and vine crops. For example, bean seedlings may take 14–21 days to emerge if seed is sown in early spring, but they will emerge in 7–10 days if the seeds are sown a few weeks later when soil is warmer.

TIME FOR GERMINATION

Quite apart from temperature, some seeds germinate much faster than others, so it is important to know when to expect seedlings from the seeds you sow. Under good conditions, seedlings of aster, marigold, zinnia, beans, peas, lettuce, vine crops and the cabbage family emerge in 6–10 days (often less). Slow starters include begonia, cineraria, coleus, cyclamen, delphinium, larkspur, pansy, polyanthus, primula, verbena, parsley and parsnip. Seedlings of these may take 21–28 days to show, so keep the soil damp, but not wet, for this length of time. Seeds of Australian native plants are notoriously slow to germinate. Some—boronia and Sturt's Desert pea—may take two or three months to emerge. The number of days for the seedlings of vegetables and flowers to emerge is given in the Sowing Guides in Chapters 8 and 9.

SEED TREATMENT

It is best to treat seeds with a fungicide dust before sowing. This protects the germinating seeds and young seedlings from a soil-borne fungus which causes the disease called 'damping off'. You can use any fungicide dust—Captan, Zineb, or copper oxy-chloride (Dry Bordox)—or a combination dust like Rose Dust or Tomato Dust, both of which contain a fungicide. Do not use Cabbage Dust which contains insecticides only. (See Chapter 17—Pests, Diseases and Weeds.) Add a small quantity of fungicide dust (enough to cover the tip of a penknife blade) to the seeds in the packet. Fold over top of packet and shake well for about 5–10 seconds. This will give each seed a coating of fungicide.

SOWING SEEDS DIRECT INTO GARDEN BEDS

Many seeds can be sown direct. Direct sowing avoids double handling. Plants are usually more vigorous, because 'transplant shock' (damage to roots in lifting seedlings followed by temporary wilting) is avoided too. If you are running late with your plantings of flowers or vegetables, direct sowing can often make up for lost time.

Most plants recommended for direct sowing have relatively large seeds, but all have vigorous seedlings which can cope with conditions in the open garden. Among popular flowers for direct sowing are alyssum, aster, balsam, calendula, celosia, larkspur, linaria, lupin, marigold, mignonette, nasturtium, nemesia, phlox, stock, sweet pea, verbena and zinnia.

Vegetables with large seeds (beans, broad beans, peas and sweet corn) and root crops such as beetroot, carrots, parsnip and turnip (all of which transplant badly) are direct sown. Many others (cabbage, Chinese cabbage, lettuce, onion, silver beet and spinach) take to this method too. This way, they avoid transplant shock and grow faster. You can sow late crops of tomato, capsicum and vine crops direct, too, but most gardeners prefer to raise early plants in punnets or pots.

PREPARING SOIL AND APPLYING FERTILIZER

When sowing direct, prepare the garden bed a week or two beforehand. Always work the soil in dark damp condition to preserve a good crumb structure.

The reason for preparing the garden bed a week or two before using it, is that the preparation stimulates the germination of weed seeds which are inevitably present in the soil. By removing these weeds the gardener is then able to sow into a relatively weed-free seed bed. It is important that the germinating seed has as little competition from weeds as is possible.

Before sowing, apply a complete fertilizer with an approximate analysis N.P.K. 5:7:4, such as Gro-Plus Complete. The high phosphorus content will ensure vigorous seedling growth immediately after seed germination. Fertilizer may be broadcast at $\frac{1}{3}$ cup per square metre and raked into the topsoil before final levelling. When sowing in rows or a definite pattern it is better (and more economical) to scatter fertilizer in a band 15–20 cm wide at $1\frac{1}{2}$ tablespoons per metre where the seed is to be sown. Rake into topsoil and level as before.

Large seeds (beans, broad beans, lupins, peas, sweet peas and sweet corn) are liable to fertilizer burn if in direct contact with it, so it is best to apply fertilizer in a band to the side or below the seed. There are two ways to do this:

1. Open a furrow 15–20 cm deep, scatter fertilizer along the bottom at 30 grams per metre and cover it with soil by raking the shoulders of the furrow. Then make a shallow furrow for seeds on top of the fertilizer band.
2. Open two furrows 10–15 cm deep on either side of a line where seed is to be sown. Scatter fertilizer at 15 g/m along the bottom of each and cover in furrows as before. Then make a shallow furrow for seeds midway between fertilizer bands.

Another method of direct sowing is to sow seeds in clumps or stations at the required distance apart. Scatter fertilizer in a band and rake in as before, or scatter the fertilizer at each position where seed is to be sown and mix it with the topsoil. Make saucer-shaped depressions at each station and sow a few seeds in each. Many bedding plants, including nemesia, phlox, stock and zinnia, can be sown in clumps about 20 cm apart. Usually all seedlings can be retained without thinning to give a denser mass of colour. Seeds of lettuce, silver beet, spinach and late sowings of tomato, capsicum and vine crops can be sown in clumps (3–5 seeds to each) in the same way. Thin each clump to the strongest seedling. With tomato and vine crops, retain two seedlings.

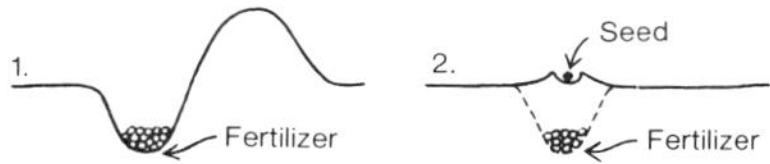

Applying fertilizer in a band, first method. Open a furrow 15–20 cm deep, scatter fertilizer along the bottom, then cover with soil. Make a shallow furrow for seeds on top of the fertilizer band

Applying fertilizer in a band, second method. Open two furrows, 10–15 cm deep, on either side of a line where seed is to be sown. Scatter fertilizer along the bottom of each and cover with soil. Make a shallow furrow for seeds midway between fertilizer bands

SOWING DEPTH

Sowing depth depends on seed size—the smaller the seeds, the shallower they are sown. When sowing direct, sow seeds of medium size in rows 12 mm deep. They can be covered with soil, but it is far better to cover with vermiculite, compost, peat moss, or Seed Raising Mixture. These light-textured materials hold moisture but have large spaces between particles to provide good aeration. Vermiculite is so light it can be used quite liberally to cover even small seeds. An easy way to apply dry vermiculite evenly is to fill a milk jug with the material and pour it along the row or over the seeds at each station. Vermiculite also acts as a marker for the position where seedlings will emerge. A very light mulch of grass clippings over the entire bed helps to retain moisture and prevent soil caking.

Large seeds (beans, peas, lupins, sweet peas, vine crops, etc.) can be sown deeper,

Seeds sown in heavy garden soil (too wet and little air) germinate poorly, whereas those sown in a friable sandy soil germinate quickly and well. These two trays of cabbage were sown at the same time, 10 days before this photograph was taken

at 25–50 mm. Always sow them into dark damp soil at the bottom of the furrow. Then cover the furrow with soil and lightly tamp down with the back of the rake. Covering with compost or vermiculite is not necessary unless the soil is extremely heavy. Lightly rake the whole bed and spread a light mulch of grass clippings. If the seeds are pressed into dark damp soil, further watering is usually not required until seedlings emerge. Too much moisture for these large seeds, especially in the first day or two, can be harmful to germination because water is trapped beneath the seed coat and excludes air for respiration. For this reason, do not soak beans, peas, lupins or sweet peas in water overnight before sowing.

SPACING SEEDS

The spacing of seeds, expecially for those sown in rows or clumps, depends on the kind of seed to be sown. This is given in the cultivation notes on the reverse side of each packet. It is also shown in the Sowing Guides in Chapters 8 and 9.

How to sow seeds thinly is a difficulty faced by many home gardeners but, like most things, it's simple once you know how! First, make a V-shaped crease in the packet as a 'track' for the seeds. Then take the packet between thumb and fingers, but leave the index finger free to tap the packet. As you tap, seeds will shuffle along the crease and fall onto the soil. With a little practice you can tap out one or two seeds at a time. Alternatively, empty a quantity of seeds into the palm of the hand. Take a pinch of seeds between thumb and forefinger and sprinkle along the row or drop a few seeds at each station.

MOISTURE CONTROL AND EARLY CARE

Apart from large seeds, which usually do not require extra water before seedlings emerge, most seeds must be kept damp—but not wet—until seedlings show through. This is most important for seeds which may take 21–28 days (or more in cool weather) to emerge. If you have used compost or vermiculite to cover the seeds, plus a grass clipping mulch on the beds, you will have ideal conditions for retaining moisture, making water penetration easy from light watering and preventing surface soil from caking.

When seedlings have emerged, continue watering, but rather more thoroughly and less frequently, to encourage deep rooting. As a general guide, a seedling 2.5 cm high may have its roots at a depth of 5–7 cm, so there is not much benefit to be had from a light sprinkling which wets the surface soil only. For watering seeds and seedlings, always use a fine hose spray, a Dram water-breaker nozzle or a watering-can with a fine rose. Slugs and snails just love young seedlings. Scatter Defender or Baysol baits over the bed a day or two after you sow the seed. Repeat the treatment when seedlings emerge, or sooner if the baits disintegrate in heavy rain.

RAISING SEEDLINGS

Seeds of many plants are best raised in seed beds, seed boxes, punnets and pots. The main reason is that the seeds are small and the seedlings lack the vigour and rapid growth of larger seeds. Begonia, cineraria, coleus, cyclamen, poppy (Iceland), petunia, polyanthus and primula are good examples of this tender group. For example, petunia seedlings may take three or four weeks to reach a height of 2.5 cm, whereas marigold seedlings grow to that height in a week. In raising seedlings, you have much better control over early growing conditions—light, temperature, soil mix, water, nutrients and so on. Some flower seeds are expensive, so it is important to raise as many seedlings as possible.

There are, of course, many flower seeds which can be sown direct or raised as seedlings, just as you choose. This gives greater flexibility to your planting programme. For example, you may have a bed of petunia or phlox still flowering well in late summer. Rather than sacrifice this colour display for an empty bed, raise seedlings of, say, nemesia, which will be ready for transplanting when the summer flowers are finished.

It is often more convenient to raise seedlings of some vegetables, especially when you need only a few plants. This is particularly true of broccoli, cabbage, cauliflower and of early plants of capsicum, eggplant, tomato and all the vine crops. Sow the seed in punnets and prick out seedlings into 10 cm plastic pots to grow on for another four weeks or more.

SOIL MIXTURES FOR SEEDLINGS

A soil mix for seedlings can be tailor-made to promote rapid germination and vigorous growth. You will need a mix which is open, friable and well drained. Sand, preferably coarse river sand, is the best material to meet this requirement. Do not use beach sand unless it has been thoroughly washed to remove salt. You will need a good moisture-holding material too. Choose from vermiculite, spent mushroom compost, garden compost, peat moss or seed raising mixture. Vermiculite is one of the best materials because the tiny root hairs of the seedlings penetrate the particles to provide a built-in moisture supply. When seedlings are transplanted, the young roots cling to the particles and so take moisture with them to the new spot in the garden.

For a start, try a soil mix containing one part garden soil, one part coarse sand and one part vermiculite, compost or peat moss (you will have to dampen the peat moss before adding to the mix). Add 30 g of a pre-sowing fertilizer such as Gro-Plus Complete and 90 g of lime to each bucket of mix. The mix must be freeflowing and should not compress in the hand when damp. If your garden soil is very sandy, you will need less sand and more moisture-holding material. If your garden soil is heavy or contains a lot of organic matter, you will need more sand. Test the mix by filling a seedling tray or punnet and watering it with a fine spray. The mix should absorb moisture quickly—in a few seconds—and drain freely. If it does not, add a little more sand.

SEED BEDS

A permanent seed bed is a good idea in a large garden or if you wish to raise large numbers of seedlings. Select a sunny, sheltered spot in the garden, build it up 15–20 cm for drainage and contain it with boards or a brick surround. An area one metre square is a convenient size. On sandy soils add liberal quantities of vermiculite or other moisture-holding material. On clay soils spread a 5 cm layer of coarse sand as well. Then add the mixed fertilizer at $\frac{1}{3}$ cup per square metre and lime at 1 cup per square metre. Dig the bed over to mix the ingredients thoroughly to a depth of about 10 cm.

You will need some protection for young seedlings in an open seed bed. A frame of 50 mm × 25 mm timber covered with flyscreen wire or 32 per cent shade cloth is suitable. Make wooden legs at each corner or rest it on upright bricks. Use the frame to protect young seedlings from direct sunlight during the hottest part of the day. Don't keep seedlings covered all the time because they become

Seedlings are easily removed from the Speedling Starta Boxes when ready to transplant

soft and lanky. Give the seedlings more sun and less shade as they grow. The frame is useful for covering the seed bed in heavy rain too.

SEED TRAYS, PUNNETS AND POTS

These are shallower than a seed bed so do not hold as much moisture, but have the advantage that they can be moved about easily to give seedlings the required conditions of shade or sunlight. After sowing, the containers can be kept indoors to provide a rather higher and more even temperature (especially in winter), and moved outside as soon as the seedlings emerge. Plastic trays, punnets and mini-punnets are already provided with good drainage, are easy to wash and clean, and can be used over and over again. You can fill these containers with soil mix, sow seeds and prick out seedlings while standing or sitting at a workbench—much easier than squatting down at the edge of the seed bed.

A recent innovation is a polystyrene tray with cells formed in the shape of an inverted cone. This product, called the

Jiffy peat pellets are used extensively to raise flower and vegetable seedlings from seed

Capsicum seedlings, undusted (left); dusted with fungicide (right)

Speedling Starta Box, is very useful for raising seedlings from seed as the seedling is easily removed from it with little or no damage to the root system of the plant, thus minimizing transplant shock.

Fill the container of your choice with dark damp soil mix to the top of the container. With a flat board, firm the mix to a level 6–12 mm below the rim of the container. (You can use the bottom of an empty tray or punnet for firming the soil, because the bottom is slightly smaller in area than the top.) Another method is to almost fill the container with dry soil mixture, dump it on the bench to level it and then water well with a fine spray until water seeps from the drainage holes. Alternatively, stand containers in shallow water until moisture seeps to the surface. With both these methods, the soil mix is often too wet for immediate use so leave for 24–48 hours before sowing.

Another method of raising seeds favoured by many home gardeners and commercial growers is planting the seed in Jiffy 7s.

Jiffy 7s are pellets of compressed peat moss which, when expanded by adding water, make a block of peat suitable for raising seeds or cuttings. When the plant is of suitable size for transplanting, the peat block complete with plant is put into the permanent position in the garden and no 'transplant shock' is encountered.

SOWING SEED

Before you sow, it is best to dust the seed in the packet with a fungicide, as described earlier. This cannot be over-emphasised. The dramatic difference between capsicum seedlings grown from dusted and undusted seeds is shown in the photograph.

When sowing in seed beds, mark out shallow rows or drills about 6 mm deep and 5–7 cm apart with edge of a flat board or dowel stick. Scatter seed thinly along the rows by tapping seeds from the packet. Do not crowd seeds in the row. For every 2.5 cm of row, sow 5–7 small seeds like primula or 3–4 medium size seeds such as aster, pansy or stock. Then sprinkle vermiculite or other moisture-holding material to a depth of 3 mm over the surface. Make sure the material you use is free-flowing and spreads evenly. This covering allows water to penetrate, forms a mulch to stop surface drying and protects seeds from washing away. Water well with a fine hose spray or watering can with a fine rose. For plastic seed trays, which measure 35 cm × 30 cm, mark out shallow rows or drills with a board or dowel stick as described for seed beds. Sow seed, cover with moisture-holding material and water gently as before.

An alternative method is to use a 'marking board'. This is a flat board slightly smaller in area than the tray with 100 evenly spaced flat-headed clouts driven

Make a shallow saucer-shaped depression and sew a few seeds in each

Form drills by using the back of a rake

into it. The heads of the clouts are approximately 12 mm in diameter and protrude from the board about 6 mm. Press the board firmly into the damp soil surface, leaving 100 shallow holes. If the soil is too dry, the holes will fill with loose soil. Tap out 2–4 seeds in each hole. Then cover the whole tray with vermiculite or compost to a depth of 3 mm. Water well with a fine spray as previously.

Plastic punnets and mini-punnets are ideal for very small seeds like petunia, poppy and primula. Soil mix must be free of lumps or clods and in dark damp condition. After dusting seed with fungicide—this also helps you to see the fine seed more clearly—sprinkle seed over the surface. Cover evenly with vermiculite or compost to a depth of 3 mm and firm down with a board or the bottom of an empty punnet. Water very gently.

Punnets and mini-punnets are good for vegetable seedlings when you only need a few plants. Scatter about 20 seeds to a punnet or about 10 seeds to a mini-punnet. This number will give you about a dozen or half a dozen seedlings respectively of broccoli, cabbage or tomato. Press into the dark damp soil with the bottom of an empty punnet, cover and water gently. For large seeds of pumpkin, cucumber, zucchini, marrow and other vine crops use eight seeds to a punnet or four seeds to a mini-punnet. Press seeds into soil

A marking board is useful for spacing seed in seedling trays

After sewing seed cover with vermiculite

point down. Vine crop seeds need to be kept on the dry side, so do not over-water.

GENERAL CARE OF SEEDLINGS

You must keep seeds moist but not wet until seedlings emerge. This may be 7–10 days for fast germinating seeds but two to four weeks or more for slow starters. Check the Sowing Guides in Chapters 8 and 9 for approximate times.

Although the covering of moisture-holding material may appear dry on top, the soil underneath can be quite damp. Test this by scraping some of the covering away with your finger and feeling the dampness of the soil underneath. The same rule applies to seedlings once they are started. Keep them damp but not wet. As they grow stronger, thorough but less frequent watering is needed. Morning, rather than evening, watering is recommended. Seedlings require some shade when very young, but as they grow they need more sunlight. This way, they become more accustomed to conditions in the open garden. Too much shade makes soft, lanky seedlings which transplant poorly.

For extra warmth, especially in winter, you can cover seed beds with glass or clear plastic until seedlings emerge. Remove the cover immediately seedlings break the surface. Don't use these coverings in direct sunlight—the temperature increases so much that the seedlings 'cook' as they break the surface. Trays and punnets may also be covered with glass or enclosed in clear plastic bags to increase temperature and prevent evaporation (both indoors and outdoors) but remove them when seedlings emerge. In very cold districts you may consider a garden frame equipped with electric heater tape to raise soil temperature.

Control slugs and snails by scattering Defender or Baysol baits on or around seed beds, trays or containers. If you don't do this you may wake up one morning to find no seedlings at all! Damping off can occur in seedlings after they have emerged—especially if the weather is humid and the seeds are sown thickly. If some of your seedlings start to topple over at soil level, immediately drench them with a solution of Captan, Zineb or Dry Bordox fungicide.

PRICKING OUT

This is a way to remove crowded seedlings when they are quite small (about 12 mm tall) to a larger container, for example, moving petunia, poppy or primula seedlings from a punnet to a seedling tray. Fill the seedling tray with soil mix as before and use the marking board to make 100 shallow holes. Then gently prise out a few seedlings with a dibble (a sharp-pointed planting stick). Separate the seedlings on newspaper covering the work bench. Take a seedling (gently but firmly) between thumb and forefinger in one hand and with the other use the dibble to make a small hole where marked on the tray. Lower the seedling roots into hole and push a little soil mix around them with the dibble to firm the seedling. When tray is full, gently water and keep in shade or under a fly screen or shade cloth frame for a few days until seedlings are well established. Grow seedlings on until 7–10 cm high and then transplant to garden beds.

TRANSPLANTING

Prepare the bed to receive seedlings two or three weeks before transplanting. During preparation add mixed fertilizer (N.P.K. 5:7:4) at $\frac{1}{3}$ cup per square metre. Level the bed and water well so the soil is dark damp. Also water the seedlings the day before transplanting.

Mark out the position for each seedling. Make a hole with a trowel or your hand 7–10 cm deep and wide enough to accommodate the seedling. Gently ease out seedlings from seed bed or container, taking as much soil as possible with each one. Lower seedling into hole and press soil firmly around it, making a small depression at the same time to direct water to the roots. For seedlings in individual pots, simply tap the pot and plant the seedling slightly deeper than it was in the pot.

If you have garden compost or dry grass clippings available, spread some around each plant in an area about 30 cm in diameter. Then water each plant well with hose or watering can to settle soil around the roots. It is best to transplant seedlings in late afternoon or evening. If planting during the day, provide protection with pieces of brush or a handful of long dry grass or straw, especially in hot weather. Scatter snail and slug baits over the bed to protect young plants from slugs and snails. Repeat if heavy rain falls.

Advancing annuals in Jiffy Peat Pots is a good way of giving quick colour to a garden

Chapter 6

NEW PLANTS FROM OLD

Vegetative propagation means increasing plants by any means other than seeds. This way of reproducing plants has many advantages. The gardener is certain that new plants will be identical with old ones and he has a usable plant in less time than one grown from seed. In some cases, plants set seed poorly or not at all, so one *must* use vegetative methods. With others, you have a choice—you can sow seed or propagate.

Examples of vegetative propagation in the garden are not hard to find. Strawberries have stolons or runners which form new shoots and roots at every node or joint. Mint (the one we use for sauce with lamb) has underground stems called *rhizomes*. These form a dense mat and each small piece can become a new plant. Potatoes have swollen underground stems or *tubers*. Each bud or eye on a potato is a potential potato plant. *Bulbs* and *corms* are other examples of this ability of plants to multiply and grow. Unfortunately, many of our most troublesome weeds also have this ability.

All forms of vegetative reproduction and the methods developed by gardeners to assist it depend on small regions of tissue which produce new plant cells. *Growing tissue*, where cells are actively dividing, is located in the tips of stems and roots and in lateral buds. Stems and roots increase in diameter too, especially in larger plants. To achieve growth in diameter there is another area of growing tissue called the *cambium layer*. It is best described as a thin, unbroken cylinder of dividing cells which connects every part of the plant. The cambium layer splits off new cells from both its inside and outside layers.

The cambium layer is the vital part of any plant when taking cuttings, or in layering, budding or grafting. For this reason, cuttings are taken just below a node or joint in the stem or below a leaf axil where one finds the greatest concentration of dividing cells which heal the wound ('form a callus') from which new roots will grow. In budding and grafting the cambium layers of the two plants are placed in contact. This allows the bud or graft (the *scion*) to draw on food from the growing plant (the *stock*) and the wound caused by the operation heals.

PLANT DIVISION

With the exception of taprooted plants, all perennials which form clumps of roots, shoots and foliage can be lifted and divided into a number of pieces for replanting. This is a quick and simple way to get new plants. Some clumps divide easily, but in others you must be more ruthless and use a spade or a sharp knife. Often you can cut off pieces with sufficient root without lifting the parent plant. Division is a useful method for herbaceous perennials (perennial roots and annual flower stems) like perennial phlox, Michaelmas daisy, gerbera, chrysanthemum and dahlia, and also for true perennials—agapanthus, iris, canna, strelitzia, violet, nandina, New Zealand flax and ornamental grasses. Many perennial plants become too large and overcrowded and benefit from division every two or three years.

LAYERING

Layering is another easy method of propagation. It has the advantage that the 'layer' is not completely cut from the parent plant and continues to draw nourishment from it. Layering is a good way to multiply many small trees and shrubs, including azalea, rhododendron, magnolia, gardenia and daphne. Carnations, rather difficult as cuttings, take to this treatment too.

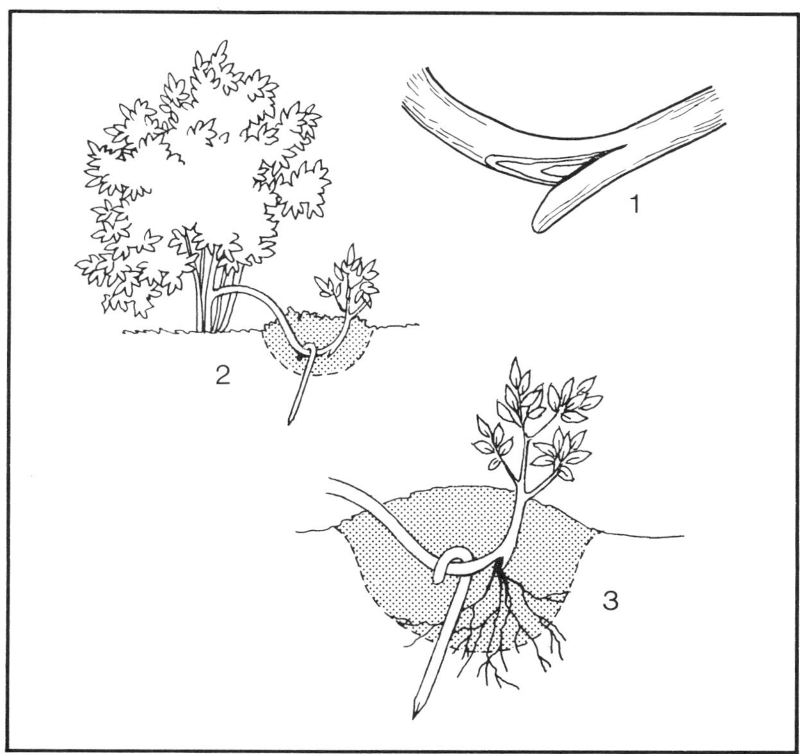

Layering is an easy method of propagation. (1) Make a slanting cut 5–7 cm long on underside of branch, finishing the cut at a node. (2) Bend branch towards ground and cover with soil. (3) Anchor the branch in the soil with wire in the shape of a U. When a good root system is formed, cut the layer from the parent plant

When layering small trees or shrubs, select one or more thin, supple branches close to the ground. Bend the branch towards the soil and make a slanting cut 5–7 cm long with a sharp knife on the underside of the branch. Finish the cut at a node, but take care not to cut more than half-way through the branch. The cut can be kept slightly open with a match or sliver of wood. Now bury the cut portion in the soil and anchor it with a piece of stout wire bent to form a U. If the soil is hard and lumpy, scrape it away and replace it with lighter soil or preferably a mixture of sand and compost or leaf mould. The end of the branch can be kept in an upright position by tying it to a small stake. You can, of course, layer a number of branches at once. Keep the soil damp, but not wet, to encourage roots to form. If the branch is layered in late winter or early spring, roots are usually formed by the following autumn. You can gently scrape away the soil occasionally to check if roots are showing. When a good root system has formed, cut the branch from the parent plant. Leave the new plant for three or four weeks to adjust to its independent status, then lift it carefully and replant in its new position.

Much the same method is used for carnations, but start layering in December

Sim carnations are reproduced by cuttings of firm laterals

or January. The rooted plants will be ready for planting in April or May. Remove the lower leaves (next to the parent plant), leaving a tuft of leaves at the end of the stem. These leaves can be shortened slightly. Many climbing and trailing plants are layered easily. Bend down well-ripened stems and peg them in the ground so that one or two nodes are covered. Cutting the stems is rarely necessary.

AIR LAYERING

If branches or stems are too large and stiff to bend to the ground, air layering is an alternative method. Again, make a slanting cut in the stem finishing at a node. Keep the cut open with a sliver of wood. Then wrap the cut with damp sphagnum moss and cover with aluminium foil or plastic tied in place with string or raffia. This keeps the cut moist. Periodic inspection will tell you when roots are well-developed and the time has come to cut it from the parent plant. Air layers need careful attention for the first few weeks after removal until the plants have become adjusted to relying on their new root system.

CUTTINGS

Many garden plants are grown from cuttings. A cutting is a piece of stem, leaf or root which, when planted under favourable conditions, produces another plant. Taking cuttings is a more artificial way of propagation but is simple and effective. You can collect, prepare and grow cuttings without too much effort and little space or green-finger skill is needed.

Cuttings fall into three main groups—stem cuttings, leaf cuttings and root cuttings. Stem cuttings are further divided into three subgroups depending on the type of wood used. These are softwood, semi-hardwood and hardwood cuttings. But before you start, there are a few other aspects to consider.

PREPARATION FOR CUTTINGS

Tools and Equipment You need a good quality knife for a start. A razor blade with one edge covered with insulating or adhesive tape is useful for small cuttings. A good pair of secateurs which cut cleanly without tearing plant tissue is essential. Both knife and secateurs should be kept in top condition by regular sharpening. A plastic collection bag for cuttings is useful to prevent them drying out and wilting, especially for softwood cuttings or if you are taking cuttings away from home.

Containers and Rooting Mixtures Best containers are plastic pots, punnets or trays. These are easy to clean so wash thoroughly before using. All containers must have free drainage. The rooting mix must be as free as possible from disease, insects and weed seeds. Leading nurseries sterilize the rooting medium and the containers they use, but this sophisticated treatment is not available to the average home gardener.

The materials recommended below are relatively sterile and relatively free of these problems. The mix must be porous, free-draining and yet hold moisture. Coarse river sand is often used but this dries out quickly. A mix containing two parts of coarse sand and one part of peat moss or vermiculite is much better. A mix which holds more moisture is one part of coarse sand and one part of peat moss or vermiculite. Charcoal, when crushed and screened, may be added to keep the mixture clean and more open. A suggested mix is two parts coarse sand, one part peat moss or vermiculite and one part charcoal. Usually rooting mixes do not contain added fertilizer, but a light dusting of superphosphate or Gro-Plus Complete, which is high in phosphorus, may assist root development. Proprietary potting soils are not recommended for cuttings.

Root-Promoting Hormones Hormone 'cutting' powders can help cuttings to form a callus and make roots quickly. Seradix, one of the most popular, is available in three grades for softwoods, semi-hardwoods and hardwoods. Dip the cut ends in the powder before planting. Full instructions are given on the label. Seradix is also useful for dusting the cut surface of stems for layering (described earlier).

Work Space A solid bench at a suitable height for standing or sitting makes the work much easier. It should be located in a corner of a garden shed made of laths or slabs of wood for shade and ventilation. Shelves behind the bench can store pots and tools, and plastic bins underneath can hold materials for the rooting mix.

Protection for Cuttings Your success in striking cuttings depends a lot on providing a suitable microclimate. High humidity to prevent moisture loss is important. An inverted glass jar or a clear plastic bag supported on a wire frame placed over a pot of cuttings offers a simple solution. Large boxes (with the bottoms knocked out) or frames with glass or plastic covers are good for a number of pots. Stand them on a 2–3 cm layer of moisture-holding material—peat moss, charcoal, pine bark—so water will evaporate continuously. Boxes and frames should not be too deep—about 50 cm is a good height. This allows about 15 cm between the top of the cuttings and the cover. Provide some shade because direct sunlight will build up too much heat. Most cuttings strike best at a temperature of 20–25°C, so keep your pots in a shady spot or place newspaper or shade cloth over the top of your box or frame for a week or so. Then harden them to direct sunlight. Ventilation is needed as well—prop a small piece of timber under the jar, open the plastic bag, or raise the frame cover slightly.

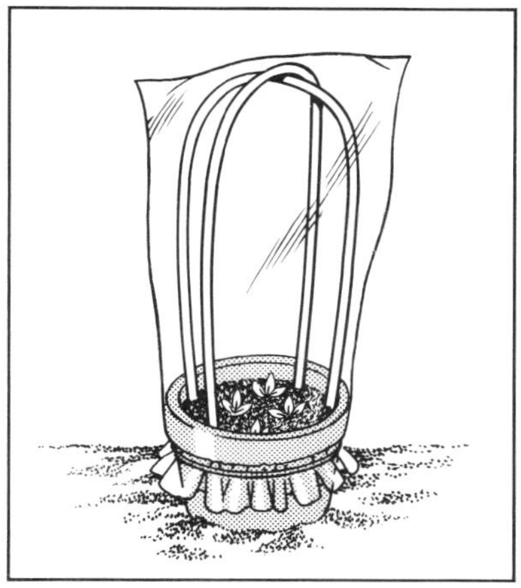

A clear plastic bag supported on a wire frame and placed over a pot of cuttings is a simple way to prevent moisture loss. Do not place the mini-glasshouse in direct sunlight

SOFTWOOD CUTTINGS

These are taken from shrubby plants with soft green shoots. Because the plants are in active growth, take care to prevent cuttings wilting before you plant them. Take softwood cuttings in early morning when the plant is full of sap. If they start to dry out cover them with moist newspaper. There are exceptions to this rule—very sappy plants like geranium and cactus will tolerate wilting even for a day or two. This gives the cut section time to dry out and start to heal.

With tip cuttings, start by pulling off leaves around the stem where the cut is to be made. Make a clean cut at a slight angle just below a node or leaf axil. For plants with large leaves, remove all but a few top ones. These can be shortened back to about half their length. Tip cuttings are usually 5–10 cm in length and have four to six nodes. This method suits azalea, gardenia and similar evergreens with large leaves. Heel cuttings are good for conifers and many other plants. Take off side shoots so that a small heel of the older branch is attached. Trim off any excess bark.

SEMI-HARDWOOD CUTTINGS

Take cuttings 10–15 cm long from evergreen trees and shrubs where the stems have started to mature into firmer brown or grey wood. These cuttings can be side shoots with a heel or a section of the lower, more mature part of the stem with the cut just below a node. Again, reduce the

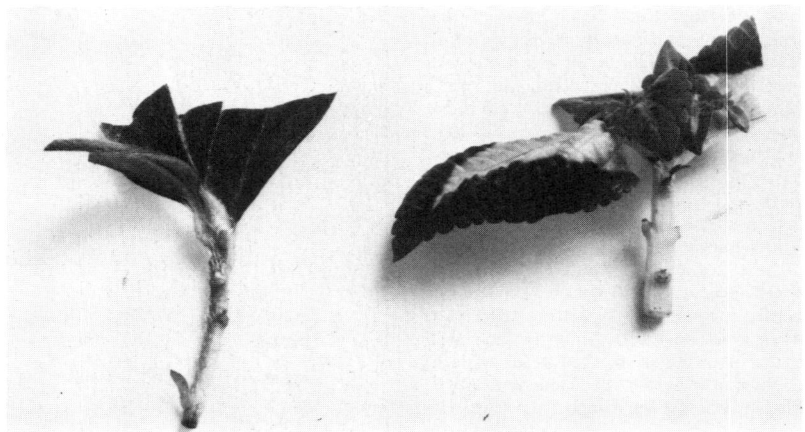

Tip cuttings of azalea and coleus ready for planting

Many conifers strike best when a heel is left on the cutting

Ficus (rubber plant) cutting prepared for planting. Leaf is rolled to save space

A typical cutting of camellia

Cinerarias grow well where trees provide a semi-shade aspect

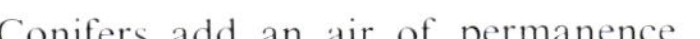

Conifers add an air of permanence

Rockeries are ideal for sloping sites

Swimming pools are an important feature of many Australian gardens

Dazzler petunia brightens summer gardens

Twinkle phlox is a free-flowering dwarf variety for edges and rockeries

number or size of the leaves to avoid excessive moisture loss.

HARDWOOD CUTTINGS

These are taken from deciduous trees or shrubs in winter when they are dormant. Select wood about 6–20 mm in diameter and about 20 cm long with four to six nodes on each cutting. These cuttings usually root easily, in the same way that branches of willows, poplars and coral trees will grow when simply pushed into the ground. To avoid planting hardwood cuttings upside down make a slanting cut at the bottom just below a node and a straight cut about 6–12 mm above the top node. This way you will be reminded to plant the cutting point down.

PLANTING CUTTINGS

Plant cuttings as soon as possible. Fill the pots with rooting medium, which should be slightly damp. Dip the cut surface into water and then into hormone powder. Make holes with a dibble or pencil about 5 cm apart in the damp mix around the edge of the pot. Set the cuttings in the holes about one-third of their length deep. Firm the mix around them and water the pots to field capacity. Protect cuttings as described above for the first week or two, and then allow more sun and ventilation. Always keep pots moist but not wet. Roots take time to develop. It may be many weeks before they will support the new plant, so do not try to move cuttings too early. It is best to transfer rooted cuttings

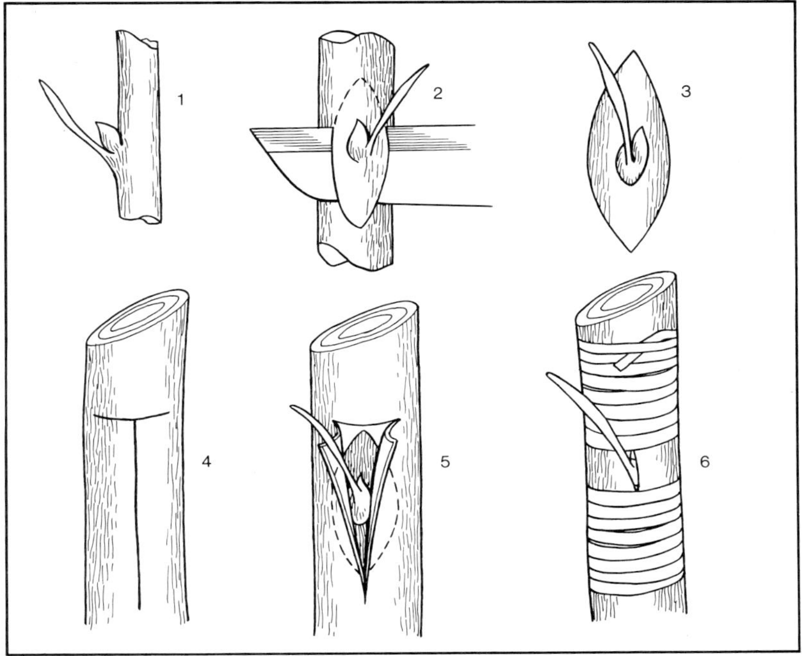

Shield budding is the method used for grafting fruit trees, citrus and roses. Diagrams 1, 2 and 3 show how the bud is sliced from the scion. 4. A T-shaped cut is made in the stock. 5. The bud is inserted in the stock, closing the bark flaps over the edges of the shield. 6. Tie the shield firmly to the stock with plastic grafting tape but take care not to cover the bud

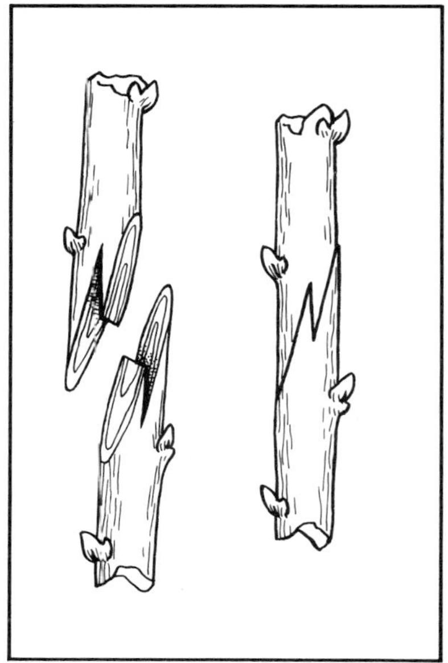

Whip-tongue grafting is useful when both scion and stock are the same thickness. The diagram shows how the cuts are made and fitted together before binding with insulating or grafting tape

to individual 7–10 cm pots and grow them on until the root system is well advanced. Make up a potting mixture similar to the one recommended for raising seedlings (one part loam, one part sand, one part compost, peat moss or vermiculite plus 30 g of Gro-Plus Complete for each bucket of mix). Proprietary potting mixtures are also suitable.

LEAF CUTTINGS

These are used for many indoor plants—for example, the leaves of Rex begonia can be cut in sections and laid on a moist rooting medium, or a single leaf can be laid flat and the veins cut with a sharp knife or razor blade. For African violet, remove a mature leaf with about 25 mm of petiole (leaf stem) and bury the petiole in the rooting mix. Shoots and roots develop from the base of the leaf. Sansevieria can be propagated from 5 cm sections of leaf. The bases of these sections are placed vertically in rooting medium.

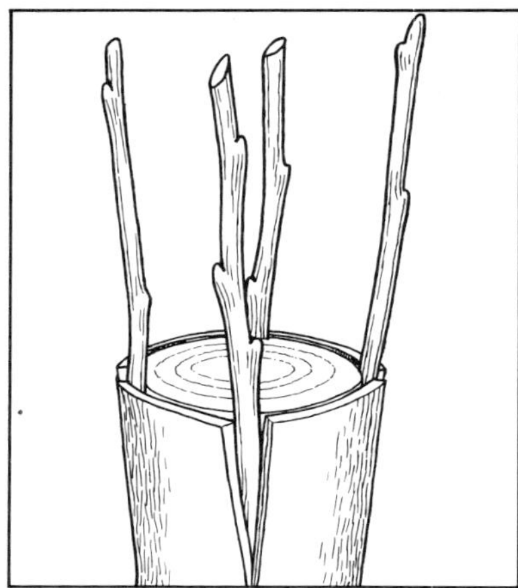

Crown or rind graft is used for grafting one or more scions into a thick trunk or branch of the stock

ROOT CUTTINGS

This is not a common method of propagation in the garden. Some plants can produce buds from their thick fleshy roots. Root sections 8–10 cm long and 12 mm in diameter are covered with 1–2 cm of rooting mix. Perennial phlox, albizia, bouvardia, lagerstroemia and wisteria have been propagated by this method.

BUDDING AND GRAFTING

These methods of propagation involve the union of a cut portion of one plant (the scion) with a stem or branch of a growing plant (the stock). In other words the cambium layers of both scion and stock must be in contact. Both methods are used by nurseries, especially when a selected variety is budded or grafted onto a more vigorous or disease-resistant root stock.

Budding and grafting are really tasks for the specialist and few home gardeners have the skill or experience to undertake such methods successfully. Only a brief summary with diagrams is given here. Those who wish for more detail should contact the Department of Agriculture in their own State or consult one of the books available on this subject.

SHIELD BUDDING

Shield budding is mostly used for propagating citrus trees, deciduous fruit trees and roses. Spring or early autumn is a suitable time for budding.

1. Preparing the stock. On one- or two-year-old wood make a T-shaped cut in the bark. The flaps should lift easily from the wood.
2. Cut the bud from a pencil-thick shoot of the scion. Use a very sharp knife to slice an oval-shaped piece of bark and wood with the bud in the centre. Make the shield 2–4 cm long. Remove the piece of wood behind the shield if it will lift out easily.
3. Place the shield into the T-shaped cut, closing the bark flaps over its edges. Trim off surplus bark at top of shield.
4. Starting at the bottom, with damp raffia or plastic grafting tape tie the shield firmly to the stock for its full length. Take care not to cover the bud.
5. After three weeks cut the raffia or tape away. If the shield is still green and the bud plump, it has taken.
6. If budded in spring, cut off stock above bud when binding is cut. If budded in autumn, the bud remains dormant until early spring when stock is cut back.

WHIP-TONGUE GRAFT

This method is used when scion and stock are of the same diameter—to 2.5 cm thick.

1. Cut top of stock and bottom of scion with a slanting cut 4 cm long. Scion should be about 10 cm long.
2. Cut along the grain on both faces, one third of length from tips to a depth of 6 mm.
3. Fit scion to stock with tongues interlocking and cambium layers matching on at least one side.
4. Bind graft carefully with insulating or grafting tape to make it firm and airtight.

CROWN OR RIND GRAFT

This method is used for one or more scions on a trunk or branch of stock over 2.5 cm in diameter.

1. Saw off stock at right angles. Smooth cut surface with sharp knife.
2. Make a vertical cut 5–7 cm long for each scion. Knife should penetrate the bark to the wood, but not further.
3. Prepare 15 cm scions with a long slanting cut to the base to form a tapered wedge.
4. Slip thin end of wedge between bark and wood of stock and push down until cut surface of scion fits snugly against exposed tissue of stock.
5. When all scions are in position bind them with raffia or grafting tape and cover the whole cut surface with grafting wax to exclude air and rain.

TISSUE CULTURE

Plant tissue culture is essentially a laboratory technique and very few home gardeners would have the necessary skills or equipment to successfully undertake such a task.

It is the vegetative propagation of plants on artificial nutrient under aseptic (sterile) conditions.

The main objectives of plant propagation under the tissue culture system are the rapid regeneration of plant specimens ensuring genetic homogeneity, the multiplication of important plant species which are difficult to propagate by conventional means (e.g. hybrids) and the elimination of virus from infected plants.

The plant parts which are suitable for plant tissue culture range from plant organs, plant tissue, diminutive shoot tips, seeds, anthers, pollen grains and plant cells.

The general method employed for cultivating larger plant tissue is to sterilize the surface of the tissue and place the sterile tissue on a sterile nutrient base under aseptic conditions for exclusion of micro-organisms, such as fungi and bacteria. The cultured tissue and nutrient are usually maintained in a flask or bottle and are incubated under artificially controlled conditions of light and temperature.

Plants which are commonly regenerated by plant tissue culture for commerical or scientific purposes are strawberries, forest trees, azaleas, ferns, rice, sugar cane, grapes, tobacco, carnations, chrysanthemums, roses and several foliage plants.

Chapter 7

LAWNS

Lawns have been a feature of home gardens in the British Isles and Europe for centuries. Australian gardeners have followed this tradition but there are many other reasons why we grow them. A lawn is the best way to cover and maintain large sections of garden easily and quickly. The initial cost of establishment is relatively inexpensive and a lawn will last indefinitely. This is a very practical consideration. But a lawn blends the house with the garden, softens harsh outlines, and complements trees, shrubs and colourful annuals. Even in the smallest garden, an attractive lawn adds a touch of spaciousness. On hot summer days, a lawn reduces temperature and glare to give a feeling of coolness. Last, but not least, it is a place for relaxation—pleasant to look at, delightful to walk or lie on and the ideal surface for children to romp and play.

LAWN PLANNING

In many respects, a lawn is an unnatural way to grow millions of grass plants which are all competing for light, water and nutrients. We cut them constantly, which makes their task of growing more difficult. We remove grass clippings for mulching garden beds or composting. This is an added drain on soil nutrients. Fortunately, lawn grasses are well adapted to this harsh treatment and will thrive if we understand their needs and go about providing them in the right way. Because a lawn is so permanent, it pays to spend some time planning it. This will help with maintenance and often makes the difference between a good lawn and a poor lawn struggling to survive.

Always shape the lawn area with flowing curves. Curves are easier to water and mow

A good lawn complements trees, shrubs and annuals

than square or sharp corners. Avoid small beds and specimen trees or shrubs—these detract from the space of the lawn and make mowing difficult. Very steep slopes are hard to mow too. Concrete mower strips adjacent to gravel paths, driveways, gardens, rockeries or other features will reduce edge clipping enormously. It is best to construct mower strips or paving after the lawn has been established. If constructed beforehand, the lawn may subside, throwing these aids to maintenance out of level.

Avoid growing grass to the building line. Pathways or paving adjacent to walls are better, because they eliminate wetting and drying effects in the foundation area when watering the lawn. Grass growing under overhanging eaves misses out on rain and will need extra watering.

Avoid heavily shaded areas for lawns—especially the southern side of buildings, fences, dense shrubs or trees. Trying to grow grass under trees is rarely successful—there is competition for moisture and nutrients as well as for light. These areas are best paved or separated from the lawn by a mower strip and covered with gravel, pebbles or pine bark. Shade-loving groundcovers (see Chapter 13) are an attractive alternative.

Avoid growing grass in situations liable to excessive wear—at the corners of buildings, outside doorways and so on. Paths or paving are the best solution here, although a few stepping stones may be enough to solve the traffic problem.

PREPARING THE SITE

SOILS AND DRAINAGE

Whether you sow lawn seed, plant sprigs of running grasses or lay turf, soil must be well prepared. The success of the lawn largely depends on the effort you put into soil preparation. Most grasses prefer well drained loam or sandy loam soils to heavier, wetter soils. Good drainage means better penetration of water, air and grass roots. On very heavy soils, it is wise to import some sandy loam soil, spread it over the area to a depth of 8–10 cm and incorporate it into the topsoil. In extremely wet or puggy situations, artificial drainage may be needed. (See Chapter 3.) Very sandy soils may drain too easily, so improve them by adding animal manure, spent mushroom compost, or other organic matter to increase their water-holding capacity.

Start preparing the soil well before the time for sowing or planting. This allows you to form the correct levels and contours, prepare a crumbly soil structure, and destroy any weed seedlings. You can dig small areas with a fork or spade, but it is worthwhile in the long run to hire a rotary hoe to save time and labour for large areas. There is no need to dig the soil deeply. On most sites, a depth of 15 cm is ample. Always dig the soil, especially heavy soil, when it is dark damp. Cultivating heavy soils when too wet (or too dry) will spoil their structure. After the initial cultivation remove large stones, gravel and other rubbish.

Allow the soil to remain in a loose or fallow condition for several weeks. Exposure to weather will help to break down large clods and form a crumbly structure. Watering (or rain) will germinate many weed seeds. Destroy them by hoeing, raking or spraying with desiccant weedicides. Track down and dig out the underground parts of persistent, perennial weeds or use weedicide. (See Chapter 17.)

SOIL pH AND USE OF LIME

Grasses grow happily in medium acid or slightly acid soils (pH 5.5–6.5). In high rainfall districts, soils may be strongly acid (less than pH 5.5). To these, apply lime at the rate of 25 kg per 100 square metres. This is equivalent to 1 cup per square metre. Lime also helps to improve the structure of heavy soils, so a good time to apply it is soon after initial cultivation. After spreading the lime, rake it into the topsoil.

LEVELLING AND GRADING

Levelling with pegs and lines is rarely necessary unless the lawn is to be used for tennis or bowls. Lawns need not be flat. Those which follow the natural slope of ground are usually more attractive. For drainage reasons, the lawn should slope away from the house. Grading the soil is necessary to fill in noticeable hollows and to scrape off high spots. You can make an improvised grader from an old wooden window frame or wooden gate, or you

An old wooden gate makes an excellent grader for scraping off high spots and filling in hollows

can nail together some 75 mm × 50 mm hardwood to make a frame. These do a good rough grading job when dragged across the surface. The amount of bite taken into the soil by the leading edge depends on the position you attach the towing rope. If it is too far forward, the grader skims over the high spots rather

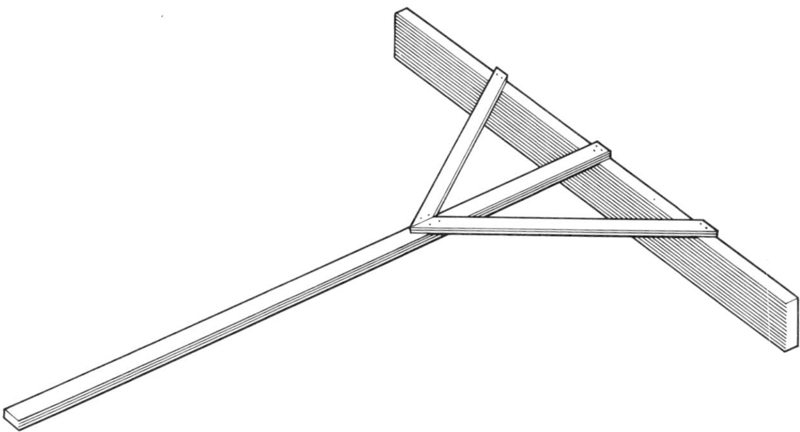

This home-made dummy rake does a good job of final levelling and is also useful for spreading topdressing soil

than levelling them. A dummy rake is useful for final grading and for spreading soil evenly. Grading must be done when soil is loose and fairly dry. When you are satisfied with the lawn's levels and contours, water the area well and leave for about a week. Then rake and cross-rake again and adjust any spot levels by further grading.

CONSOLIDATING

During the fallow period, natural settling or consolidation of soil below the crumbly surface will occur. Walking over the area for grading, raking or destroying weeds (plus occasional watering or rain) does this for you. If soil is still soft and spongy all over, rolling may be necessary. If soft in spots, tread these firmly one by one. But only roll or tread when soil is dark damp to avoid compacting it. Rake the area again after rolling or treading.

FERTILIZER

Your soil is now in good condition for sowing seed, planting sprigs or laying turf —a 12–20 mm layer of loose, crumbly, surface soil and firm soil below. This is the time to apply a base or pre-planting fertilizer. All soils for lawns need fertilizer. Having a dark, rich-looking loam is no guarantee of its fertility. Soils in high rainfall districts are deficient in phosphorus which is the nutrient most needed by developing seedlings. Phosphorus is important for new root growth of sprigs and turf too. Use a mixed fertilizer with 7 per cent phosphorus or more—Gro-Plus Complete, Gro-Plus Lawn Starter, Shirley's G5 or Q5 (Queensland). All have an approximate analysis N.P.K. 5:7:4. Avoid using lawn foods at pre-sowing or pre-planting. These are low in phosphorus and high in nitrogen, being intended for regular feeding of established lawns only.

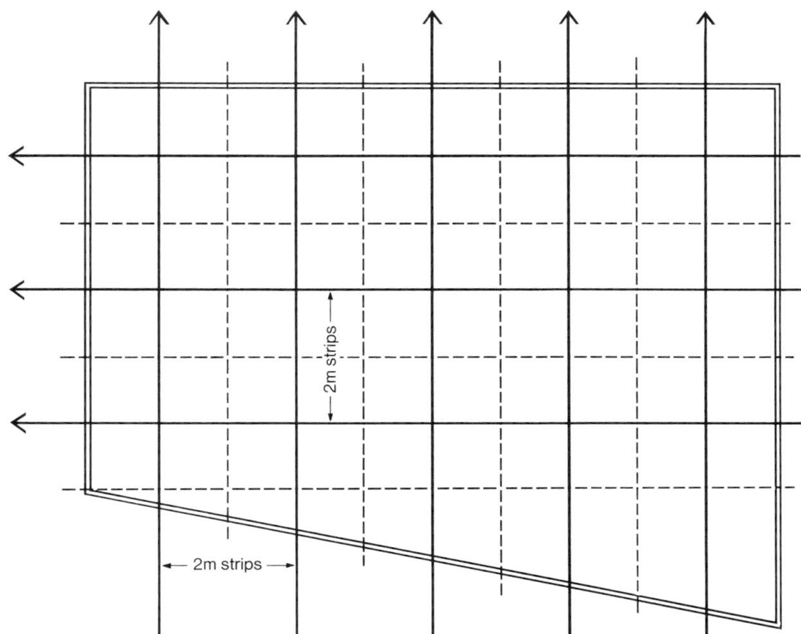

Always broadcast fertilizer or seed in two directions for even coverage. Divide the area into strips 2 m wide with garden twine and calculate the quantity of fertilizer or seed for each strip

A recommended rate for a pre-sowing or pre-planting fertilizer is 10 kg per 100 square metres. This is equivalent to $\frac{1}{3}$ cup per square metre.

SPREADING FERTILIZER

It is essential to spread fertilizer evenly. Mechanical spreaders are available but you can broadcast it very effectively by hand. Probably the best method is to divide the total quantity of fertilizer into two halves. Then spread the first half in one direction (say north/south) and the other half at right angles to it (east/west). This way you cover the area twice. You can be even more accurate if you divide the area into strips with garden twine stretched between pegs at 2-metre intervals. This is a convenient width for scattering fertilizer by hand. Then calculate the quantity of fertilizer for each strip. After spreading the first half of the fertilizer, move the twine and pegs to make strips at right angles for spreading the second lot. After spreading, water well and leave for 5–7 days. Any weed seedlings can be killed easily by raking and cross-raking.

SOWING SEED

Lawn grass seed is small and light, so choose a calm day for sowing. A good time of the day is early morning when there is generally little or no wind. Only sow an area which you can water effectively. It is better to sow small sections over a few weeks than to have patchy germination of the whole lawn because of inadequate watering. For even sowing, use the same method as suggested for spreading fertilizer. That is, sow the seed in two directions. Best sowing rates and sowing times are given in the section 'Selecting your Grass' on page 73.

After sowing, lightly rake the crumbly surface. A light roller may be used to press the small seeds into contact with the soil particles, but this is not essential. On sandy soils, a very light covering of dry, sieved mushroom compost retains moisture and helps to prevent soil washing away during rain or watering. On sloping sites, hessian spread on the surface and anchored with pieces of light timber will prevent soil wash, but remove it as soon as seeds start to germinate. Keep the surface soil moist with light watering until seeds germinate and seedlings are well established. As they grow stronger, give the lawn rather heavier but less frequent watering to encourage roots to go deeper.

PLANTING SPRIGS OR RUNNERS

This method is often used to establish creeping grasses such as couch, Queensland blue couch, buffalo grass and kikuyu. A sprig or runner is a small piece 7–15 cm long, with one or more nodes (joints) from which leaves and roots grow. Planting runners does not require as close attention as sowing seed. It is a fairly inexpensive method but time-consuming.

Prepare the soil in the same way as for sowing seed. Plant runners into dark damp soil about 30 cm apart by hand or with a trowel. Alternatively, make a furrow 5–7 cm deep, lay runners in the furrow and rake back the soil over them. In both cases, firm the soil around the roots and make sure some of the leaves are above ground. On very large areas, runners can be 'chaffed' (chopped into small pieces), broadcast and covered with a light sandy soil. Some landscape contractors will quote for the complete operation.

Only plant sections you can handle and water. Cover unplanted runners with a wet bag or hessian to prevent drying out. Keep soil moist for about a week, especially in hot dry weather. Best time to plant runners is spring and early summer, but autumn is satisfactory in subtropical and tropical areas.

LAYING TURF

Laying turf is an expensive method but provides an instant lawn with little chance of erosion or loss of planting material. It is a good method for sloping sites. Soil preparation need not be so thorough as for sowing seed or planting runners, but the surface should be crumbly and well graded. Early maintenance, especially watering, is less critical. Turfing is confined to the creeping grasses—couch, Queensland blue couch, buffalo and kikuyu—and turf supplies are available from turf farms, contractors and nurseries. Turf for Kentucky bluegrass/fescue and bentgrass/fescue mixtures is available in cooler climates like Canberra and Melbourne. Obtain your turf from a reliable supplier of 'cultivated' turf. The turf must be free of weeds such as nut grass and onion weed, and also of the 'grassy' weeds (paspalum, water couch and Mullumbimby couch).

Cutivated turf is machine-harvested and cut to an even thickness. It comes sometimes in 30 cm squares, more often in rolls 30 cm wide, which are laid in place like a carpet. Place individual squares or edges of rolls to fit snugly against each other and fill cracks between them with dry, sandy soil. Tramp or roll the turf after laying and water well. Turfing can be done at almost any time of the year but late winter or early spring is probably best. It is important to lay turf as soon as possible after delivery. The turf tends to deteriorate —especially in warm weather—if stored for more than a few days.

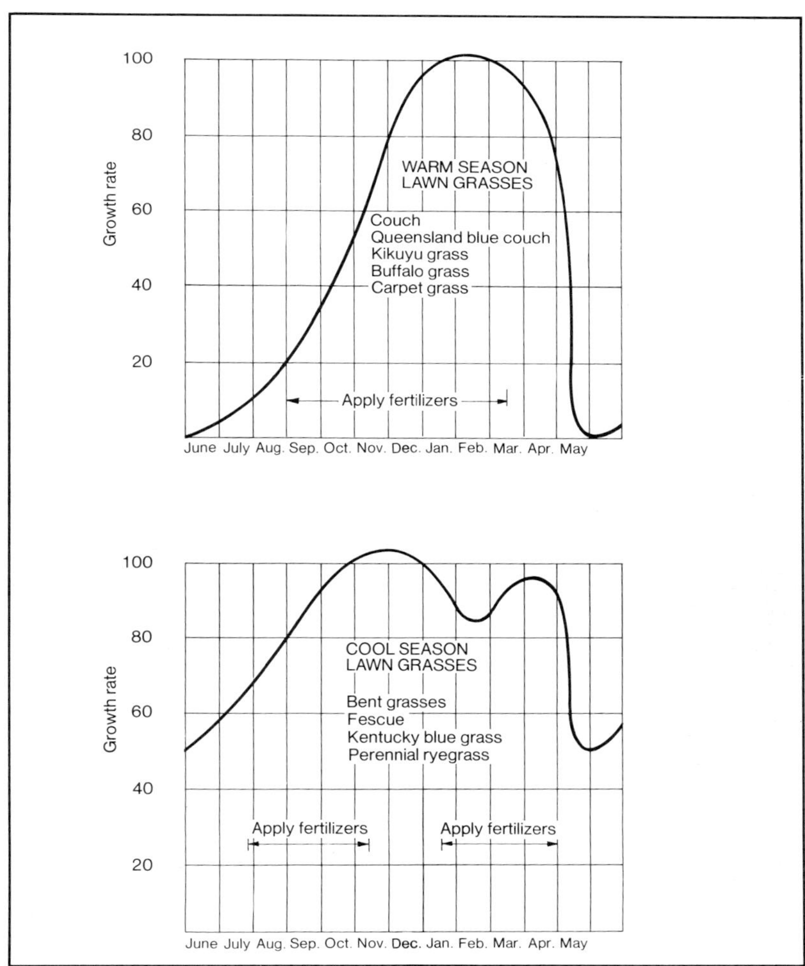

Growth pattern of popular lawn grasses

SELECTING YOUR GRASS

Lawn grasses can be divided into two groups depending on their growth pattern. Warm season grasses grow well from spring to autumn with peak growth in summer. They grow slowly or become dormant in winter. At temperatures below freezing they are frosted (but not killed) and recover in spring. Generally, grasses in this group are best suited to warm coastal and hot inland districts.

Cool season grasses are cold-tolerant, frost-resistant and stay green all year round. Peak growth is in spring with a smaller flush of growth in autumn. They are best suited to cool temperate and cold climates but are successful in warmer areas provided they are watered well during summer.

The growth pattern of these two groups of lawn grasses is shown in the diagram which tells us how they grow each month in a temperate climate like Sydney, Adelaide or Perth. This is important because it determines the time of the year when each group needs water, fertilizer and other care.

WARM SEASON GRASSES

Couch grass Couch (known as Bermuda grass in the United States) is the most popular and widely-grown lawn grass in Australia. It is not successful in cold climates such as the Eastern Highlands of New South Wales and colder parts of Victoria and Tasmania. It is an aggressive 'running' grass with strong stolons which form a dense mat of hard-wearing turf. It is heat-resistant and tolerates dry weather better than most grasses. Couch grows on a wide range of soils but responds dramatically to nitrogen fertilizers applied during the growing season. It needs full sunlight and will not grow in shade. It is dormant in winter. A couch lawn can be grown from seed, runners or turf. Sow seed in late spring or early summer at 2 kg per 100 square metres. Seedlings emerge in 10–15 days in warm weather. Sprigs or runners of couch are planted at the same time but couch turf can be laid in late winter or early spring.

Speedy Couch This is the same grass species, but the seeds of Speedy couch have had the outside husk removed with

Couch makes a hardy lawn in warm climates

the result that the number of seeds per unit weight is almost doubled and germination is faster. Sow seed at the same time of the year (late spring or early summer) but reduce the seeding rate to 1 kg per 100 square metres.

Queensland Blue Couch This lawn grass is very similar in appearance to couch grass but the stolons and leaves are rather finer and blue-green in colour. It is slightly more shade-tolerant than couch but is not as resistant to wear, close cutting and dry weather. Blue couch makes a beautiful lawn and is best suited to southern Queensland, the coast of New South Wales and southern parts of Western Australia. Seed is now available, providing a quicker method of establishment than sprigs; turf is also available. Best time to put down is spring or early summer.

Buffalo grass Buffalo grass has broad, light green leaves and forms a dense, coarse-textured lawn which resists wear and weeds. Buffalo was previously one of the most popular lawn grasses but now appears to be losing favour. It does not set fertile seed and must be established from turf or runners. Buffalo prefers full sunlight but tolerates shade rather better than most warm season grasses. It tends to become dense and spongy after a few years and requires frequent cutting and renovation to preserve a well-kept appearance. Generally suited to warm coastal climates, it tolerates heat and dry weather well. Plant runners or lay turf in spring or early summer.

Kikuyu grass Kikuyu, a native of the highlands of East Africa, is now naturalized in many coastal and inland regions of Australia. It is the most vigorous of all lawn grasses, with stout stolons and rhizomes, and for this reason has often been regarded as undesirable in home gardens. But when kept within bounds—by mower strips and the use of desiccant weedkillers—kikuyu makes an attractive, hard-wearing lawn. It stays greener in winter than other warm season grasses and tolerates partial shade, growing well to the base of trees. It revels in warm weather, tolerates dry spells but needs watering in very hot conditions. Kikuyu responds dramatically to nitrogen fertilizers and the dense turf resists weeds, insects and disease.

Previously, kikuyu was established from runners or turf but seeds of two cultivars—named Whittet and Breakwell—are now available. Sow seed at 125 g per 100 square metres in late spring or early summer. The vigorous seedlings emerge in 10–15 days. Runners can be planted in spring or summer too and turf can be laid almost any time of the year.

Carpet grass Carpet grass has become naturalized in southern Queensland (where it is often called mat grass) and coastal New South Wales. It is a vigorous grass, spreading by stolons to form a hard-wearing, weed-resistant but rather coarse-textured lawn. Like other warm season grasses it is dormant in winter. It tolerates shade fairly well and does not become as spongy as buffalo or kikuyu. Generally, carpet grass is suited to warm coastal and subtropical climates. Carpet grass is established from seed sown in late spring or early summer at 2.5 kg per 100 square metres. Runners can be planted at this time of the year too.

Carpet grass is more correctly known as Narrowleaf carpet grass to distinguish it from Broadleaf carpet grass, a related species which is widely used for lawns in tropical regions, including North Queensland and Darwin. The latter is established from runners as no seed is available.

COOL SEASON GRASSES

Bentgrass Bentgrass (also known as New Zealand bent, browntop or highland bent) is the most widely grown cool season grass, either alone or in grass-seed mixtures. It makes a beautiful, evergreen, fine-textured lawn and is preferred for top-class bowling and golf greens. It is best suited to cool temperate or cold climates but will grow in warmer areas with plenty of water through summer. Bent grass prefers full sunlight but will tolerate partial shade. It is resistant to wear when well established. Sow seed in early spring or autumn at 1.5–2 kg per 100 square metres. Seedlings emerge in 7–10 days—sometimes sooner in favourable weather. You can over-sow couch lawns with seed of bentgrass to provide winter green. Give a light top-dressing and a ration of mixed fertilizer in autumn. Sow seed at 1 kg per 100 square metres and water regularly until seedlings are established.

A soft lawn is a perfect place for relaxation

Kentucky Bluegrass Kentucky bluegrass is hardier than bentgrass and makes an attractive lawn. It withstands extreme cold and frosts in climates like Canberra, where it is often sown with white clover, strawberry clover or other cool season grasses. It also does well in warmer districts of south-western New South Wales, northern Victoria and South Australia, where it is used in mixtures with couch grass and strawberry clover. Kentucky bluegrass lawns are more resistant to wear than most cool season grasses, but thin out when poorly grown or cut too closely. Apply fertilizer in both spring and autumn (but not in summer when it is dormant). Lenient mowing at a cutting height of 2.5–4.0 cm is strongly recommended. Sow seed in spring or autumn at 2 kg per 100 square metres. Seedlings emerge in 15–20 days. In Kentucky mixtures which contain couch grass too, make spring sowings only.

Chewings Fescue This grass is seldom grown alone as a lawn species but it is widely used in lawn grass mixtures with bentgrass, Kentucky bluegrass and couch. Seeds germinate in 7–10 days and seedlings grow quickly. This gives good protection to slower seedlings in the mixture. Fescue prefers full sunlight but will tolerate partial shade. It needs lenient mowing and is a good companion grass with Kentucky bluegrass for this reason.

Perennial Ryegrass Perennial ryegrass forms a rather open turf when sown alone. For this reason it is usually combined with bentgrass and fescue. Ryegrass mixtures are sown at heavier rates to give an attractive, hard-wearing lawn. Ryegrass mixtures are less expensive than those containing 'fine' turf grasses only. They are popular for play areas, swimming pool surrounds and nature strips. They are widely used to give quick cover for large-scale planting on playing fields and parks. Perennial ryegrass has deep roots and withstands heat and dry weather better than bentgrass and fescue, but requires lenient mowing—4 cm cut—for best results.

LAWN GRASS MIXTURES

A wide range of lawn seed mixtures are available in home garden packs or in bulk for large areas. It is important to select a mixture containing grasses best suited to your climate.

For hot, dry climates mixtures containing couch are best. Florida, the most popular blend, contains couch and bentgrass for an evergreen lawn the year round. Landscape, a special blend for dry summer conditions in South Australia, contains couch, Kentucky bluegrass and a small proportion of strawberry clover. Mixtures which contain mostly bentgrass are best for cool temperate and cold conditions. Canberra blend contains a high proportion of bentgrass with some Kentucky bluegrass. Kentucky blend and Kentucky Strawberry blend both contain less bentgrass and more bluegrass. Fescue is a small component of all three. For areas with wet winters and hot dry summers, like Adelaide and Perth, mixtures such as S.A. Hardwearing and Westralia are best. Both contain couch blended with Kentucky bluegrass. Mixtures containing perennial ryegrass are economical and give a quick cover for less formal lawns. Green Valley, the most popular, contains bentgrass and ryegrass. Park Mix contains a large proportion of ryegrass with small amounts of couch and bent.

A new and interesting addition to the lawn grass mixtures range currently available is 'Sun and Shade'. Its constituents make it suitable for growing in full sun or full shade and it will solve many a garden area problem.

Two of the major constituents of the mixture are Derby fine leafed perennial Ryegrass and Poa trivialis. Derby Ryegrass copes well with a wide range of temperatures, from cold to hot, shade to full sun. Poa trivialis does well in damp shady conditions and makes a dense deep green turf.

Seed rates for lawn grass mixtures vary slightly according to seed size of the components. Usually this is 3–4 kg per 100 square metres.

CARE OF NEW LAWNS

WATERING

Lawns started from seed need frequent watering until grass is growing strongly. When the grass is about 2 cm high, allow the soil to dry out for a few days to encourage roots to go deeper. Then give a good watering and repeat the process. Unless the weather is very hot, you will find that watering may now be reduced to one good soaking each week. Always aim to wet the soil thoroughly to a depth of 15–20 cm with each watering. Lawns from planted runners need care in watering, too, because there is a lot of bare soil between individual plants. Again, encourage deep rooting by a few days' spell between waterings. Lawns from turf get by with a good soaking each week after laying.

MOWING

Don't allow grass to grow too high before you cut it. When the grass is 4–5 cm tall, mow it with the blades set at 2.5 cm above the soil. This will encourage new shoots at the base of each plant. The next three or four cuts should be repeated at the same height—a light 'topping' only. Then reduce the cutting height gradually to 1 cm for once-a-week mowing.

FERTILIZER

The pre-sowing or pre-planting fertilizer will keep the new lawn green and vigorous for eight to ten weeks after sowing or planting. At this stage, the need for nitrogen will be increasing. Lack of this nutrient will cause yellowing of leaves and a generally unthrifty appearance. Nitrogen fertilizers, such as sulphate of ammonia or Nitram, can be used but a lawn food such as Gro-Plus for Lawns is recommended for regular application. Lawn foods contain about 17 per cent nitrogen. Should young grass seedlings be stunted and show a reddish-brown or purple pigment in the leaves, this points to a phosphorus deficiency. It may occur only in patches due to uneven spreading of the base fertilizer. Correct it by watering each patch with one of the soluble fertilizers—Thrive, Aquasol or Zest.

LAWN MAINTENANCE

For good maintenance, you must consider the kind of grass and its growing season. Refer to the diagrams of warm season and cool season grasses earlier in the chapter.

WATERING

There are no set rules on when to water or how much water to use. This will depend on the kind of grass, soil type and weather conditions. Bentgrass and other cool season grasses need more water than tougher summer growers like couch, especially in hot weather. Heavy soils hold moisture well and may only need a good soaking once a week in summer, but sandy soils

Concrete blocks make an effective margin between lawn and flower beds

may need watering every day or two under the same conditions.

Always encourage deep rooting by thorough watering. Frequent surface watering makes for soft, sappy growth which is more prone to disease. Generally, it is best to have a fairly dry surface soil and dark damp soil below. Check this by removing a square or plug of lawn to see how far water has penetrated. The plug can be replaced without damage to the lawn.

Early morning is the best time to water lawns because the sun will dry out the surface during the day. Watering in the evening or at night creates a high humidity layer in the lawn. This also favours the spread of disease. A very common mistake is too much water, especially with couch and other summer growers which are dormant in winter. Watering at this time of the year does nothing for the lawn and very often encourages annual weeds like winter grass to germinate and grow.

FERTILIZERS AND LIME

Only fertilize lawns when they are growing or starting to grow actively. This is the time they make best use of it. Use lawn foods which are high in nitrogen, because this is the nutrient removed in greatest amount in grass clippings. Apply them little and often at 2–2.5 kg per 100 square metres every five or six weeks during the growing season: for summer growers, from spring through to autumn; for cool season grasses, from late winter to early summer and again from late summer to autumn. Except in tropical and subtropical climates with very mild winters, fertilizing lawns in colder months is unnecessary and wasteful.

When applying fertilizers make sure to spread them evenly from end to end and again from side to side, as suggested previously. Water them into the lawn as soon as possible to avoid fertilizer burn. Some lawn foods are soluble and may be applied with a watering-can or through hose sprayers. Continuous use of sulphate of ammonia, or of lawn foods containing it, increases soil acidity. Lawns of couch and Queensland blue couch do not thrive on acid soils, although bentgrass is less particular. The degree of acidity can only be determined accurately by a soil test but in high rainfall districts, a general recommendation is to apply lime at 25 kg per 100 square metres every second or third year. Avoid using too much lime, however, as this encourages clover, weeds and earthworms.

MOWING

Always mow lawns regularly at a constant cutting height. This develops a balance between the shoot and root system. Mowing too closely results in shallow roots and weakens the turf, which in turn encourages weeds. Close mowing of couch in autumn causes premature browning and interferes with food storage in the stolons on which spring growth depends.

For most home lawns, unless they are used for bowls or tennis, a cutting height of 1–2 cm is suitable. A slightly higher cut of 2.5 cm helps to lower soil temperatures in summer. Kentucky bluegrass needs this higher cut at all times. In the growing season, mow once each week, but during winter mowing every four or five weeks is sufficient. Remove the clippings to improve the appearance of the lawn and provide material for compost. Removing the clippings also helps to prevent a spongy surface developing—especially in running grasses.

RENOVATION

Renovation is any mechanical treatment to improve the surface of lawns and to allow free entry of air, water and nutrients. Always renovate lawns when they will recover quickly—in spring for summer growers, in late winter or early autumn for the cool season grasses. 'Mat' or 'thatch' is an undecomposed layer of old roots and runners which acts very much like a thatched roof and prevents the entry of air, water and fertilizers. To get rid of it, cut the lawn closely and rake severely. Then cut again and repeat the raking until there is a bare cover of grass over the soil. Follow with a ration of fertilizer and water well.

Bare patches may occur in compacted heavy soil, or perhaps as a result of severe traffic. Compacted soil, like thatch, prevents the entry of air, water and nutrients. Give the soil a thorough soaking and use a garden fork to penetrate the soil 10–15 cm deep at the same distance apart. Work the fork back and forth to enlarge the holes. Special hollow-tine forks can be bought for this coring treatment. Spiked rollers are best for very large areas.

TOPDRESSING

Fortunately, topdressing is no longer the annual ritual advocated by many people in the past. The main purpose of topdressing is to correct any unevenness in the lawn surface, so it is still very important in turf used for cricket, tennis, bowls and golf putting greens. It has little place in home lawns once the correct levels and grades are made. Topdressing is useful as a light soil covering after de-thatching or coring, and also for over-sowing bentgrass on couch lawns in autumn.

Topdressing soils are usually light sandy loams and provide little in the way of nutrients, that is, they are no substitute for fertilizers. If topdressing is used for any reason, spread it thinly with the back of a rake or a dummy rake. Do not bury the grass completely, as deep covering will retard it. After rubbing-in, the tips of the grass should show through the soil. Usually, a bucket of topdressing soil per square metre is sufficient. Apply a mixed fertilizer at 3 kg per 100 square metres. This is equivalent to $1\frac{1}{2}$ tablespoons per square metre.

WEED CONTROL

A healthy, vigorous turf which is difficult for weeds to invade is the first line of attack in weed control. This depends largely on good maintenance. Adequate and regular fertilization is essential. A poor soil means a poor, open turf which invites weed invasion. Fertilizers, especially sulphate of ammonia and lawn foods containing it, discourage weeds, especially clovers and flatweeds. On the other hand, fertilizers high in phosphorus and organic fertilizers like blood and bone favour clover and weed growth. Correct watering and regular mowing are important factors in having a weed-free lawn. A wide range of weedicides can be applied to lawns as selective sprays. These chemicals, together with those to control insects and diseases in lawns, are discussed in Chapter 17.

Chapter 8

HOW TO GROW VEGETABLES AND HERBS

The benefits of growing your own vegetables need no emphasizing. Of all gardening activities, to be able to harvest fresh, vitamin-packed vegetables for salads or the kitchen pot is one of the most rewarding. Growing vegetables is an intensive form of cultivation, but it is very satisfying to plan a programme of small, successive sowings and plantings for a continuous harvest throughout the year. To do this, you should know when and how to grow different vegetables, how long each kind will take to mature and what yield to expect.

Vegetables are often divided into three groups depending on the part of the plants we eat:

1. *Fruit and seed vegetables*—beans, peas, capsicum, eggplant, tomato, sweet corn and vine crops (cucurbits).
2. *Leaf and stem vegetables*—cabbage, celery, lettuce, rhubarb, silver beet and spinach. Broccoli and cauliflower are included in this group, too, although the part we eat is the flower bud and not the leaves or stems.
3. *Root and bulb vegetables*—beetroot, carrots, onions, parsnips, potatoes, radish and turnips.

But whatever kind you grow, you should aim to promote young, growing tissue. To do this, all vegetables must be grown quickly—so there is not a great deal of difference in the way you grow them. You do not need different soil for different vegetables. If you can grow good tomatoes there is no reason why you cannot grow good beans, cabbages or carrots, too. However, the grouping of vegetables into fruit, leaf and root plants does give good guidelines for fertilizer use. Fruit and root vegetables need large quantities of phosphorus in fertilizer, because this element stimulates flowers, fruits and seeds and root development. Fertilizers high in nitrogen may produce too much leaf growth and reduce yields of fruits and seeds. On the other hand, nitrogen fertilizers are needed in greater quantities by leaf vegetables like broccoli, cabbage, cauliflower, lettuce, silver beet and spinach. The fertilizer needs of vegetables are partly dictated by the part of the plant we eat.

COOL AND WARM SEASON VEGETABLES

A most important point in growing vegetables is to know their requirements for cold and warmth. Some vegetables are disappointing if they are planted out of season.

Cool season vegetables grow best at low temperatures of 10–20°C. They tolerate even colder conditions and usually are frost-resistant. This group includes broad beans, broccoli, Brussels sprouts, cauliflower, onions, peas, spinach and turnips, which are sown to grow during the cooler months of the year.

Warm season vegetables grow best at a temperature of 20°C or above. They grow poorly in cool weather and are frost susceptible. This group includes beans, capsicum, eggplant, potato, sweet corn, sweet potato, tomato and all the vine crops. They are sown in spring or early summer to grow during warmer months.

A third group have intermediate temperature requirements and grow best at 15–25°C. This group includes beetroot, cabbage, carrot, celery, leek, lettuce, parsnip, radish and silver beet. It is important to sow them at the correct time of the year for your climate because they tend to 'bolt', or run to seed, if they are sown too early or too late. Root crops like beetroot, carrot and parsnip may run to seed if sown too late in autumn or winter. Silver beet, a close relative of beetroot, may do this too. Some varieties of lettuce run to seed if sown in warm weather, so choose those varieties which have been selected for sure-heading in summer. The best months for sowing each kind of vegetable in each climate zone—tropical/subtropical, temperate and cold—are shown in the Sowing Guide on pages 116–121. A map of climate zones in Australia is shown in Chapter 1.

NEED FOR SUNLIGHT

An open, sunny site is a must for your vegetable garden. To grow quickly, vegetables need as much sunlight as possible, especially in winter when days are shorter. If possible, select a part of the garden facing north to north-east to catch the morning sun and at least four or five hours direct sunlight each day. Make allowance for longer shadows in winter. Try to avoid shade from buildings, fences, or large trees and shrubs. Trees and shrubs with large root systems will compete for moisture and plant nutrients as well as light. A level site is best and easy to manage. Both beds and rows are best running north-south. This way each plant in the row receives maximum sunlight. On sloping sites, garden beds should run across the slope with a retaining wall of timber, brick or concrete on the downhill side to prevent erosion and loss of soil. In this case it may be best for rows to run across the beds instead of along them.

SHELTER FROM WINDS

Vegetables need some protection from wind. Cold winds slow down growth and hot, dry winds evaporate enormous amounts of moisture from soil and plants. Strong winds also damage leaves and stems and may loosen and weaken roots. All these effects of wind interfere with rapid and continuous growth. Windbreaks of trees, shrubs or hedges growing on the south or west of the vegetable garden do not create shade problems and can be planted well back from the garden to give good wind protection. Artificial windbreaks of lattice, slatted timber or light brush fences make an excellent windbarrier. They reduce wind velocity without creating turbulence. Brick or concrete walls (with about 50 per cent opening) have this effect too (see diagrams showing effects of windbreaks in Chapter 1). Windbarriers on the south offer a bonus, too, by providing support for climbing beans and cucumbers in summer and climbing peas in winter.

SIZE AND LAYOUT

The space available and your own enthusiasm will determine the size of your vegetable plot. Generally, it is best to have beds 150 cm wide with paths 30–40 cm between each bed. This allows you to work from both sides. This width is doubly convenient because it will accommodate three plants of broccoli, cabbage or tomato across the bed or three rows of beans, peas or sweet corn. With smaller, more upright growers such as beetroot, carrots, lettuce and silver beet you can fit in five or six rows across the bed. Beds may be of any length but 9 m is a convenient length which can be divided into 3 m sections for successive plantings. To provide year-round vegetables for a family of four you need about six beds, each 9 m by 150 cm—a total area of about 100 square metres. For new gardeners, a smaller area of three beds is suggested. It is better to look after a small garden well than to have a large garden which may get too hard to manage. You can always increase the area by adding extra beds or making them longer.

SOIL FOR VEGETABLES

The main requirement of a soil for vegetables is a good physical condition. It should have a loose, crumbly structure which is capable of absorbing and holding water and nutrients but it should be well aerated and drain easily.

As described in Chapter 3, you can improve sandy soils easily by adding moisture-holding materials—animal manure, mushroom compost, garden compost, vermiculite or peat moss. The crumb structure of heavy soils also benefits from organic matter and their texture is improved by adding coarse sand. Always dig heavy soil when dark damp. For most vegetables, including root crops, dig soil to 25–30 cm deep but do not bring subsoil to the surface. Drainage is usually necessary on heavy soils too. Raise each bed 15–20 cm by shovelling soil from the pathway onto beds on either side. Slope the edge of the beds to 45–60 degrees. This allows downward and lateral movement of water. Beds can also be contained by flat boards held in position with stout pegs or surrounded by bricks or concrete —but make provision for weep holes for drainage.

FERTILIZERS AND LIME

It is best to apply a pre-sowing or pre-planting mixed fertilizer (such as Gro-Plus Complete, with an approximate analysis N.P.K. 5:7:4) for each vegetable crop you grow. The high phosphorus content will ensure vigorous seedlings and good root development. Broadcast the fertilizer, apply in bands or scatter in furrows as described in Chapter 5. Root and bulb vegetables will produce a good crop without side dressings of additional fertilizer, but fruit and seed vegetables benefit from a side dressing of mixed fertilizer when flowering commences. Leaf vegetables demand extra nitrogen in the form of sulphate of ammonia or liquid feeds of Thrive, Aquasol or Zest every 10–14 days while they are growing.

What about lime for vegetables? Most vegetables and herbs grow successfully

Patio Pik cucumber. Good in the garden or in tubs

Herbs are popular and fun to grow

in soil which is slightly or very slightly acid (pH 6.0–7.0). A few (potatoes, sweet potatoes and watermelon) will tolerate a strongly acid soil. In high rainfall districts near the coast it is a good idea to add agricultural lime to vegetable beds every year or two, but there is no need to apply lime before every crop. Too much lime can be harmful (see section on pH and Lime, Chapter 4). A good time to apply lime is after summer crops have finished and before cool season crops are established.

CROP ROTATION

The only reason for crop rotation in the vegetable garden is to prevent the spread of diseases (or insect pests) which may attack vegetables belonging to the same family or group. If disease in any crop is serious, it would be unwise to plant a closely related crop in the same bed. The theory that different vegetables absorb different kinds and amounts of nutrients was started by Viscount Townshend in the eighteenth century and has been perpetuated ever since. The availability of modern fertilizers and regular application of organic matter to the vegetable garden has made this theory out of date.

Complicated charts and diagrams for crop rotation are rather confusing and seldom work out as planned. In Australia, there is a natural rotation between warm and cool season crops anyway. Providing you apply fertilizer and organic matter and are aware of the possibility of disease carry-over to related plants, crop rotation can be forgotten.

WATERING AND MULCHING

The general principles of watering and mulching have been discussed in Chapter 3. To grow quickly, vegetables need adequate water at all times. Most of the water absorbed by vegetables passes through the plant and is evaporated or transpired by the leaves. In hot dry weather, leaf vegetables may lose several times their weight in water each day. Vegetables use water most easily when the soil is at or near field capacity, which means that sandy soils need more frequent watering than heavy soils. Once vegetables are established, a good soaking will encourage deep roots. This helps them to withstand dry conditions for longer periods.

Sprinklers with a slow application rate and fine droplets are the best for vegetables. For watering the whole garden, Selecto sprays, which can be varied to throw from a few degrees to a full circle, are useful. A vegetable garden can be watered by placing the spray (set at 90 degrees) at each corner in succession. More often, single beds or sections of beds will need water, so a hose with adjustable nozzle is preferable. Soakit hoses are also useful for slow watering and can be laid along the rows. Furrow irrigation is also suitable for flat beds, or for those with a gentle slope, especially for watering tomatoes, potatoes and other crops where it is desirable to keep the leaves dry to lessen the spread of leaf diseases. Make a deep furrow, with a gradual fall to one end, between the rows of plants. Water with a slow hose until the furrow is completely filled with water.

Mulching vegetable beds, especially in summer, can greatly reduce loss of soil moisture. It also gives a more even temperature and discourages weeds. Of the many mulching materials, the best for vegetable beds are garden compost, well rotted animal manure or lawn clippings. The mulch can be dug into the soil when preparing for the next crop.

CHOOSING YOUR CROPS

Naturally you will choose vegetables which you and your family like. But some take up more space for the yield they give. Yield for the area occupied is a good reason for growing a particular vegetable. Climbing beans, climbing peas and cucumbers grown on a trellis use vertical space but little soil space. Tomatoes and capsicum give high yields for space occupied.

The cut-and-come-again vegetables—broccoli, celery (pick green outside leaves progressively), rhubarb and silver beet—are good value for continuous harvesting. Salad vegetables and leaf crops, which lose quality quickly after harvest, are excellent in the home garden because of their extra flavour and freshness.

New dwarf or bush varieties of some vegetables are now available—dwarf or mini tomatoes, bush marrow (zucchini), bush pumpkin and bush squash. They take less garden space and many can be grown in tubs or large pots on a sunny balcony or patio.

A selection of the best home garden vegetables—yield for space occupied—is given below:

- Beans (dwarf and climbing)
- Broccoli
- Brussels sprouts*
- Cabbage
- Capsicum
- Carrot
- Cauliflower*
- Cucumber (on trellis)
- Lettuce†
- Marrow (bush)
- Onion
- Peas (climbing)
- Parsnips
- Pumpkin (bush)
- Radish
- Rhubarb
- Silver beet
- Spinach
- Tomato
- Turnip

* for temperate and cold climates
† select suitable varieties for summer and winter

GENERAL HINTS FOR VEGETABLE GROWING

SUCCESSIVE SOWINGS

Some home gardeners, particularly new ones, sow or plant too much at the one time. Beds are quickly filled and when harvest arrives there are too many vegetables to eat. For a continuous satisfactory supply, make small successive sowings. Always have an empty bed or section of bed in preparation for the next sowing or planting.

TALL VEGETABLES

Plant tall crops (tomatoes, sweet corn, broad beans) on the southern end of beds to prevent shading of low-growing vegetables. If making successive sowings of any crop in the one bed, say, dwarf beans or sweet corn, use the southern end first so that subsequent sowings get full sunlight.

GROUPING VEGETABLES

If possible try to group together vegetables which grow to the same height and mature about the same time (carrots, leeks and parsnips). This gives each a fair share of sunlight and after harvest the whole section can be dug and prepared for the next crops. Root and bulb crops are usually easier to handle in long rows of 2–3 m.

LONG-STANDING CROPS

Plant perennial crops, e.g. asparagus and rhubarb, in a separate section or at one end of a bed where they can grow undisturbed. Crops with an extended harvest period (capsicum, celery (picked green) and silver beet) will also occupy space for two or three months after your first picking.

MAKE WAY FOR NEW CROPS

Always pull out and compost crops which have passed their best as soon as possible. Why keep a whole row of beans or peas for just a few pods? It is best to use this unproductive space by planting something else.

HOW TO GROW INDIVIDUAL VEGETABLES

ARTICHOKES

The globe artichoke is a grey-green thistle-like plant which grows to a height of 1 m or more. It takes up a lot of space in the vegetable garden but you may find an odd corner where it can flourish undisturbed for three or four years. Often a few plants can be grown in a sunny spot in the flower garden. The best climate is one with a mild winter (no frosts) and a cool summer.

Seeds of globe artichoke may be sown in spring. The seedlings are very variable, however, so it is best to start plants from shoots or suckers. Plant these in late winter in cool districts or in autumn where it is warmer. Shoots should be about 30 cm in length with well-developed roots. Plant them 1 m apart each way and rake in a small handful of pre-planting fertilizer in a circle around each plant.

The globe-shaped buds appear in early spring and plants keep bearing until November. After flowering, keep plants watered and mulched through summer. In autumn cut plants back to 30 cm high and apply a side dressing of mixed fertilizer. Add animal manure if available or compost to maintain good soil structure. Prune back to the four or five strongest shoots in winter for buds next spring. Plants will bear well for three or four years, after which they should be divided and replanted. Harvest buds when 5–10 cm in diameter and still tight and tender. For larger main buds, prune out lateral buds when about the size of a golf ball. Three to five plants should be sufficient for an average family.

The Jerusalem artichoke is really a large sunflower with tuberous roots like a potato. It is a perennial, but is grown as an annual from tubers planted in late winter or early spring. Plant tubers 10–15 cm deep and 50–60 cm apart each way with a pre-planting fertilizer scattered in a circle around each. The plants produce yellow flowers in summer, but pinch them out in the bud stage. Tubers are ready four to six weeks after buds appear, but can be left in the ground until winter if necessary. Nine to twelve plants should be sufficient for the average family.

ASPARAGUS

Asparagus is not a difficult crop to grow. Once established, the plants are very long-lived and keep producing for twenty years or more. Asparagus is best suited to mild or cold climates. Frosts are no problem because the plant (often called asparagus fern) dies off each winter to produce new shoots or spears in spring. Light soils, through which the spears can push easily, are preferred.

Sow asparagus seed in spring in a seed bed in a corner of the vegetable garden because seedlings must be two years old before planting in their permanent position. Asparagus is the only vegetable with male and female flowers on separate plants. Male plants produce bigger and better spears, so female plants with red berries in the second autumn from sowing should be discarded. Some gardeners prefer to buy two-year-old crowns from nurseries in winter. Mary Washington is the standard variety.

Before planting crowns prepare the bed to spade depth and add a pre-planting fertilizer plus liberal quantities of organic matter for good soil structure. Set crowns 15–20 cm deep and 30–50 cm apart at the bottom of a trench along the row. Cover with about 5 cm of soil, filling in the remainder as the fern grows. Do not cover new shoots. Water regularly and give liberal dressings of high-nitrogen fertilizer (N.P.K. 10:4:6) in summer to encourage vigorous top growth. Cut down the dry, yellow fern in winter to ground level and rake it up for burning. Give another fertilizer ration in late winter to encourage spears in spring. Do not cut any spears in the first spring after planting. Cutting can increase each year as plants grow older and reach full bearing in four or five years. Each year start cutting when spears appear (August or September, depending on district). Harvest every day or two and continue cutting for eight to ten weeks.

'Green' asparagus is cut from level beds when spears are 15–20 cm long and before the tips open. For 'white' (or blanched*) asparagus, hill the soil over the row to a depth of 25–30 cm in late winter. As the tip of the spear breaks the surface, push a sharp knife through the soil to cut the spear about 15 cm below. Hills may be levelled when fern is removed in winter. Cultivate and fertilize, then rebuild the hills before spring. Whether you grow green or white asparagus, 20–25 plants are ample for the average family. Spears keep well for several days in the crisper tray of the refrigerator or, after washing and blanching* for three minutes, can be packaged for the home freezer.

BEANS

Packed with vitamins and easy to grow, beans are among the most popular of all home garden vegetables. They produce outstanding yields for the space occupied. Beans are warm season vegetables and susceptible to frost. They can be grown all year round in warm northern regions. In temperate areas the growing season is five to six months, but in cold climates only three or four months. Beans do best on well drained soils, but may suffer on very sandy soils, so add organic matter to improve water-holding capacity. On most soils, mulching with compost or grass clippings is recommended in very hot weather. Beans are also susceptible to wind damage so protect them with wind breaks. Dwarf beans are ready to pick in 8–10 weeks. Climbing beans are great space-savers, and yield more pods over a longer period, but take 10–12 weeks to picking. The cultivation of both dwarf and climbing beans is the same.

Dusting seed with a fungicide before sowing in recommended. Sow dwarf beans in rows 50–60 cm apart, spacing seeds 7–10 cm along the row. A row 3–5 m long at each sowing is suitable for the average family. Make the next sowing when plants in the previous crop develop their first true leaf.

Sow climbing beans to grow on a fence or trellis, spacing seeds 10–15 cm apart. You can make a simple trellis by stretching two wires between steel droppers or timber posts—bottom wire at 15 cm and top wire at 2 m. Use garden twine between

*The word 'blanching' has two separate meanings. First, it is used for keeping the stems of vegetables such as asparagus and celery white by excluding sunlight. Second, it is used in quick freezing for 'fixing' the colour of vegetables with boiling water or steam before they are frozen.

the bottom and top wires for the bean plants to climb. A row 1–3 m is suitable for the average family. Another way is to form a tripod or A-frame of garden stakes and sow 2–3 seeds at the bottom of each stake.

When sowing dwarf or climbing beans, always apply a pre-planting fertilizer in a band, as suggested for sowing large seeds direct in Chapter 5. Be careful not to place bean seeds in direct contact with fertilizer because they are susceptible to fertilizer burn. If you press bean seeds into dark damp soil they absorb sufficient moisture to germinate. Avoid watering for a day or two after sowing and do not soak seeds in water overnight. This may hinder rather than help germination.

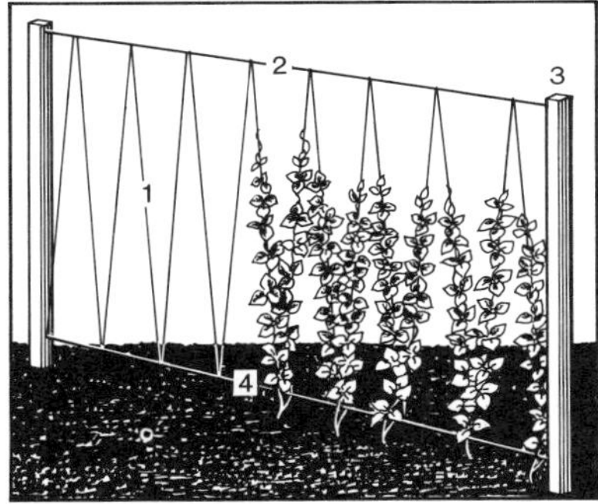

1. Garden twine. 2. Top wire 2m from ground.
3. Hardwood post or steel dropper.
4. Bottom wire 15cm from ground.

A simple trellis for climbing beans

A tripod of garden stakes for climbing beans. Sow 2–3 seeds at the bottom of each stake

Bean plants usually give good yields without additional fertilizer as they grow, but a side dressing of nitrogen fertilizer or liquid feeds of water soluble fertilizers will increase yield and quality if applied when flowering commences. Always pick beans when young and tender—before the seeds swell to make the pods lumpy. After the first picking the pods mature quickly, so pick them at least every 5–6 days. This gives better quality and prolongs flowering, too. Use beans when fresh. If possible pick and cook the same day, although they keep well in the refrigerator for a few days. For quick freezing, wash and prepare pods as for cooking, blanch for three minutes, drain and package.

There are many varieties of beans—dwarf, climbing, string, stringless and wax-podded (butterbeans).

Dwarf string beans Royal Windsor, a selection from Windsor Longpod, is one of the best string varieties for the home garden. It bears prolific crops of long, straight, fleshy pods. Brown Beauty is a good hot weather variety and bears a heavy crop over a period of three weeks or more when picked regularly. Hawkesbury Wonder is also popular. It sets pods in cooler weather and is excellent for sowing late in the season. All varieties of string beans are almost stringless when picked young—the pods need 'topping and tailing' only.

Dwarf stringless beans Pioneer, previously known as Redlands Pioneer, is an excellent quality, fleshy bean which is recommended for successive sowings throughout the season. It is resistant to bean rust, a fungus disease which attacks susceptible varieties in hot, humid weather. Snapbean has round stringless pods even when mature. Pods are often cross-cut rather than sliced. Gourmet's Delight (also rust resistant) is a new stringless variety similar to Pioneer but better suited to sowing late in the season.

Wholepod is a sensational new bean which produces a prolific crop of slim,

Wholepod is a new bean which may be cooked whole

10 cm long pods. The bean is of top quality, stringless, and does not need to be sliced. To prepare for cooking simply top and tail it and cook whole. Ideal for freezing. The dark green pod contains small seeds and is borne on strong compact bushes. Pick them young to maintain continuity of crop.

Dwarf Snake is another stringless bean. Snake beans belong to a separate species closely related to cow peas. They are often called Yard Long or Asparagus Beans because the light green, rounded pods grow to 30 cm long or more. Pods are stringless but must be picked young. There is also a climbing variety of snake bean. Both dwarf and climbing types are best suited to tropical or warm temperate conditions.

Bountiful is the best of the dwarf wax-pod (butter) beans. The tender, fleshy, stringless pods are rich yellow in colour.

Climbing beans Purple King (Blue Coco) is an old garden favourite. The long, dark purple, rather flat pods turn green when cooked. It bears over a long period but is rather susceptible to rust in warm humid weather. A recommended home garden variety. Epicure is practically stringless and has been popular for its flavour and well-shaped pods over many years. Westralia (rust resistant) has a strong vine and dense foliage. It has dark green, flattened pods up to 25 cm in length. Blue Lake is the best stringless climbing bean and renowned for its flavour. It bears a heavy crop of long, round pods and is especially suited to cool, highland climates and southern states.

Climbing runner beans Runner beans are perennial plants which die back after cropping in summer but grow again from the crown the following spring. The pods are broader and shorter with a rough texture but are still tender when cooked. They are popular in cool districts and are grown extensively in Victorian and Tasmanian home gardens. The flowers are large and ornamental. The most popular varieties are Scarlet Runner (brilliant red flowers) and White Dutch (white flowers).

Dried beans Most varieties of beans can be used as dried beans. Allow pods to ripen on the bush or vine. When dry, shell the seeds which can be used in soups, stews and bean salads. Two varieties of French beans grown specifically for this purpose are Borlotti (Pinto, Italian or Cooker bean) with speckled yellow and red seeds, and Canellini (Red Kidney bean). Other beans in this dried bean group are lima bean (harvested before the seeds ripen and shelled like peas) and soybeans, which have been used in Asian countries for thousands of years to make sauces and beverages, and as a substitute for milk and cheese. Soybeans have a very high protein content.

BEETROOT

Beetroot is an attractive and tasty cooked vegetable for salads and for pickling. Beetroot is adaptable to all climate zones but plants may bolt (run to seed) if sown out of season. Sow seed from July to March in temperate climates but only from September to February where colder. In warm tropical areas sow beetroot in most months of the year, although sowing during the wet season can be risky.

The 'seed' is a cluster of 2–4 true seeds in a corky cluster which absorbs water slowly, so it's a good idea to soak seed for a few hours before sowing. Apply a pre-planting fertilizer in a band where the seed is to be sown and rake into the soil. Dust seed with fungicide and sow thinly in drills 20–30 cm apart and 12 mm deep. Cover seeds with compost or vermiculite and spread a light mulch of grass clippings over the bed. Keep damp with light watering until seedlings emerge in 10–14 days. Thin seedlings (two or three seed-

Beetroot is best when home grown

lings may emerge from each cluster) early to 3–5 cm apart and later, when roots start to form, to 7–10 cm apart. Roots grow at or slightly above soil level so do not cover them with soil when cultivating between rows. Beetroot is best when grown quickly and responds to liquid feeds of Thrive, Aquasol or Zest. Roots are ready to pull about ten weeks after sowing. Start pulling alternate roots early. This spreads the harvest and roots left in the soil gain in size. Sow successive rows 2–3 m long every 4–6 weeks during the season. Beetroot keeps well in open storage for a few weeks but it will keep for two or three months when stored in the refrigerator.

Varieties Derwent Globe is a deep, round beet with flesh of good texture and flavour. Early Wonder matures more quickly but may develop white rings (zoning) if checked in growth. The roots are wide and rather flat. Golden Apollo is a novelty variety with deep yellow roots but a similar texture and flavour to red beet.

BORECOLE

Borecole (also called Scotch Kale or Curly Greens) is a loose-leaf member of the cabbage group of plants. It is widely used in the British Isles and Northern Europe. It is very hardy and withstands severe winter conditions, but is not grown to any extent in Australia. Borecole is grown in exactly the same way as cabbage. It can be used by harvesting the outside leaves progressively, like spinach or silver beet, or by cutting the whole plant at once.

BROAD BEANS

Unlike other beans, broad beans are a cool season vegetable and provide useful meals in spring and early summer when other vegetables are often scarce. It is a useful legume for soil improvement too. Broad beans are best suited to mild temperate and cool climates and are sown from early autumn to late winter in most districts.

Broad beans are a tall leafy crop and needs plenty of space to grow. Sow seeds in rows 60–75 cm apart after banding a pre-planting fertilizer alongside. Another way is to sow double rows 25–30 cm apart with 75–90 cm between each pair of rows. Dusting seed with fungicide before sowing is recommended. Press the large seeds into 'dark damp' soil at the bottom of a furrow about 5 cm deep, spacing them 15–20 cm apart. Cover with soil, tamp down and rake surface. If sown in damp soil, extra water is not needed until seedlings emerge in 10–14 days.

Extra fertilizer is usually unnecessary while the crop is growing. Too much fertilizer—especially fertilizer high in nitrogen—promotes leaf growth at the expense of flowers and pods. Plants of

Superette hybrid cabbage

broad beans may need some support. An easy way is to use garden twine stretched between stakes at each end of the row. Flower-drop is a common problem in early spring due to low temperatures. Pod setting improves with warmer weather and greater activity of bees. You can pick pods when young and slice them like French beans or leave them to fill and then shell the half-ripe seeds. A 5–6 m row is usually sufficient for one sowing for the average family.

Varieties Early Long Pod grows to 2 m tall with pods 20–25 cm long. Roma is a very early maturing variety which bears a heavy crop of tender pods. Coles Dwarf or Dwarf Prolific grows to 1 m with smaller pods. A dwarf variety is often preferred by home gardeners as the plants are more compact and less liable to wind damage.

BROCCOLI

Broccoli, more often called Sprouting Broccoli, is an excellent cool season vegetable for use in late autumn, winter and early spring. It is a close relative of cauliflower but the tightly-packed heads are green. The large centre head may reach 20 cm in diameter. When this is cut, new shoots with smaller heads form in the leaf axils, so a single plant will bear for many weeks. Broccoli does best in temperate and cold climates but is adaptable to all climates with a cool winter. In cold districts, sow seed in December or even November. Early sowing allows plants to grow a large frame before cold weather. In temperate and warmer climates, sow from late summer to autumn. Successive sowings may be made. Usually 9–12 plants at each sowing, 4–6 weeks apart, are sufficient for the average family.

Broccoli is usually raised as seedlings, because only a few plants are required. Sow seed in punnets, prick out seedlings early into 10 cm pots and grow on until 7–10 cm high. Transfer plants to a well-prepared bed to which plenty of organic matter and a pre-planting fertilizer has been added. Seeds may also be sown direct in clumps or stations spaced 45–60 cm apart. Thin each station to one seedling. Broccoli, like all leafy crops, needs to be grown quickly and responds to side dressing with nitrogen fertilizers or liquid feeds of Thrive. Use insecticide dusts or sprays to control caterpillars and aphids. (See Chapter 17.)

Cut the centre head when tightly packed and before the individual flower buds open. Take about 10 cm of main stem with a slanting cut. This prevents water lodging in the stem and causing rotting. Cut side shoots as they develop—again taking 10 cm of stem. Centre heads and side shoots store well in the crisper tray of the refrigerator for several days, or can be prepared as for cooking, blanched for 3–4 minutes, packaged and deep frozen.

Varieties Greensprout—a Calabrese type of Italian origin—has been the main variety for many years. Green Duke Hybrid, a uniform and vigorous variety, has been recently introduced. It is very adaptable to a wide range of conditions and can be sown well into winter or even in early spring. Skiff sets fruit better under cold weather conditions.

BRUSSELS SPROUTS

Brussels sprouts is another cool season crop but it is less adaptable than broccoli. It does best in mild temperate and cold districts and is not really suited to warm northern climates. Make sowings from October to early February in cold districts and from December to late February in mild temperate areas. Brussels sprouts are usually raised as seedlings in the same way as broccoli. Transplant seedlings at the same size into a prepared, well-manured and fertilized bed. The plants need more space to grow than broccoli so allow 60–75 cm each way. Apply side dressings of nitrogen fertilizer or liquid feeds regularly from transplanting onwards. Hill soil around plants as they grow to lessen wind damage. Dust or spray to control caterpillars and aphids. For the average family, 6–9 plants are sufficient for one sowing.

Brussels sprouts are ready for harvest 16–20 weeks after transplanting. The cabbage-shaped sprouts form in the leaf axils of the main stem and mature progressively from bottom to top. When bottom sprouts are quite small, start stripping lower leaves with a sideways pull. This allows sprouts to develop more easily. Pick the sprouts with a downward and sideways action. Discard any fluffy sprouts at the bottom of the stem. Continue stripping leaves and pick sprouts as they become usable. Modern varieties—especially hybrids—tend to form sprouts from bottom to top about the same time. If plants are well-grown, cut the main stem at ground level and pick all sprouts at once. This makes harvesting easier but you may have to sacrifice a few small sprouts at the top of the stem. Sprouts store well in the refrigerator for 7–10 days or they can be washed, trimmed of loose outside leaves, blanched for 4–5 minutes, packaged and quick frozen.

Varieties Top Score Hybrid is adaptable to a wide range of climatic conditions, is vigorous and very uniform. If well grown, the hard, tight sprouts can all be harvested at the one time.

Red Hybrid cabbage—an interesting change and easy to grow

CABBAGE

Cabbages are very adaptable to climatic conditions. In warm northern areas they are sown during most months of the year, although they may be difficult to grow well in the wet season. In temperate and cold districts they can be sown from early spring to autumn. Some old varieties tended to 'burst' or run to seed if sown out of season, but modern hybrid varieties are more reliable. Like other leaf crops, cabbages like good going, so the soil must have good structure (apply plenty of organic matter). Also use a pre-planting fertilizer when preparing the bed.

Cabbage plants can be raised as seedlings in the same way as broccoli or sown direct in clumps in the garden bed and later thinned to one seedling. Plant spacing depends on variety. Space small varieties 40–50 cm apart each way but give more room (60–75 cm) for large varieties.

Plenty of water (but good drainage) and regular side dressings of nitrogen fertilizer or liquid feeds of water-soluble fertilizer every 2–3 weeks will promote quick growth and crisp, solid heads. Dust or spray regularly to control caterpillars and aphids. (See Chapter 17.)

You can make successive sowings of cabbage over a long period. Usually 9–12 plants are sufficient at each sowing for the average family. Make next sowing when seedlings of previous batch are 15–20 cm tall. Start cutting the first plants from each sowing when the heads are firm but quite small. Hybrid varieties have an advantage here because they 'hold' well in the garden and are slower to burst or run to seed. Cabbages store well in the refrigerator for a week or you can slice them, blanch for 3–4 minutes, package and quick freeze.

Varieties Small early varieties mature in 8–10 weeks and weigh 1–2 kg. Earliball (Golden Acre) is very early with small, solid round heads. Sugarloaf has conical heads weighing up to 2–3 kg. Both are recommended home garden varieties. Two new early maturing varieties are Cape Horn Sugarloaf Hybrid with attractive bright green conical heads weighing 1–2 kg, and Superette Hybrid with uniform round heads weighing 2–3 kg. Large late varieties mature in 12–16 weeks and weigh 3–4 kg. In the past, the open-pollinated varieties like Succession and Early Drumhead were the main late types, both with large, flat heads. These have been replaced by hybrids such as Greengold Hybrid and Green Coronet. Greengold Hybrid is very uniform and has bright green solid heads. Green Coronet is a newer variety with an attractive mid-green colour. This flattened ball-shape cabbage grows to about 4 kg. Heads are tightly packed, mid-season maturing, with a short core which holds extremely well. Good tolerance to black rot. Also in the large cabbage group are Savoy Hybrid, with dark green, blistered leaves, and Red Hybrid, with firm, round, deep red heads which are popular for coleslaw and pickling.

CAPE GOOSEBERRY

Cape gooseberry, also known as Ground Cherry, Husk Tomato or Chinese Lantern plant, is grown for its small, globe-shaped, yellow or red fruits which are enclosed in papery husks. The fruit may be eaten fresh but is more often used to make jams or jellies. Cape gooseberry is a warm season plant. In frost-free, warm and tropical climates, the bushes are perennial and reach a height of 1 m or more. In cooler temperate climates it is grown as an annual during summer. It needs a warm, sheltered position. In suitable areas it may be grown as an ornamental shrub.

Generally, the cultivation of Cape gooseberry is very similar to that of capsicum. Sow seeds in punnets in spring when the weather is warm. Transplant seedlings to the garden when 8–10 cm tall, spacing them 100 cm in the row. For the average family 2–3 plants would be ample. Plants take five or six months before fruit is ready for picking. In warm, frost-free districts cut plants back hard after fruiting to induce new growth for next year's crop. Plants may bear well for three or four years. Golden Nugget (Golden Gem) is the best-known variety.

CAPSICUM

There are two kinds of capsicum or pepper. The sweet (mild) ones are eaten raw in salads or used in cooked dishes, soups and stews. The hot pepper or chilli is used fresh or dried as a flavouring and for sauces and pickles. Whether sweet or hot, capsicums are warm season plants like tomatoes. In warm tropical/subtropical climates you can grow capsicums almost all year round. In temperate zones sow seed from August to December and in cold climates September to November only. In frost-free areas, capsicums will die back over winter and shoot again in spring, but they are best grown as annuals.

You can sow seed direct but, because of the short growing season in most districts, it is best to raise seedlings for transplanting to the open garden as soon as it is warm enough. You only need a small number of plants too—4–6 well-grown plants are sufficient for the average family. There is hardly time, even in a mild climate like Sydney, for successive sowings. Seedlings transplanted in mid-September will bear fruit about Christmas and keep bearing until autumn. In warm northern areas, providing there are no frosts, successive sowings every 8–10 weeks will give a continuous supply. It is easy to raise seedlings in plastic punnets. Keep the punnets indoors until seedlings emerge, then prick them out into 10 cm pots to grow on in a sunny, sheltered spot until about 15 cm tall. Transfer them to a bed prepared beforehand with organic matter and a pre-planting fertilizer added. Space plants 50–60 cm apart each way.

Capsicums have a fairly deep root system and are adaptable to both heavy and light soils, but they need regular watering. Do not force plants, especially with nitrogen fertilizers, in the early stages. This makes too much leaf growth. After flowering has started, a side dressing of mixed fertilizer, such as Gro-Plus Complete, scattered around each plant will promote good fruiting. Repeat every 4–5 weeks while plants are bearing. Capsicums rarely need staking but well-grown plants may need support if carrying a heavy crop and exposed to wind. Drive in a stake close to each plant and tie the main stems to it with garden twine. Capsicums can be useful and ornamental when grown in large pots or tubs on a sunny terrace or patio. A pot 40 cm in diameter and the same depth will grow a good-sized plant—but pay special care to watering because pots dry out quickly.

You can pick sweet capsicums at any stage; there is no need to wait till they are full size. Frequent picking encourages more flowers and fruits. You can pick hot capsicums (chillies) when immature or leave them on the bush until full-coloured and shrivelled.

Varieties Giant Bell (Californian Wonder type) is a popular variety of sweet capsicum with large, bell-shaped, dark green fruit turning red as they mature. Long Sweet Yellow (Sweet Banana) has attractive, tapered fruit. Skin colour is lime-green, turning yellow then red with splashes of purple. Sweet Mixed contains both green and yellow varieties. Hot capsicums have rather smaller, green, tapered fruit turning red at full maturity. Hot Pepper or Hot Chilli (Long Red Cayenne) is the most popular variety. Bar-B-Q is a compact hybrid producing a heavy crop of small, round, sweet fruit. May be eaten whole, pickled or shishkebabed. Festival produces neatly shaped fruit on dwarf bushes. Sweet to eat and attractive display in the garden.

CARROTS

Carrots are an adaptable crop in the home garden and yield well for the space occupied. In warm northern zones, you can sow seed almost any month of the year although midsummer sowings are often avoided. Best months to sow in temperate zones are July through to March and in cold districts August to February. Sowings in late autumn or winter may run to seed without forming roots.

Deep sandy soils or heavy soils with good structure allow roots to grow and expand quickly. On clay soils, add coarse sand to improve texture and organic matter to improve structure. The myth that if you add organic matter, the roots will be fanged and horribly misshapen, is not true. You may have to discard a few split or malformed roots but the overall yield will outweigh this slight disadvantage. For carrots, you need a well-prepared bed for direct sowing—firm soil below and a loose crumbly surface. After scattering pre-planting fertilizer in a band where seed is to be sown, mark out shallow furrows 20–30 cm apart and sow seeds 6 mm deep by tapping the seeds from the packet as described in Chapter 5. Cover with compost, vermiculite or seed raising mixture and water gently. Seedlings may take 10–21 days to emerge so keep the bed damp until they do. A light sprinkling of grass clippings makes a good mulch through which seedlings can push easily.

Western Red Carrot

When seedlings are 5 cm high, thin them to 2–3 cm apart. Later, when 15 cm high, thin again to 5 cm apart, The seedlings removed will have tender roots large enough to eat. While thinning, remove weed seedlings too. If space between rows is then cultivated and mulched, further weeding is seldom needed. The base fertilizer applied before sowing may be sufficient to grow a good crop, but liquid feeds every few weeks will promote faster growth. Do not overfeed, especially with fertilizers high in nitrogen.

Most varieties take 12–16 weeks from sowing to harvest. For the average family a row 4–6 m long is sufficient for each sowing. Make further sowings at 4–6 week intervals during the season. Pests and diseases are usually not a serious problem. Start pulling early to spread the harvest and allow remaining roots to grow larger. Carrots keep well in open storage (remove tops) but even better in the refrigerator crisper tray. For freezing, slice or dice carrots as for cooking, blanch for 5 minutes, package and freeze.

Varieties Topweight has been the leading carrot variety in Australia for many years. It has a strong top with long, tapering roots of good colour. Other long-root varieties are Western Red, with a tall, strong top and smooth, good-quality roots, and All Seasons, similar to Topweight but a deeper colour. All three varieties are resistant to virus disease.

Shorter, stump-root varieties are often preferred in the home garden and are better suited to shallow soils. King Chantenay has excellent colour and has been a favoured home garden variety for a number of years. Early Horn is an early maturing, short stumpy variety for shallow soils. Roots have a sweet flavour but are rather paler in colour than other varieties.

Baby Carrot is a sweet, tender, finger-size carrot which is ready for harvest in 10–12 weeks from sowing. It can be sown more thickly in rows 10–15 cm apart and rarely needs thinning when grown in light, friable soil. Because of early maturity it can be sown later in the autumn than other varieties. Very suitable for growing in pots or troughs in a sunny, sheltered position.

CAULIFLOWER

Cauliflower, like broccoli, is a valuable winter vegetable but is not so adaptable to climate. Cauliflower takes longer to grow—14–24 weeks depending on variety—and there is only one head per plant. Cauliflowers are best grown in cool to cold climates but are also successful in mild temperate areas on the coast. Varieties available in Australia are a doubtful proposition in tropical zones. Cauliflowers need low temperatures for flower heads (called 'curds') to form, so they must be sown from mid-summer to autumn to grow a large frame before cold weather sets in. The best months to sow them are shown in the Sowing Guide in this chapter. You can have an extended harvest of cauliflower by sowing two varieties of different maturity at the one time or you can make successive sowings of the one variety. For the average family, plants of varieties in successive sewings will give a continuous harvest for 12–13 weeks.

Cauliflower seeds, like those of broccoli, brussels sprouts and cabbage, are usually sown in boxes or punnets and the seedlings transplanted when 7–10 cm high. Space plants 50–75 cm apart each way—late varieties need rather more space than early varieties. Like other leafy plants in this group, cauliflowers are hungry plants. Prepare the bed with plenty of organic matter and a ration of pre-planting fertilizer as described previously for broccoli. Give regular side dressings of nitrogen fertilizer every 2–3 weeks, or liquid feeds every 10–14 days. Dust or spray regularly to control caterpillars and aphids.

Cut the curds when tight and solid for best quality—do not wait until they become soft and fuzzy. Protect the white curds from direct sunlight (and yellow discoloration) by breaking outside leaves inward or by tying the ends of longer leaves together with raffia or string to form a shady tent over the centre. Start cutting early to extend the harvest. Curds store well in the refrigerator crisper for up to a week, but you can break them into serving-size pieces, blanch for 4–5 minutes, package and freeze.

Varieties All varieties produce top quality curds when grown quickly and harvested at the correct time. The main difference in varieties is the time they take to mature. The most popular cauliflower varieties (maturity in weeks from transplanting to harvest is shown in brackets) are Phenomenal Early (14–18 weeks); Deepheart (18–22) weeks).

In a climate like Sydney, three sowings of Phenomenal Early and Deepheart in early January, mid February and early March would be ready for transplanting (9–12 seedlings of each variety) in mid February, end of March and mid April. These successive plantings would be harvested from mid June to mid September or about three months.

There are, of course, other varieties which may be available as seedlings from nurseries. You may be able to buy seedlings of Snowball (12 weeks), South Australian Early (14 weeks), Sharpe's Shorts (16 weeks) or Westralia (24 weeks). Snow-

Cauliflower is a valuable winter vegetable. Ideal for freezing

ball, an early variety, is only suitable for cold climates.

CELERY

Celery is a very good home garden vegetable, especially if the outside stems are picked like silver beet. These green stems give a continuous harvest over two or three months for use in salads, soups, stews or as a cooked vegetable. If you prefer white celery, the stems must be blanched by excluding sunlight. Celery prefers a mild to cool climate but grows well in warmer areas in late summer and autumn. Raise seedlings in boxes or punnets in much the same way as other vegetable seedlings. Seeds are small and it may take 14–21 days for seedlings to emerge. Seedlings grow slowly too and it is best to prick them out into small pots or mini-punnets to grow on until large enough (8–10 weeks) to plant out.

Prepare the bed with liberal amounts of compost or animal manure, if available, and add a ration of pre-planting fertilizer as well. Space plants 30–40 cm each way and water well. For the average family 16–20 plants for each sowing are sufficient. Celery is shallow-rooted and regular watering is needed—every day or two in hot weather. Plants need generous feeding to grow quickly, otherwise stems become coarse and stringy. Give nitrogen side dressings or liquid feeds of Thrive every two weeks or so. Leaf spot disease may be troublesome but can be controlled with a fungicide spray. (See Chapter 17.)

For green celery, simply pick outside stems with a sideways pull to break them off at ground level, but leave as many younger leaves as possible for re-growth. You can blanch celery plants 3–4 weeks before harvesting by wrapping black polythene, Kraft paper or even a few thicknesses of newspaper around the stems from ground level to about 40 cm high, and tying loosely with string or raffia. A scattering of snailbait will deter snails and slugs from sheltering inside the cover. Blanching by setting seedlings in a trench and covering with soil as they grow is not recommended because of disease problems.

Celery keeps well in the refrigerator crisper for up to a week, or you can trim off leaves, chop stems into 5 cm lengths, blanch for 3 minutes and quick-freeze for later use. Leaves can be dried until brittle, chopped or crushed into small pieces to use for flavouring in the same way as dried herbs.

The vigorous Crisp Salad is a popular long-stemmed celery

Varieties Crisp Salad is stringless, has good flavour and a crisp texture. It is the standard variety.

CELERIAC

Celeriac (Turnip Rooted Celery) is grown in the same way as celery. The tops may be used as green celery. The tuberous roots, which may reach a diameter of 5–8 cm and the same length, can be grated for salads or used in soups and stews.

CHICORY

Chicory, or Witloof, is not widely grown in home gardens in Australia. The Asparagus chicory is grown for its dandelion-like leaves which provide useful greens. Chicory also produces roots which can be dried, ground and used as a coffee substitute or supplement. Roots can be dug up in late summer and buried upright in damp sand, peat moss or vermiculite with an 8–10 cm covering of the material on top. New growth forms plump white shoots called ‘chicons’ which can be used for winter salads.

Sow chicory in the same way as carrots and thin seedlings to the same distance.

Fertilizer and cultivation requirements are almost identical. Red Heart is the most popular variety of chicory. Long Greenleaf is an annual chicory of the asparagus type and grown for its upright stems and leaves.

CHINESE CABBAGE

Chinese cabbage is a close relation of cabbage but is a different species. It is widely grown in Asian countries where it is called Pe-tsai, Pak Choy, Bok Choy, Wong Bok, Kim Chee and other names. There are many different varieties available. Generally, plants of Chinese cabbages are smaller than ordinary cabbages. Plants are more upright, with loose heads, and the leaves have a texture resembling lettuce, but with a mustard-like flavour.

Chinese cabbage is grown in the same way as cabbage, but is best sown in late summer and autumn. Late winter and early spring sowings tend to run to seed in warm weather. Seed is best sown direct in clumps 30–40 cm apart and seedlings thinned to the strongest. For the average family 6–9 plants are sufficient for each sowing. Protect plants by dusting or spraying against caterpillars and aphids. (See Chapter 17.) Chinese cabbages, like other leaf crops, need generous feeding, so follow the same programme as for cabbages. Plants grow quickly and are ready for harvest 8–10 weeks from sowing. Chinese cabbage can be used in salad or coleslaw or cooked in the same way as cabbage.

Varieties Seeds of Chinese cabbage have been sold under many of the Chinese names given to it—Pe-tsai, Wong Bok or Chinese Great Luck. A recent introduction is a hybrid variety called Hong Kong Hybrid. This is more productive and adaptable than open-pollinated varieties. Another variety is Chinese Spinach, which does not form a head. It has dark green leaves on tall white stalks, not unlike silver beet. The outside leaves can be picked separately or the whole plant cut at ground level.

CHOKO

Choko, also known as Chayote or Alligator Pear, is a vigorous perennial vine crop which is adapted to mild temperate and subtropical zones. It is frost susceptible and needs a warm growing season of five to six months. The pear-shaped fruits have a texture and flavour rather like marrow or summer squash. The choko vine is best grown on a fence or trellis in an out-of-the way part of the garden where it can run wild. The vine is started from a single sprouted fruit but it is important to select a well-matured fruit with a smooth skin free of prickles. Keep the fruit indoors until it sprouts.

Prepare soil well by adding organic matter and a pre-planting fertilizer. Plant the choko into damp soil with the shoot and top of the fruit just above soil level. One well-grown vine is sufficient for the average family but plant two or three fruits about 100 cm apart in case of failure. While the vine grows, give side dressings of high nitrogen fertilizer such as Gro-Plus for Citrus every 5–6 weeks. Plants started in spring will flower in late summer to bear fruit in autumn. In winter, cut the old vines down, leaving 2–4 young shoots for the next crop. Cultivate around the plants in early spring and work in organic matter and fertilizer in the same way for starting a new vine.

Pick the fruit when lime green and 5–7 cm long. If left on the vine too long, fruit becomes coarse and loses flavour. Chokos store well in the refrigerator crisper for a week or more. They are not suitable for freezing as the flesh becomes soggy on thawing out.

CRESS AND MUSTARD

Garden cress and mustard greens make tasty additions to salads and are used for sandwiches and garnishes. They are grown in all climate zones and seed can be sown at any time of the year. Seeds are best sown in separate boxes, pots or punnets. Prepare a good seed-raising soil plus a mixed fertilizer as described in Chapter 5. Broadcast seed freely on the surface to have plants spaced 1–2 cm apart. Cover seeds lightly with compost or vermiculite and water gently. Keep damp until seedlings emerge, usually in 6–10 days but often less in warm weather. Thinning is seldom necessary. Plants are ready to harvest in about four weeks' time but may take longer in winter. When plants are 10–15 cm tall cut them with scissors just above ground level. Sow

Supersprint is a new, early-maturing rockmelon with excellent flavour

Hybrid Sugarloaf cabbage

Golden zucchini hybrid produces a heavy crop

Chinese spinach

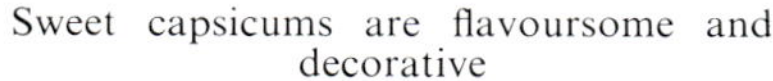

Sweet capsicums are flavoursome and decorative

Ha-ogen is an unusual and interesting melon for desserts or salads

successively for a continuous supply.

Varieties Curled Cress and Finest White Mustard are the standard varieties of garden cress and mustard greens. Water-cress, a close relative of nasturtium, grows in running streams. You can grow it in pots standing in water, preferably placed in the shade. It is a perennial plant and is started from cuttings, root divisions or seed. American or Land Cress is another kind which prefers wet shady conditions, but plants or seed may be difficult to obtain.

CUCUMBER

Cucumber is a warm season vegetable but adaptable to all climate zones. In tropical/subtropical areas, sow any month from July to March. In temperate areas, best months for sowing are September to January but in cold districts, with a short growing season, October to December.

You can sow seed direct in a well-prepared bed with added compost or animal manure plus a pre-planting fertilizer. For direct sowing, soil must be warm —20°C or above—for good germination. Dusting seed with fungicide is recommended as cucumber seeds (and those of all vine crops) are liable to damp off. Press 4–5 seeds into dark damp soil at each clump or station, spaced about 100 cm between rows and 40–50 cm between clumps. Thin seedlings to the two strongest. Although clumps are often called 'hills', they should really be saucer-shaped depressions so that water is directed to plant roots. To save space, cucumbers can be grown on a fence or trellis. The vines need some help to climb when young, so tie the stems to the wire support. Later, tendrils cling to the wire quite well. 'High-rise' cucumbers not only save garden space, but fruits are not in contact with the ground so shape and quality is improved.

For early cucumbers, it is best to sow seeds in punnets. Prick out seedlings into 10 cm plastic pots to grow on before planting in the garden. This method is described in Chapter 5. The secret of success is to prick out seedlings at the cotyledon stage before they form the first true leaf—they can be easily handled then without root damage.

Cucumbers and other vine crops like

Crystal Apple cucumber

'good going'. Good soil preparation with organic matter and pre-planting fertilizer will see the plants through to flowering; after flowering commences, scatter a mixed fertilizer around the base of the plants. Repeat this side dressing at 4–5 week intervals while plants are bearing. For the average family 4–6 plants are sufficient. These will continue to fruit well into autumn. The worst enemies of cucumbers and other vine crops are mildews. Some varieties are mildew-resistant but refer to Chapter 17 for control of these diseases.

You can pick long green varieties as gherkins when 5–10 cm long, to use fresh or for pickling. For high quality salad use, pick green varieties when 15–20 cm long, or round (apple-shaped) cucumbers when no larger than a cricket ball. Early and regular picking promotes further flowering and fruit setting. Like all vine crops, female flowers are pollinated by bees after visiting the male (pollen) flowers. Failure to set fruit is often due to cold weather or the absence of bees. Fruit setting improves in warm sunny weather.

Varieties Green cucumbers are long and thin, sometimes rather oval-shaped. They start bearing 8–10 weeks from sowing. The best green varieties are Green Gem, Burpless Hybrid and Sweet Slice. All three are tolerant to downy and powdery mildew. Burpless has long, thin fruit which may grow to 40 cm but are still fleshy and tender. This is an excellent home garden variety especially suitable for trellis growing. Supermarket (Long Green) has dark green fruit with pointed ends and is resistant to downy but not powdery mildew. Patio Pik, a recent introduction, has a small vine and short fruit about 10 cm long. This variety can be grown in a large pot or 2–3 plants to a garden tub. Provide short stakes for support. Sweet Slice is a hybrid with long, straight fruit and a unique sweet taste. It is burpless and does not require skinning. Pickling has tender crisp fruit which may be picked when less than 5 cm in length to produce a heavy crop or left to reach 15 cm to 18 cm if larger fruit is preferred.

Dominus is a Lebanese-type cucumber. It is recommended that the fruit should be picked and eaten when small—approximately 10 cm in length. Sweetly flavoured burpless fruit may be eaten whole or sliced for salads.

Round or apple-shaped cucumbers have a lime-green, cream or white skin. They take 10–12 weeks to fruiting but are extremely prolific. Crystal Apple, with a cream to white smooth skin, is the most popular of the round varieties. Richmond Green Apple has larger, oval-shaped fruit with a light green skin. S.A. Large Apple is similar but has white skin. None of the round or apple-shaped varieties are mildew resistant.

There are many other varieties. White Spiney is a popular variety on the north coast of North South Wales. It is late in maturity but a prolific bearer. Fruit is creamy-white with small black spines, and turns a deep yellow at maturity. African Horned is a novelty variety which has almost oval fruit with prominent spines on the greenish skin.

EGGPLANT

Eggplant or Aubergine is closely related to potato but is grown for its purple egg-shaped or pear-shaped fruits which vary in length from 10 to 25 cm. It is a native of Africa and southern Asia so needs a long, warm growing season. In tropical/subtropical climates, sow seed from September to March. In mild temperate climates like Sydney sow from September to December. In cold districts, sow in October and November only because plants take 14–16 weeks to bear.

You can grow eggplants in the same way as capsicums. Because of the short growing season, it is best to raise plants in punnets. Prick out seedlings into pots and transfer them to the garden bed when weather is warm. Plants grow 60–90 cm tall so space them 60–75 cm apart. A short stake to tie the plants to may be needed in exposed situations. For the average family, 4–6 plants are usually sufficient. Harvest fruit when the skin is smooth and rich purple in colour. If the skin has started to wrinkle with maturity, the flesh will be coarse and tough. The fruit stalks are hard and woody so cut them with a pair of secateurs to avoid damaging the plants. When well grown, you can expect 6–8 fruits on each plant.

Eggplant (Aubergine)

Varieties The best varieties are Long Tom Hybrid with small slender fruit, and Yates Supreme, with large pear-shaped fruit but a week or two later in maturity.

ENDIVE

Endive is closely related to chicory but is grown for its serrated, frilled leaves which form a loose heart and add an interesting taste to salads. It is similar in appearance to lettuce and is grown in much the same way. Endive is usually sown in late summer and early autumn for winter harvest. In warm northern zones it can be sown from autumn to spring. For the average family

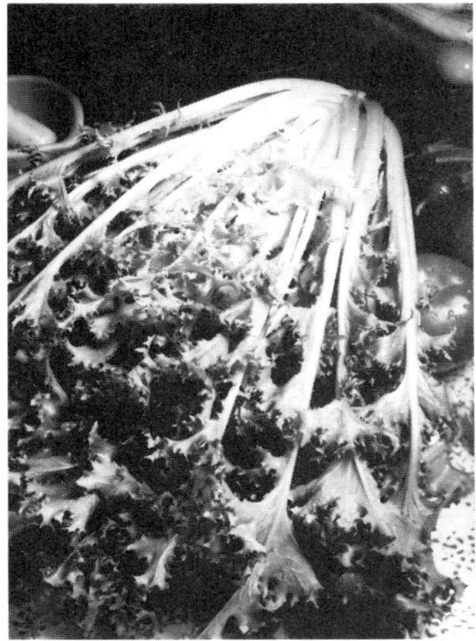

Green Curled endive

6–9 plants are sufficient. The leaves may have a slightly bitter taste which can be removed by blanching. Cover plants with large plastic pots or a thick layer of straw about three weeks before cutting.

Varieties Green Curled, Full Heart and Salad King are the most popular varieties. Another variety is Batavian, sometimes called Escarole, which has broad lettuce-like leaves and more compact heads.

HERBS

Herbs, fresh and dried, are becoming more popular in Australian kitchens to add piquant and interesting flavours to a wide variety of dishes. No garden is complete without some of the dozens of kinds of herbs which can be grown. Most herbs need a friable, well drained soil. Some need full sunlight, others grow well in partial shade. You can grow them in a separate herb garden but many smaller kinds can be grown as rockery plants or borders to flower gardens. All but very tall herbs can be grown in pots, tubs or troughs.

Many herbs are annuals but others are biennials or perennials. A large number can be started from seed sown in spring or autumn in temperate and cold zones but in almost any month of the year in

Versatile herbs can be grown in a variety of locations.

warm northern areas. Seeds can be sown direct but some are best raised in boxes or punnets for transplanting when large enough to handle. Always keep soil damp, but not wet, until seedlings emerge. Once established, herbs need little attention apart from regular watering and weeding. Herbs grown in pots benefit from liquid fertilizer feeds as they grow.

Harvest fresh herbs as required. For drying, cut off stems when plants are well grown, tie them with string and hang upside down in a dry but shady place. When completely dry, store leaves and stems—uncrushed or crushed—in air-tight bottles or jars. There are well over fifty different species of herbs. Here are brief notes on the most popular kinds.

Angelica A tall 1.5–2.4 m biennial with bright green serrated leaves and branching hollow stems, and a celery like texture. The hollow stems and stalks may be crystalized, and used for decorating cakes and pastry, whilst the leaves may be added to salads. A tea can be made from either the leaves, stems, seeds, or the dried roots. Grow from seed.

Basil (or Sweet Basil) An attractive annual plant, up to 40 cm tall, with shiny oval leaves and white flowers. It prefers full sun but tolerates semi-shade. It is a useful border plant and grows well in pots or tubs. Leaves have a clove-like flavour and can be used fresh or dried. Sow seeds in spring and space plants 20 cm apart.

Bay Tree A large tree, slow growing to about 12 m. It is the Laurel tree of ancient Greece and Rome. It has glossy dark green leaves which are narrow and about 4 cm long. The leaves are used extensively in many different types of cooking. Propagation is by seed or cuttings.

Borage A tall annual herb, up to 60–90 cm, with purple-blue flowers. Both leaves and flowers are used for flavouring soups and stews. Sow seeds in spring and summer and space plants 30 cm apart.

Caraway A 60 cm biennial, requiring a sheltered sunny position. It has finely cut and frond-like foliage with white flowers in summer.

Chamomile (Anthemis nobilis) see page 203. An ancient herb and a traditional ground cover for around garden paths and walks. The flowers may be used to flavour dry sherry, and a tea may be brewed from the blossoms.

Chervil Best treated as an annual, chervil will grow for eighteen months or more under good conditions. Plants grow to 30–60 cm tall and the divided parsley-like leaves have a mild anise flavour. Sow seed in spring, summer or early autumn, spacing plants 30 cm apart. Chervil grows well in partial shade and prefers a moist situation.

Chives Close relatives of onion, chives are perennial plants which grow in grass-like clumps 20–30 cm tall. They grow well in sun or semi-shade and are ideal plants for pot culture. Sow seeds in spring, summer or early autumn and space plants 20 cm apart. New plants can be started by dividing clumps if they become over-crowded. Chopped leaves are useful in salads, soups, stews and egg dishes.

Coriander (or Chinese Parsley) An annual herb to 40 cm tall. Plants prefer full sun but will tolerate semi-shade. Sow seeds in spring, summer or early autumn and space plants 20 cm apart. The parsley-like leaves have a sharp taste and the mature seeds are used for flavouring salads, bread and confectionery.

Dill An annual herb to 90 cm tall with light green, feathery leaves and umbrella-shaped flower heads. It prefers full sun-light. Sow seeds direct in spring or early summer in clumps spaced 30 cm apart. Leaves have a pungent, bitter-sweet taste. The seed can also be used for flavouring.

Fennel (or Sweet Fennel) A perennial herb but usually grown as an annual.

Coriander or Chinese parsley

Dill

Plants resemble those of Dill but are taller and coarser. Sow seeds direct in spring or early summer, spacing plants 50–60 cm apart. Both leaves and seeds have an anise flavour. The variety called Florence Fennel is not so tall and has a bulbous base which is eaten as a vegetable called Finocchio.

Garlic Another close relative of onion, garlic is a bulbous perennial with a clump of flat leaves to 60–90 cm tall. Plants die back after flowering. Plants are grown from separate bulblets or cloves which make up the compound bulb. Dig the bulbs for use as flavouring or for replanting cloves in late winter or early spring.

Horseradish A perennial plant with large spinach-like leaves and a long white taproot. Start plants in late winter or early spring from sections of root 15 cm long. Space them in well-prepared soil 30 cm apart at an angle. Cover the thick end with 2.5 cm of soil. When shoots appear, reduce their number to two or three. Although a perennial, horseradish is best grown as an annual. Take care to dig up all the pieces of root as the smallest piece will regrow. Well-grown roots are 5 cm in diameter and can be freshly grated or dried for used in spreads, dressings and horseradish sauce.

Hyssop A strongly-flavoured, perennial herb growing to a height of 50 cm. Sow seed from spring to early autumn. The leaves have a minty taste and are particularly useful for flavouring salads and soups. This herb is reputed to have the ability to make rich food easily digestible.

Lavender (Lavendula) see page 198. This traditional herb is a hardy perennial plant and is widely grown. The English Lavender (*L. spica*) is the most widely planted and the French Lavender (*L. dentata*) is also well known. Both are compact bushes growing in excess of 50 cm. Both are used extensively to perfume soaps, linen and toilet waters.

Lemon Balm A bushy, perennial herb which grows to a height of 50 cm. The crushed leaves have a delightful lemon fragrance. Sow seed from spring to early autumn. A useful herb for adding to sauces with poultry, fish and pork. Also used with fruit jellies, tarts and custard.

Marjoram (or Sweet Marjoram) A perennial herb to 30–40 cm tall, but often grown as an annual. Sow seed in spring or autumn and space plants 20 cm apart. Plants need full sun and rather moist conditions. The oval leaves and small white flowers are used fresh or dried for flavouring meat dishes.

Mint A rambling perennial, spreading by means of rhizomes. It can often spread too well in the garden so it is a good idea to grow it in a large pot or tub. Seed of some varieties is available or you can start plants from pieces of stem at any time of the year. It is one of the few herbs which prefer shade and very damp conditions. Leaves have a strong aroma and flavour and are used in mint sauce, mint jelly and for garnish. Peppermint, Spearmint and Curled mint are the most common varieties but there is also Applemint, Golden Applemint (variegated leaves), Eau de Cologne mint, Orange or Bergarmot mint and Pennyroyal. All have distinct flavours.

Oregano (also known as Wild Marjoram or Pot Marjoram). A perennial herb very similar to sweet marjoram but with a distinct aroma and flavour. Sowing, spacing and cultivation are the same as for marjoram. Leaves have a sharper flavour than marjoram and are used fresh or dried in Italian, Spanish and Mexican dishes, especially those with spaghetti and tomatoes.

Parsley The best known of all herbs. It is a biennial plant to 30 cm tall, but much taller when it runs to seed. It is best grown as an annual. Sow seeds direct in spring, summer and early autumn, either in clumps spaced 15–20 cm apart in the garden or broadcast in large pots or tubs. Seedlings may take 21–28 days to emerge so keep the bed or container damp for this length of time after sowing. Parsley grows well in sun or semi-shade and prefers high fertility soil and rather damp conditions. When established, plants respond to regular side dressings of nitrogen fertilizer or liquid feeds. Leaves are used fresh as a garnish and fresh or dried for flavouring salads, vegetables, meats, stews, soups and egg dishes. The most common variety is Curled parsley but another variety with stronger flavour is Italian Plainleaf parsley.

Rosemary An attractive perennial shrub to 60–150 cm with dark green, needle-like leaves and lavender blue flowers. It grows well in sun or semi-shade but soil must be well-drained. It is very suitable for growing in large pots or tubs. Sow seeds in boxes or punnets in spring, summer and early autumn or start plants from cuttings in late winter. Leaves have a pine-like flavour and are used, fresh or dried, with chicken or meat dishes and stews.

Rosemary

Sage A perennial herb to 60 cm tall with long grey-green leaves and tall spikes of violet-blue flowers. Sow seeds direct in spring, summer and early autumn in clumps 30 cm apart or raise seedlings for transplanting at this distance. Plants need full sunlight and will tolerate quite dry conditions. Leaves, fresh or dried, are traditional for flavouring seasonings for poultry, pork, lamb and beef.

Sorrel A perennial plant to 60 cm tall, but flower stalks may reach 120 cm. Sow seeds direct in clumps 20 cm apart in spring or early summer. Sorrel needs full sunlight and prefers rather damp conditions. Leaves have a sharp acid taste and are used as an addition to salads or cooked like silver beet or spinach. Also for flavouring meat dishes, stews and soups.

Summer Savory An annual herb to 30 cm tall. Sow seeds direct in clumps 15 cm apart in spring and early summer. It needs full sunlight and a well-drained soil. Suitable for container growing. Leaves have a peppery flavour and are used with meats, fish, eggs, beans, stews and soups.

Tarragon A perennial herb to 60–90 cm tall, with dark green, pointed leaves with a liquorice flavour. Sow seeds direct in clumps 60 cm apart in spring or early summer. Tarragon needs full sunlight. Leaves are used in salads, egg and cheese dishes, with fish and in sauces.

Thyme A small, prostrate, perennial herb to 20–30 cm tall. Sow seeds in spring, summer and early autumn in clumps spaced 30 cm apart. It is a useful plant for ground cover and for rockeries and garden borders, and is also suitable for container growing. Plants prefer full sunlight and will tolerate rather dry conditions. There are many different varieties of thyme and all are suitable for use as seasoning. Use the fresh or dried leaves in soups and stews or in seasoning for poultry, meat and fish dishes.

KOHL RABI

Kohl rabi is easy to grow and delicious to eat. The plant forms a swollen stem above the ground, so it is really a turnip-rooted cabbage. It can be sown in all climates from late summer to autumn.

In temperate and cold districts, early spring sowings are also successful.

Prepare soil well for direct sowing, adding plenty of organic matter. Scatter a mixed fertilizer in a band where seeds will be sown and rake into topsoil. Sow seeds in clumps 10–15 cm apart with 30–40 cm between rows. Cover seeds with compost or vermiculite, spread a light grass clipping mulch over each clump and water gently. Thin seedlings at each position to the strongest. Kohl rabi is best when grown quickly with regular watering and side dressings of nitrogen fertilizer or liquid feeds. Do not hill plants; prevent weeds by shallow cultivation between rows. Control caterpillars and aphids as for cabbage. For the average family a 1–2 m row is sufficient for each sowing at 4–5 week intervals from mid-summer to autumn.

Kohl rabi is ready to pick in 8–10 weeks. Start pulling the 'bulbs' early to spread the harvest. For top quality, bulbs should not exceed 5–7 cm in diameter. They store well in the refrigerator for a week or two. For freezing, select young bulbs, peel and dice or slice, blanch for 2 minutes, package and freeze. Early Purple is the standard variety, with flattened, globe-shaped bulbs and purple stems.

LEEKS

Leeks are close relatives of onions but are grown for their long, white (blanched) stem and bulbous base. They are more adaptable to climate than onions and grow more quickly. In temperate and cold climates you can sow seeds from spring to autumn but in warm tropical areas the best sowing period is late summer and autumn for plants to grow during the cooler months.

Leeks are best raised as seedlings in boxes or punnets. Grow them on to 20 cm tall before transplanting into a bed well prepared with organic matter and mixed fertilizer added. The easiest method of planting is to make holes with a dibble or rake handle, 2–3 cm wide and 15 cm deep. Drop seedling into hole so that roots rest on the bottom. When watered, enough soil will wash into the hole to cover roots. As plants grow regular watering will fill the hole with soil. Another method is to set seedlings at the bottom of a trench 20 cm deep. Fill in the trench with soil as plants grow. With either method space plants 15–20 cm apart each way.

Leeks need regular watering and respond to side dressings of nitrogen fertilizer or liquid feeds every 2–3 weeks. Generous feeding promotes quick growth and plump, tender stems. For the average family 40–50 plants are sufficient for each sowing. Successive sowings every 4–6 weeks can be made.

Start harvesting when stems are 2 cm thick—usually 12–14 weeks after transplanting. This way you can harvest 3–4 leeks each week for several weeks. On heavier soils, dig each plant with a long trowel so that stems are not damaged. Some plants may form small stems around the main one. Separate these carefully and replant to grow on. Leeks store well in the refrigerator crisper for several weeks. Musselburg is the most widely grown variety.

LETTUCE

Lettuce is not difficult to grow but, of all vegetables, needs to be grown quickly for crisp, tender hearts. The main requirements are:

1. Friable, well prepared soil which absorbs and holds moisture but drains readily.
2. The right varieties for the time of the year.
3. Regular and thorough watering.
4. Generous feeding.

Lettuce, by nature, is a cool season crop, but plant breeders have evolved sure-hearting varieties which can be grown in summer. You can sow lettuce all year round in each climate zone. Lettuce is the mainstay of summer salads so it is best to make successive sowings from early spring to January or early February.

Prepare the bed with plenty of organic matter for good soil structure and add a pre-planting fertilizer. Broadcast fertilizer at $\frac{1}{3}$ cup per square metre or scatter in a band where plants are to grow and rake it into the top soil.

You can raise seedlings in a good seed-raising soil for transplanting, but direct sowing in clumps or stations is more reliable, especially in warm weather. Make shallow, saucer-shaped depressions 20 cm

Imperial Triumph lettuce is an ideal variety for winter cutting

apart for small varieties or 30 cm apart for large varieties. Tap out several seeds at each, cover with compost or vermiculite and water gently. Keep the bed moist (a light mulch of grass clippings helps tremendously) until seedlings emerge, usually in 6–7 days but sometimes sooner. Scatter snail baits to protect the seedlings from snails and slugs. Thin each clump to the strongest seedling. Poor germination of lettuce seeds may be a problem when sown direct in very hot weather. When the soil temperature is 30°C or above lettuce seeds have trouble in germinating. You can overcome this by moistening the seeds—spread them on a piece of damp flannel or blotting paper—and keeping them in the refrigerator for a day or two. Then sow them direct as before. For the average family 9–12 plants should be sufficient for each sowing. Make successive sowings every 3–4 weeks.

Lettuce plants have shallow roots so they need plenty of water—every day in summer. Mulch plants with grass clippings or compost too. Give light side dressings of nitrogen fertilizer or liquid feeds of Thrive every 10–14 days while the plants grow. Start picking lettuce early when hearts are just forming—the young plants are crisp and tender and this spreads your harvest. Each sowing should give a 2–3 week harvest if you start picking early. Lettuce keeps well in the refrigerator crisper for about a week. Lettuce cannot be quick frozen because leaves become soggy when thawed out.

Varieties Large, sure-hearting varieties for growing in summer are Great Lakes and Pennlake. Both have crisp, solid hearts and will not run to seed when well-grown. Large, cool weather varieties for winter-cutting are Imperial Triumph and Winterlake. Yatesdale, Imperial 847 and Sunnylake are good between-season varieties for spring, early summer and autumn cutting. They are all excellent lettuces with solid hearts.

Small varieties are often preferred in the home garden. They can be planted closer but do not form hearts as solid as the large types. They may run to seed in hot weather, too. Mignonette is an old garden favourite. Green Mignonette has pale green loose hearts and Brown Mignonette is a brighter green tinged with reddish-brown. Buttercrunch (Butterhead type) is another small variety with waxy, light green outer leaves and yellowish green hearts, an excellent variety for tossed salads. Cos or Romaine lettuce has rather upright leaves forming a tall, rather loose, heart. All small varieties are excellent for growing in large pots, tubs or troughs—but keep them well-watered.

MARROWS

Marrows belong to the warm season group of vine crops—often referred to as 'cucurbits' after the family name. This family includes cucumbers, marrows, melons, pumpkins and squash. The climatic requirements, time of sowing, soil preparation and fertilizer, cultivation and pest control for each of these vegetables is much the same. This has been described for cucumbers and the reader is referred to this section. There are a few minor differences in detail, such as sowing depth, plant spacing, time to maturity and number of plants for the average family. These figures are given for marrow in the Sowing Guide.

Varieties In the past many varieties of marrow were 'running' types. These take a lot of room in the home garden so plant breeders have developed bush varieties. Long Green Bush and Long White Bush are both good home garden varieties. They are best picked when the fruit is 30–40 cm long. The best home garden varieties are zucchini marrows. President Hybrid Zucchini is a prolific bush variety with very dark-green fruit. Fruit can be picked at almost any size but most people prefer them about 15 cm long. Greyzini Hybrid is another excellent bush variety. It is similar to Blackjack Hybrid but fruits have grey-green, mottled skin. Golden Zucchini Hybrid is a yellow-skinned variety. All varieties can be grown in a tub on terrace or patio.

MELONS

Melons are another member of the vine crop group. For details of sowing and growing see the section on cucumbers. Refer also to melons in the Sowing Guide.

Rockmelon or Cantaloupe Hales Best is the leading variety of rockmelon. It is powdery mildew-resistant, and is also known as PMR 45. Fruit weighs about 1 kg with a netted yellow skin and sweet salmon-coloured flesh. Dixie Jumbo is a new large-fruited variety of good quality deep-coloured flesh and excellent flavour. Supersprint is a new hybrid which gives hybrid vigour and very even, productive crops. It is a high-yielding early-maturing type with round fruit of ideal table size. Salmon-coloured flesh sweetly flavoured. Honey Dew has oval fruit to 3 kg in weight with smooth white skin and sweet green flesh. Ogen is an interesting hybrid melon with a greenish yellow skin and green flesh, taste similar to Honey Dew. It is medium-sized globe to flattened globe shape, mid-season maturing. An unusual and interesting melon for desserts or salads. Casaba Golden Beauty is rather similar to Honey Dew but with yellow skin and a distinctive flavour. Rockmelons are ready to harvest when the stem pulls easily from fruit. Ripen indoors for a day or two for full flavour. Bush Star is an outstanding new non-running variety carrying 2–3 fruits per bush. Fruit is netted with flavoursome flesh.

Hales Best rockmelon

Watermelon Candy Red Hawkesbury is the leading large-fruited variety. It has large, oblong fruit to 14 kg in weight with grey-green skin and deep red flesh. It is resistant to *Fusarium* wilt and to a disease called Anthracnose. Sugar Baby matures at the same time but has small round fruit to 4 kg in weight. Garden Baby is a new bush-type similar to Sugar Baby. It bears 1–2 dark-skinned, round fruit per bush. Close planting is recommended. Sunnyboy is another new introduction—a striped watermelon of excellent quality and flavour. Harvest watermelons when the underside, in contact with the soil, turns yellow and the fruit gives a dull, hollow sound when tapped. All varieties of melons are running types so need plenty of room in the garden.

MUSHROOMS

The mushroom is not a vegetable but a fungus which has a fleshy, fruiting body arising from the underground web of hair-like filaments known as the mycelium. The

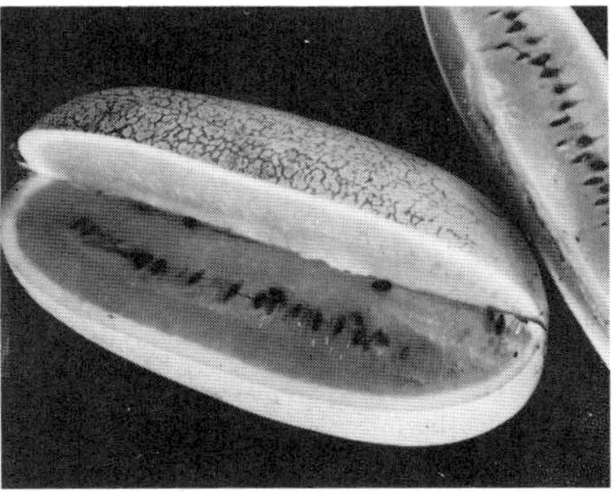

Candy Red Hawkesbury watermelon

stalk of the mushroom is topped by the pileus or cap underneath which are the gills containing the spores or reproductive cells. Mushrooms are prized for their delicate flavour and are used as a vegetable in cooking, either alone or combined with many dishes.

In their natural habitat, mushrooms and their inedible and sometimes poisonous relatives called toadstools occur in open grassland in autumn or spring when favourable conditions of moisture, temperature and humidity are present. Care must be taken when gathering field mushrooms because edible and inedible species are similar in appearance. The shape and texture of the cap, the colour (pink turning brown) of the gills underneath and their characteristic smell are good guides for identifying field mushrooms.

Mushrooms can be grown artificially on compost and those bought in shops are all produced in this way. Commercial mushroom growing is a highly specialized process which involves careful preparation and pasteurization of the compost, the addition of the mushroom inoculum (called spawn) and finally covering the compost with a layer of peat or soil (the casing layer). The commercial crop is grown in trays or plastic bags which are placed in specially constructed growing houses in which the temperature, humidity and ventilation are controlled. Similar but less sophisticated methods were used by home gardeners in the past but preparing the compost from fresh animal manure, straw, soil and lime is a laborious task and accurate control of moisture and temperature during the composting process is not easy.

However, an offshoot of commercial mushroom growing is the marketing of 'mushroom farm' compost in large plastic bags for home growers. The compost is already inoculated with spawn and a smaller plastic bag containing the casing layer is also supplied, together with instructions. For best results, the bags should be placed in a well-shaded location with still, fresh air, a high humidity and a temperature of 15–18°C. A good spot for the 'mushroom farm' is a corner of the garage or garden shed, in the cellar or underneath the house. Complete darkness is not necessary.

After spreading the casing layer of peat, sprinkle the surface with water to keep it damp but not wet—a light sprinkle two or three times each week is usually sufficient. The whitish-gray strands of the mycelium will cover the surface of the peat in ten to fifteen days and a few days later the filaments will clump together to form 'pin heads' which then develop into the first flush of mushrooms. After picking the first crop, sprinkle again regularly to keep the surface damp and new 'pin heads' will form for the next flush. A well-grown bag of mushrooms should produce a crop about every ten days over a period of two or three months. If, after several crops, no mushrooms appear for two or three weeks, the compost is exhausted and a new bag should be started. Use the spent compost for mulching or digging into the vegetable or flower garden.

Mushrooms can be picked in the button, cup or flat stage, whichever you prefer. Hold the cap of the mushroom in your fingers and gently twist the stalk from the casing layer. It is a good idea to keep a small amount of peat in reserve for filling in small holes which may occur on the surface when removing the mushrooms.

Success with mushrooms grown in plastic bags depends largely on their location, which must be free of draughts, and on very careful attention to watering. Growth rate depends on temperature, so the flushes of mushrooms take longer in cold weather. 'Mushroom farms' usually do best in autumn and spring when temperatures are close to the optimum of 15–18°C.

Keep the area around the bags clean at all times, especially before starting a new bag. Pests such as slugs, slaters, cockroaches and mice may be troublesome so scatter appropriate baits on the floor around the bags.

MUSTARD

(*See Cress and Mustard*)

OKRA

Okra or Gumbo is an annual plant related to hibiscus. It is most suited to tropical/subtropical climates or warm temperate climates with a long growing season. Plants grow to 90 cm tall and have large, hibiscus-like yellow flowers with red centres. These form the edible pods 7–10 cm long.

The edible seed pods of okra are delicious for flavouring soups and stews

Raise seedlings in late spring or early summer in a similar way to capsicum and transplant when 10 cm high to a well-prepared bed, spacing plants 50–60 cm apart. For an average family 4–5 plants are sufficient. Plants are grown on in the same way as capsicum with a side dressing of mixed fertilizer when flowering commences. The tender pods are ready to pick 4–5 days after the flowers have opened. They become very tough if left on the bush, and plants stop flowering. The pods, fresh or dried, are used for flavouring soups and stews.

ONIONS

Onions are a very good winter crop in the home garden. For best results, it is important to choose the right variety for sowing at the right time of the year in different climate zones. Generally, onions are classified as early, mid-season and late maturing types. In warm northern areas, early onions are sown from February until May. In temperate climates like Sydney, sow early onions from March to May and mid-season onions June to July. In cold southern areas sow early, mid-season and late onions in succession from April to August or September. It is important to sow early onions first, mid-season onions next and late onions last. Premature bolting (running to seed) may occur if maturity groups are sown out of sequence or season.

You can raise onion seedlings in beds, boxes or punnets for transplanting when 10–15 cm tall, spacing plants 7–10 cm apart in rows 20–30 cm apart. Do not plant deeply—just cover roots and base of stem. Direct sowing saves double handling and is less trouble. Prepare the bed well, as for sowing carrots, and scatter a pre-planting fertilizer in a band where seed is to be sown and rake in. Make a shallow furrow 6 mm deep and tap out seeds thinly on to the dark damp soil. Cover with compost or vermiculite and water gently. Seedlings generally emerge in 10–14 days, but may take longer in colder weather. Thin seedlings early to 2–3 cm and later to 7–10 cm. For the average family a 4–6 m row is sufficient. Make successive sowings with varieties of different maturity.

If the bed is well-prepared and fertilized, additional fertilizer is seldom necessary. A light side dressing of mixed fertilizer or liquid feeds when bulbs start to form will boost plants along if they are not growing strongly. Control weeds by hand weeding between plants and shallow cultivation between rows. Do not hill the plants—bulbs sit on the soil surface, not below it. For control of diseases and pests see Chapter 17. Onions take 24–32 weeks to picking. Bulbs are ready to pull when tops dry and fall over. After pulling, leave them in the sun for a few days to cure. When outside skin is quite dry, screw off tops and rub off remaining roots. Select sound bulbs without blemishes for storage in a cool, dry place. Wire baskets or plastic mesh bags give good ventilation.

Varieties Onions not only vary in maturity but come in different shapes, colours and degree of onion flavour (strong or mild). Generally early maturing onions do not keep as well as mid-season and late onions.

Early Barletta is the earliest onion with flat white bulbs. It is a favourite for early sowing. Hunter River White, Hunter River Brown and Gladallan Brown are also early onions. All three have small to medium sized, globe-shaped bulbs with fair keeping quality. They are the main varieties for early sowings in warm northern and mild temperate climates.

Odourless is a mid-season, large flat onion with light brown skin and mild cream flesh and is popular with home gardeners. Mild Red (Red Skin) is also mid-season in maturity with purple-red, globe-shaped bulbs which keep fairly well. Creamgold is an excellent mid-season onion and is a very good keeper. It has light brown, globe-shaped bulbs with pungent cream flesh.

Late maturing varieties are White Imperial Spanish and Brown Spanish. Both have medium sized, globe-shaped bulbs which have excellent keeping quality.

Many varieties of onion can be used as green salad onions or spring onions, but it is best to sow seed of the Welsh or Japanese Bunching type of onion for this purpose. These are available as Spring Onion or Shallot Bunching Onions. You can sow both varieties direct at almost any time of the year from spring through to autumn. Sow seed more thickly with rows 5–10 cm apart. Thinning is seldom necessary. Spring or bunching onions are ready to harvest in 8–12 weeks. Make successive sowings every 4–6 weeks as required. Both varieties are ideal for growing in pots or troughs.

Pickling onions are easy to grow

Potato Onion (Multiplier Onion) is grown from small sets or bulblets planted in autumn. Groups of bulbs are formed below ground. Tree Onion (Egyptian Onion) is grown from sets in the same way. Bulbs are formed below ground but also at the top of the flowering stem.

PARSNIPS

Parsnips, like carrots, yield well for the space they occupy. They can be grown in all climate zones. In warm tropical/subtropical areas, sow seed from February to September and so avoid the hot, wet season. In temperate districts, sow from July to March and in cold districts from August to February. Late sowings in autumn and winter may produce small roots and plants may run to seed prematurely.

Parsnips like a friable, open soil and good drainage for best root development. Compost or animal manure, if added well before sowing, will not result in forked or misshapen roots. Dig the bed to spade depth and prepare the bed for direct sowing in the same way as for carrots. Add mixed fertilizer in a band and rake into the soil. Sow seed thinly in a furrow 6 mm deep. Rows should be 30–40 cm apart. Cover seed with compost or vermiculite and water gently. Parsnip seeds are slow to germinate (21–28 days) so keep the bed damp with light watering until seedlings emerge. A light grass clipping mulch helps to control moisture loss in hot weather. For the average family 3–5 m of row is sufficient. Late summer and early autumn sowings are the most useful for harvesting the roots for baking, stews and soups in winter.

Thin seedlings to 5–7 cm apart and control weeds by hand weeding and cultivation. If parsnips are grown on a well-prepared and fertilized bed, extra fertilizer is rarely needed, but liquid feeds, when roots start to form, will promote faster growth. Do not over-feed, especially with nitrogen fertilizers.

Parsnips take 18–20 weeks to grow.

Start pulling roots early to spread the harvest. The remainder keep well in the soil, especially in winter when growth is slow. Roots store well for a week or two in an airy cupboard (remove tops) but for several weeks in the refrigerator crisper. Roots can also be washed, sliced or diced, blanched for 2–3 minutes, packaged and frozen for later use.

PEANUTS

Peanuts are sometimes called ground nuts, earth nuts or monkey nuts. Americans call them 'goobers'. Commercially, peanuts are grown on a large scale for eating raw or roasted, or to be ground for peanut butter or crushed for peanut oil. They are very nutritious and contain 50–55 per cent oil and 40–45 per cent protein. Peanuts are not commonly found in the home vegetable garden but they are an interesting crop to grow. The plants are semi-erect, annual legumes which add nitrogen to the soil in the same way as beans and peas. They are natives of Brazil and need a long, warm growing season of about five months to mature. For this reason, they are best adapted to tropical, subtropical and warm temperate climates. The plants are very susceptible to frost damage.

Peanuts are grown in much the same way as dwarf beans but peanuts are strange plants. After the small, yellow flowers are pollinated, the flower stalks (called 'pegs' by peanut growers) lengthen and push downwards into the soil. The pods, containing one to four kernels, develop underground. The crop is dug when top growth begins to yellow and die down.

Peanuts prefer a well-drained, sandy soil through which the 'pegs' can penetrate easily. Heavy soils are less suitable but satisfactory if they contain plenty of organic matter to give them a friable structure. When preparing the soil, add a dressing of lime to raise the pH level to 6.5 or 7.0 (see Chapter 4). Seeds (raw peanuts—roasted ones will not germinate) are usually available from health food stores and are sown direct in the garden when soil temperatures reach about 20°C. Like bean seeds, peanuts are susceptible to fertilizer burn, so band a pre-planting fertilizer such as Gro-Plus Complete or superphosphate alongside the line where the seeds are to be sown. Also dust the seeds with fungicide before sowing (see Chapter 5). Press the seeds into 'dark damp' soil at the bottom of a furrow about 50 mm deep, spacing the seeds 10–15 cm apart. If sowing more than one row, allow 60–75 cm between each. Cover the seeds with soil and stamp down firmly with the back of the rake. Then level the surface and scatter a mulch of dry grass clippings over the whole bed to retain moisture and prevent the soil caking. If seeds are sown in 'dark damp' soil, there is usually no need for extra watering until the seedlings emerge in 7–10 days.

Cultivate between the rows to destroy weeds and water the plants regularly, especially in hot weather. As the plants grow, hill the soil slightly against them for support. Alternatively, apply a mulch of compost or grass clippings between the rows. If a pre-planting fertilizer has been used, additional fertilizer is rarely needed. However, if the plants lack vigour at any stage, apply a complete fertilizer with about 10 per cent nitrogen as a side dressing or give liquid feeds of Thrive, Aquasol or Zest. Well-grown peanut plants should reach a height of 30–40 cm at flowering time.

When the foliage turns yellow and starts to die down, this is the time to dig the plants—usually 16–22 weeks after sowing the seed. If dug too early, yields will be reduced and the kernels may be shrivelled. If left too late, the pods may break off the 'pegs' and remain in the soil. Lift each plant with a large fork and turn it upside-down to dry in the sun for a few days. After the plants have wilted, shake the roots and pods free of soil and dry them further under cover. When the pods are quite dry, strip them from the plants and store them in bags or boxes. After shelling, the kernels can be eaten raw, or roasted on a shallow tray in the oven.

Varieties Virginia Bunch (dark foliage) and Red Spanish (light foliage) are the two most widely grown varieties in Australia. Virginia Bunch is more resistant to stem rot, a fungus disease which attacks the lower stems and crown of the plants near soil level.

PEAS

Peas are one of the best cool season crops for the home garden. The yield for space occupied is not as high as some other vegetables, but peas are easy to grow and space is usually available in winter to sow them. You can grow peas in all climates. In warm northern zones, sow seed from March to July. In temperate climates sow from February to August, and in cold climates sow from June to September or early October. In any district where frosts are likely, make sowings so that the crop is not in flower during the frost period. Frost will damage both flowers and young pods.

Peas are adaptable to heavy and light soils but need good drainage and a friable, well-structured soil. If your soil is acid, apply lime as described in Chapter 4. Prepare the bed free of clods and in dark damp condition for direct sowing. Apply pre-planting fertilizer such as Gro-Plus Complete in furrows alongside where the seed is to be sown to avoid fertilizer burn. (See Chapter 5 for direct sowing large seeds.) Dusting seeds with fungicide before sowing is strongly recommended. Mark out a furrow 25 mm deep and press the seeds into the soil 3–5 cm apart. For dwarf peas allow 40–50 cm between rows. For climbing peas, space seeds at the same distance against a fence or trellis on which the plants can climb. If making large sowings of climbing peas, allow 100 cm between rows. For the average family a 3–5 m row is sufficient for each sowing of dwarf peas. A smaller row of 1–3 m is usually ample for a sowing of climbing peas because they yield more pods over a longer period.

After sowing pea seeds, cover in the furrow with soil, tamp down and rake the bed. If seeds are pressed into dark damp soil, no further watering is required until seedlings emerge. Too much moisture, especially in the first 36 hours after sowing, may do more harm than good (see Chapter 5). Birds can be a problem to emerging seedlings too. If you have bird trouble, cover the rows or the whole bed with pieces of wire netting or black cotton thread stretched between short stakes. Remove these bird deterrents when the seedlings are 10 cm high. The birds have usually lost interest at this stage. Peas usually crop well without extra fertilizer while they grow. Yellow, stunted plants are more often the result of wet soil and poor drainage rather than lack of nutrients. If drainage is adequate and the plants still lack colour and vigour, apply a mixed fertilizer as a side dressing or give liquid feeds.

Climbing peas need support but dwarf peas yield better if their tendrils can cling to brush, twigs or wire netting. This keeps the bushes upright and pods are easier to pick. For all pea crops, cultivate regularly (a day or two after watering) to destroy weeds. Plants can be slightly hilled at the same time. This discourages weeds close to the row and gives the stems more support. Pests and diseases of peas are usually not a problem. (See Ch. 17.)

Pick well-filled pods before etching of veins shows on the surface. Pick every few days for high quality and to prolong flowering. Peas in the pod keep well in the refrigerator crisper for a week or two—or you can shell the peas, blanch for one minute, package and freeze them.

Varieties Melbourne Market and Earlicrop are two early dwarf varieties. Both are ready to pick 12–14 weeks from sowing. Greenfeast is the main crop variety. It is rather taller and bears masses of pods in 14–16 weeks after sowing. Telephone is the standard climbing variety. Pods are ready about the same time as Greenfeast but plants bear for much longer (3–5 weeks). Snow Pea (Chinese Pea) is an edible, podded variety. You can pick very young pods for cooking without shelling or you can slice them like beans. Snow Pea is a climbing variety with purple flowers but there are dwarf varieties too. Sugarsnap is the sensational new edible podded pea that has enjoyed wide acceptance. First available as a vigorously growing, heavy cropping climber, it is now also offered as a dwarf variety. Produces delicious pods over a long period. Highly recommended.

POPCORN

(*See Sweet Corn.*)

POTATOES

Potatoes are warm season plants and very susceptible to frost. Severe frosts will kill the tops completely. In tropical/subtropical zones—providing the district is frost-free—potatoes can be grown all year round, but the most suitable months are

January to August. This period avoids the wet season. In southern Queensland and warm to mild temperate districts, most gardeners prefer to grow a spring crop (planted July to September) and an autumn crop (planted January or February). In cold districts the planting season is restricted to warmer months only from August to December. The time of spring sowings depends on late frosts and soil temperatures.

Potatoes are adaptable to light and heavy soils but good drainage is essential. They do best on friable soils with good crumb structure. Potatoes are grown from tubers and not from true seeds. Tubers (or seed potatoes) are available from nurseries and garden stores at planting time in spring. When buying seed potatoes, look for Government Certified tubers as these are free of virus disease. If you want to grow an autumn crop too, you can save some seed potatoes from healthy, high yielding plants in your spring crop.

For planting, tubers should be 30–60 g in weight. Cut large tubers into chunky pieces with at least one eye or sprout on each. Do not rub the cut surface in ashes or similar material—just let it dry out naturally. Spread tubers out in a shady spot for a week or two before planting. This allows the young sprouts to 'green' or harden. As a guide, a 3 kg bag of certified seed potatoes should provide 50–60 plants, which is sufficient for an average family.

Prepare the bed to spade depth well beforehand to have the soil in friable dark damp condition at planting. Mark out furrows 15 cm deep and 75 cm apart. Scatter fertilizer along the bottom of the furrow at $\frac{1}{4}$ cup per metre and cover with about 5 cm of soil from the sides. Place tubers at 30–40 cm apart, cover with soil and rake the surface level. Sprouts emerge in 3–4 weeks.

Cultivate between rows to keep down weeds and gradually hill the plants to form a furrow between the rows. Hilling supports the plants, protects new potatoes from exposure to light and prevents them being attacked by caterpillars of potato moth (see Chapter 17). On level or slightly sloping beds, the furrow between rows can be used for irrigating. On well-prepared and fertilized soil, no extra fertilizer is needed, but water regularly to promote smooth, well-developed potatoes. You can start digging 'new' potatoes about 3–4 weeks after plants have flowered and the lower leaves turn yellow. If potatoes are to be stored, allow the tops to die off completely before digging. Remove soil from potatoes, discard any damaged or blemished ones and store in a cool dry place, which must be dark to prevent the skin from 'greening'. Wooden crates, wire baskets or hessian sacks are good containers for storage.

Varieties Sebago is the main variety in most states. Other good white-skinned varieties are Sequoia, Kennebec, Exton and Katahdin. Delaware is popular in Western Australia and Brownell (red skin) is an old favourite in Tasmania.

PUMPKINS

Pumpkins, are grown in the same way as cucumbers. Refer also to the Sowing Guide.

Varieties Most pumpkins have large running vines and may take a lot of space. If you can grow them in an odd corner where they can scramble over a fence or garden shed, this is an ideal place. One or two varieties have smaller vines and fruit and one is a bush pumpkin. These are excellent for small gardens. Generally,

Crown Prince pumpkin

all pumpkins are harvested when the vine dies and the fruit stalk is dry and brittle. Fruit is then fully mature with best flavour. Mature fruit stores well for many months, although some varieties keep better than others. Store fruits in a cool airy cupboard or in cardboard cartons. Fruit for storage must be free of blemishes or broken skin through which storage rots can invade. Inspect stored pumpkins periodically for signs of damage by rots, or rats and mice.

Of the large pumpkins, Queensland Blue is an old favourite with green, turning grey, creased fruit and deep orange flesh. It is a good long-keeping variety. Triamble has grey, three-lobed fruit with golden flesh and rather dry texture. It is an excellent keeper. Crown Prince has smooth, whitish-grey, drum-shaped fruit with bright orange-red flesh and keeps well. Butter Pumpkin has a cream to yellow skin and deep yellow flesh. It is earlier to mature than other large varieties but does not keep as well. Gramma (Trombone) is another large variety with horse-shoe shaped, orange fruit and sweet, yellow, rather dry flesh. A favourite for gramma or pumpkin pie. Jarrahdale is a grey-skinned variety with deep orange flesh and a similar shape to Queensland Blue. The fruit quality is first class, cuts well, and the flavour is sweet. Good storage qualities.

Of the smaller pumpkins, Butternut has yellow, pear-shaped fruit 1–2 kg in weight and with deep orange flesh. It grows well on a fence or trellis to save space. It is a good keeper if fully mature when picked. Baby Blue has grey-blue skin with yellow-orange flesh. Vines are fairly compact and fruit sets close to the centre. Both Butternut and Baby Blue are good home garden varieties.

Two bush pumpkins are Golden Nugget and Butterbush. Both can be set about 100 cm apart each way or you can grow a single plant in a large tub. They are ideal home garden pumpkins and can be picked when fully coloured, but are best left until the bush dies for better keeping quality. In warm and mild temperate climates two, or perhaps three, sowings are possible during spring and summer. Golden Nugget bears 6–10 small, round, orange fruit with deep yellow flesh. Butterbush produces 4–5 pear-shaped fruit each weighing about 500 g. The fruit, which has a good quality, deep orange flesh and a delicious flavour, should be ready to pick about 14–15 weeks after sowing.

RADISH

Radish can be sown almost the year round in all climate zones. Successive sowings from early spring to late summer will give crisp roots for the salad season. Radish is the quickest and easiest crop to grow. Seeds germinate in 5–8 days and roots are ready to harvest in 6–8 weeks.

Sow seeds direct in a well-prepared bed as for carrots. Broadcast fertilizer or scatter in a band where seed is to be sown. Space rows 10–15 cm apart and tap out seeds in a furrow 6 mm deep. Cover with compost or vermiculite and water gently. Thin seedlings to 3–5 cm apart when they have grown their second leaf. Water regularly and give liquid feeds every 7–10 days. Mulch between rows to keep soil moist in hot weather. Make sowings every 2–3 weeks as required—a short row 50–100 cm long is usually sufficient for the average family. Start picking roots early because they get old and tough quickly. Roots keep well in the refrigerator crisper for a week or so.

Varieties Radishes come in different shapes and sizes. French Breakfast (red with white top) is tankard-shaped. Round Red and Ruby Red are the best of the globe-shaped (turnip-rooted) varieties. Long Scarlet and Long White Icicle have tapering roots to 15 cm. All varieties are ideal for growing in pots and troughs.

RHUBARB

Rhubarb—one of the few perennial vegetables—is best grown in a separate bed to stay undisturbed for three or four years. It tolerates shade and stems are usually longer when plants are grown in semi-shade. Rhubarb is adapted to all climate zones and a variety of soils but needs good drainage, regular watering and generous feeding.

Rhubarb can be established from seed sown in spring or early summer in temperate and cold districts but late summer and autumn sowings are successful in warm northern areas too. Beds must be well-prepared with plenty of organic mat-

Tomatoes are the most popular home garden vegetable. Five of the more popular varieties are: *(above left)* Grosse Lisse, a prolific cropper with medium-sized, even fruit; *(above right)* Egg Tomato, for bottling or fresh salads; *(right)* Tiny Tim has a heavy crop of cherry-sized fruit on small plants; *(below left)* Whopper, for large fleshy fruit: *(below right)* Apollo Hybrid, ideal for early or late season sowing

Crisp, tender hearts of lettuce are available for much of the year by successive sowings

Greenfeast peas are easy to grow and crop heavily

All Seasons carrots yield well and have good colour and flavour

Radishes are quick and easy to grow. Seed may be sown at any time

Dominus Lebanese type 'burpless' cucumbers are sweetly flavoured

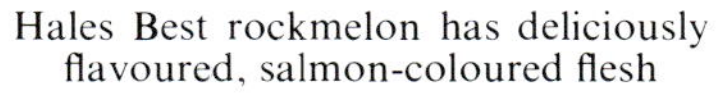

Hales Best rockmelon has deliciously flavoured, salmon-coloured flesh

Golden Nugget pumpkin is ideal for gardeners with little space

Greyzini, an excellent hybrid zucchini bush marrow, grows well in a tub

Top Score hybrid brussels sprouts mature early and are vigorous and uniform

Broccoli Green Duke Hybrid is adaptable to most climatic conditions

Queensland Blue is an excellent quality, long-keeping pumpkin

ter and a pre-planting fertilizer added. Sow a few seeds in clumps or stations, spaced 40–50 cm apart each way and thin to 2–3 seedlings. Plants from seedlings vary in colour of stems, stalk length and vigour, so select the best to grow on and discard the remainder, leaving one seedling at each clump.

Another method is to grow rhubarb from crowns or sets by dividing established plants in late winter or early spring. This way you can select the best-yielding plants with certainty. Rhubarb crowns or sets are usually available from nurseries during winter months.

When plants are established apply a nitrogen fertilizer or liquid feeds every 4–5 weeks during the main growing season—from early spring to autumn. For the average family, 12–15 well-grown plants will provide a good harvest. In winter each year, loosen soil around the plants and fork in compost or animal manure plus another ration of mixed fertilizer to give them a good start for spring growth.

Pick stalks (outside ones first) as required with a downward and sideways action so they pull away cleanly from the crown. Always leave the youngest stalks in the centre of each plant to promote new growth. If flowering stems appear, cut these off at the base and apply a nitrogen fertilizer or liquid feeds to encourage more stalks and leaves. After picking, cut the leaves from the stems. Do not use the leaves, as cases of rhubarb poisoning have been reported. Stalks keep well in the refrigerator crisper for a week or two but it is better to cook them immediately. Cooked rhubarb can be kept in screwtop jars in the refrigerator to serve chilled as required.

Varieties Most varieties are ever-bearing and grow all year round, but yields are better in warmer months. Sydney Crimson is the most widely grown.

ROCKMELON

(*See Melons.*)

ROSELLA

Rosella or Jamaica Sorrel is grown for its red fleshy fruit which are used to make sauce or jelly with a cranberry flavour. It is an annual plant closely related to hibiscus and grows to 2 m tall. Rosella is a warm season plant which needs a growing season of at least six months, so is only suitable for tropical/subtropical or warm temperate climates.

Seedlings are raised in the same way as capsicum and transplanted to the garden when 15–20 cm tall. Space plants 150 cm apart.

Grow plants on as for other fruit vegetables with a side dressing of mixed fertilizer when flowering commences. Pick fruit, which consist of red fleshy scales surrounding the green centre, when fully ripe. Two or three plants will provide plenty of fruit.

SALSIFY

Salsify, often referred to as Oyster Plant or Vegetable Oyster, is grown for its cream-coloured roots which are used in the same way as parsnips. It is not widely grown in Australia but is adapted to all climates. Time of sowing, soil preparation, fertilization and cultivation are the same as for carrots or parsnips. Roots are long and thin and ready to dig in 20–22 weeks from sowing. Wash, boil and scrape roots before baking or frying. For those who like the flavour of this vegetable, a row 1–2 m is sufficient for one sowing.

SHALLOTS

Shallots or Eschalots are a different species from spring onions or bunching onions, which are often referred to as shallots. True shallots are grown from bulblets or cloves like garlic. The 'mother' bulbs are usually planted 5–7 cm deep in autumn or early winter. Space plants 15–25 cm apart. As plants and 'daughter' bulbs develop, push soil around them to blanch the stems. Grow them quickly with generous feeding as for leeks or spring onions. You can harvest plants as chopped leaves (like chives), as green onions for salads, or as dry bulbs (like garlic) for flavouring. Small bulblets or cloves can, of course, be replanted. If you start with a few plants, you can grow shallots for ever.

SILVER BEET

Silver Beet or Swiss Chard is a close relative of beetroot—in fact, a variety of the same species. It is not a true spinach but it is often called spinach in New South

Wales and Queensland. Silver beet, with its large crinkly leaves and cream-coloured stalks, is an excellent, cut-and-come-again vegetable and it is easy to grow in the home garden. It is adapted to all climate zones. In warm northern areas it is sown almost any month of the year; in temperate and cold districts from early spring to early autumn. Late autumn and winter sowings may run to seed. (See Sowing Guide for best months to sow.) Silver beet, like other leaf vegetables, needs well drained soils with plenty of organic matter for good structure and generous feeding.

Silver beet can be raised as seedlings in boxes or punnets but direct sowing avoids double handling and transplant check. After scattering pre-planting fertilizer in a band where seed is to be sown, tap out a few seeds in clumps or stations 30–40 cm apart. It is a good idea to soak seeds for a few hours and then to dust seed with fungicide before sowing. Cover seeds with compost or vermiculite 12 mm deep and water gently. Seedlings emerge in 10–14 days and can be thinned to the strongest when 10 cm high. For the average family 9–12 plants are sufficient for one sowing. You can have a year round supply with 2–3 sowings between spring and autumn.

Grow silver beet quickly, like lettuce, with regular watering and side dressings of nitrogen fertilizer or liquid feeds every 2–3 weeks. Cultivate to control weeds and mulch around plants in hot weather. Diseases and pests are not serious but leaf spot may be troublesome. (See Chapter 17 for control.)

Start picking outside stalks and leaves when large enough. Break them off at the base with a downward and sideways action. Always leave 4–5 centre stalks for quick regrowth. Cut off any flower stems which appear—but once flowering commences plants become unproductive very quickly. Silver beet keeps well in the refrigerator crisper for up to a week but for best flavour, cook leaves immediately after picking. Silver beet, like lettuce, does not freeze well.

Varieties Fordhook Giant has been selected over many years for quality and high yield to become the leading variety for home garden and commercial growing. It has dark green, blistered leaves and creamy-white stalks. Novelty varieties are Rainbow Chard with leaf stalks in purple, red, pink and yellow, and Ruby Chard with bright crimson stalks.

SPINACH

Spinach is a cool season, short-day crop which tends to run to seed in warm weather with long days. For this reason it is widely grown in Victoria and Tasmania and is why home gardeners in New South Wales and Queensland prefer to grow the more adaptable silver beet as 'spinach'. In warm northern areas, sow spinach in winter months only. For temperate climates, sow from late summer to early winter. In cold climates sow from late summer right through to early spring.

Sow seeds direct in well-prepared beds with plenty of organic matter plus mixed fertilizer added. Plants prefer well drained, fertile soil with good structure similar to that for lettuce. Sow a few seeds in clumps 30–40 cm apart, cover with compost or vermiculite and water. Thin seedlings in each clump to the strongest. For the average family 12–15 plants are sufficient. Plants grow quickly so make successive sowings every 3–4 weeks for a continuous supply.

Like lettuce, spinach needs regular watering and side dressings of nitrogen fertilizer or liquid feeds every 10–14 days. A grass or straw mulch tucked around each plant will keep the leaves free of dirt. Major pests are leaf miner and mites. (See Chapter 17 for control.) When plants

English Hybrid spinach is a cool season crop and easy to grow

are large enough, pick outside leaves individually like silver beet. Each plant will keep producing for about four weeks or so. Leaves keep well in the refrigerator crisper for up to a week but are best cooked immediately after picking. Like lettuce and silver beet, spinach leaves tend to collapse when frozen.

Varieties Hybrid varieties have largely replaced open-pollinated varieties such as Prickly Seeded. Hybrid 102, sold as English Hybrid or Winter Hybrid, is the most widely-grown variety. It has rather upright, plain, medium green leaves with excellent flavour.

SQUASH

Squash is another warm season vine crop grown in the same way as cucumber. Refer also to the Sowing Guide.

Varieties There are two types of squash. Summer squash are picked when immature, like cucumbers or marrows. Winter squash are picked when fully mature, like pumpkins, and keep well in storage. Both types are, of course, grown during the spring, summer, autumn period.

Early White Bush is the most widely known variety of summer squash. The fruits are round, 15 cm in diameter with scalloped edges. Both skin and flesh are creamy-white. Each fruit weighs 1–1.5 kg. Green Buttons Hybrid is a recent introduction. It has lime-green fruit with scalloped edges. They are best picked when 5–10 cm in diameter. Pick regularly to encourage further flowers and fruit. Green Buttons Hybrid can be grown in a large pot or tub about 40 cm in diameter and the same depth. Another interesting new variety is Scallopini Hybrid Squash with a prolific crop of shiny, dark green fruit.

Of the winter squash, Green Warted Hubbard is a running vine with round, dark green fruit with pointed edges. The fruit weighs about 5 kg and contains deep orange flesh. Table Queen (Acorn squash) has a smaller grey-green, pear-shaped fruit to 1 kg in weight. Both varieties of winter squash are good keepers.

SWEDES AND TURNIPS

Swedes and turnips are cool season root crops which are grouped together because their climatic requirements and cultivation are almost identical. Both vegetables can be grown in all three climate zones. In all climates, sowing takes place in late summer and autumn, but in cold districts both swedes and turnips can be sown in late winter or early spring as well (see Sowing Guide for best months to sow in each zone). Swedes take 12–16 weeks to grow and have large roots with yellow or buff-coloured flesh. The roots store well. Turnips take less time (10–12 weeks) to grow, have smaller globe-shaped roots with white flesh. Turnips do not store as well as swedes.

Prepare the soil well for direct sowing in the same way as for carrots. Both crops respond to liberal quantities of organic

White Bush squash is a widely grown summer variety. Ready to pick about ten weeks after sowing

The large yellow roots of Champion Purple Top swedes are tender and palatable

matter and a scattering of pre-planting fertilizer in a band where seed is to be sown. Mark out shallow drills 6 mm deep and 20–30 cm apart. Tap out seeds thinly along the drill—4–5 seeds to each 5 cm—cover with compost or vermiculite and water gently. Seedlings emerge in 6–10 days, depending on temperature. Thin seedlings of both crops to 7–10 cm apart. For the average family a 3–5 m row is sufficient at each sowing. Make successive sowings every 3–4 weeks.

Water regularly and if plants are backward give side dressings of nitrogen fertilizer or liquid feeds. Cultivate between rows to keep down weeds but do not hill the plants. The roots are really swollen stems which sit on the soil surface, not below it. Control caterpillars and aphids as for other crops of the cabbage group. (See Chapter 17.)

Start harvesting early for best quality and to give the remaining roots more space to grow. Dig all roots before they become coarse and stringy. Swedes store better than turnips at normal temperatures but both keep well in the refrigerator crisper for 8–10 weeks.

Varieties Best varieties of swedes are Champion Purple Top and Royal Rose. Purple Top White Globe is the most popular turnip. Another good turnip variety, especially for baking, is White Stone.

SWEET CORN

Sweet corn is a very popular vegetable, looks good in the garden and is easy to grow. Home grown cobs are much tastier than the ones you buy, so if you like sweet corn it is worth that bit of extra space. A warm season crop, sweet corn is adapted to all climates. In warm northern areas the best months to sow are July to February, but almost any month (providing there are no frosts) in tropical parts of Queensland and the Top End. In temperate climates, you can sow from August to January and in cold climates from October to December. Sweet corn grows well on both light and heavy soils, providing drainage and soil structure are good. Plenty of fertilizer and water are needed for a bumper crop.

Prepare soil well with organic matter added to have the bed in friable dark damp condition for direct sowing. It is best to grow sweet corn in blocks of short rows rather than one long row. This way pollen from the male flowers or tassels at the top of the plants has the best chance of falling on the female flowers or silks halfway up the stems. You can grow three rows spaced 50–60 cm apart in a bed 150 cm wide.

Apply the pre-planting fertilizer in furrows where each row is to be sown so that seed is not in direct contact. It is best to dust seed with fungicide before sowing to protect against 'damping off'. Mark out the seed furrow 25 mm deep and press seeds into dark damp soil. Space seeds 15 cm apart. This allows for some misses when seedlings are thinned to 20–30 cm apart. Another method is to sow two seeds close together, with 30 cm between each pair, and thin to one seedling. Cover seeds with soil, tamp down, rake the bed and scatter a light grass mulch on the surface. If seed is sown in dark damp soil seedlings will emerge without further watering. Scatter snail baits to protect seedlings from slugs and snails. For the average family a sowing of three rows 2 m long will provide 20–24 plants after thinning out. Make the next sowing when the previous plants are 15–20 cm tall.

After thinning, scatter a side dressing of nitrogen fertilizer around each plant. Repeat this treatment when the tassel first appears between the top leaves. Regular watering is needed while the crops grow, especially in hot weather. A good soaking to field capacity once or twice a week is better than a light sprinkle every day. Cultivate between rows to control weeds and draw soil around the stems to hill the plants at the same time. Corn earworm and aphids are serious pests. (See Chapter 17 for control.) When tassels open out fully they are ready to shed pollen. Overhead watering in early morning will create a humid atmosphere in the crop to promote good pollination of the silks. Pollen is shed—usually about mid-morning—for several days.

Cobs must be harvested at the right time. They are ready to pick when the silks have turned brown and cobs stand out from stem at about a 30 degree angle. Make a further check by pulling open the husk from the top and pressing the grains with the thumb nail. If grain is soft and exudes juice with a creamy consistency, the cob is ready to pick. In over-ripe cobs, sugar quickly turns to starch and the grains are tough and doughy. Pick cobs with a downward and twisting action.

For top quality, remove husks and cook as soon as possible. Water must be boiling before cobs are put in saucepan. Cobs in-the-husk keep well in the refrigerator crisper for 3–4 days. To freeze corn on the cob, remove husks, blanch cobs for eight minutes, cool quickly and package each cob separately. For corn off the cob, blanch cobs for five minutes only, cool quickly, cut grains from cob, package and freeze.

Varieties Most sweet corn varieties are now F1 hybrids. The original hybrid, Golden Cross Bantam, has been replaced by superior, high-yielding varieties such as Iochief Hybrid and NK 195 Hybrid. Popcorn is different from sweet corn but it is grown in the same way. The cobs are harvested when the grain is hard and fully mature. The very high starch content in the grains makes them explode or pop when heated.

SWEET POTATOES

Sweet potato, a close relative of convolvulus, is a warm season, frost susceptible vegetable which needs a growing season of five months. It is suitable for tropical/subtropical or very warm temperature areas. The plant is a vigorous, rather untidy vine and takes quite a lot of garden space. It prefers light soils for good tuber development.

Start plants from cuttings, which may be available from nurseries in spring, or buy a few sweet potatoes and bury them in a box of moist sand placed in a warm spot. When tubers shoot, divide them up for planting. Prepare soil as for potatoes and apply a pre-planting fertilizer in a band 30 cm wide at 3 tablespoons per metre where plants are to grow. Planting on a raised ridge makes for better drainage and easier harvesting. Set cuttings 40–50 cm apart and 5–7 cm deep with 100 cm between rows. For an ample supply of tubers, 18–24 plants are sufficient.

Lift vines occasionally as they grow to prevent rooting at the nodes along the stems. Do not give additional fertilizer, especially not nitrogen, which promotes top growth at the expense of tubers. Diseases and pests are not a problem. It is best to wait until plants are completely yellow and tubers fully mature before digging. Mature tubers have firm skin and when cut dry quickly to a creamy white colour. After digging, leave tubers in the sun for a few days to cure. Discard any diseased or damaged ones and store remainder in sacks in a cool, dry, airy place.

TOMATOES

In a popularity poll for home-grown vegetables, tomatoes would top the list. Tomatoes give a higher yield for space occupied than any other vegetable. A good average yield is 3–5 kg (7–11 lb) per plant but when well grown, each plant can yield 10 kg (22 lb) of fruit or more.

Tomatoes are warm season, frost susceptible plants which need a growing season of about three months, so they are adapted to all climate zones in Australia. In frost-free, warm northern areas, tomatoes are grown throughout the year. In temperate climates the best months to start tomatoes are August to December and in cold districts September to November. In mild districts like Sydney, an early crop (August sowing) and a late crop (November sowing) will supply tomatoes for about five months of the year (mid-December to mid-April). Tomatoes grow well on light and heavy soils but the usual rules for vegetable soils apply—good drainage, organic matter for soil structure and adequate water and fertilizer. Phosphorus is a most important nutrient for tomatoes and lack of it, especially in the seedling stage, will reduce yields of fruit. Nitrogen is needed too but not in excessive quantities as for leaf vegetables.

Seeds can be sown direct (especially for the late crop) but it is more usual to raise seedlings in boxes or punnets for transplanting. Seedlings for the early crop can be pricked out into 10 cm plastic pots and grown on for several weeks in a warm sunny spot. If the garden bed is not ready, transfer the seedlings to 15–20 cm pots. Spacing plants is an important consideration. For tall, staking tomatoes, set seedlings 50–60 cm apart each way. This relatively close spacing (much closer than commercial crops) gives quite enough light to each plant, but plenty of watering and fertilizer is needed. For an average family, 12–15 plants are sufficient for one sowing. Cultivate around plants and between rows to destroy weeds. On flat or sloping beds you can make furrows between the rows for furrow irrigation. This is a useful method as many leaf spot and leaf blight diseases are spread by overhead watering.

With adequate base fertilizer (in seed bed soil or applied before direct sowing) extra fertilizer is not needed until plants have set their first truss of fruit. At this stage, scatter a tablespoon of mixed fertilizer around each plant and water

Pruning tomato plants. Break off laterals (shown by arrows) which grow from the leaf axils. If laterals are well developed cut them off with a sharp knife or pair of scissors. Tie the leader(s) to the stake with garden twine at about 30 cm intervals just above a leaf. Make a figure-eight tie which allows the leader(s) to increase in size. Take care that flower trusses are not squeezed between the leader and the stake

in. Repeat treatment every 4–5 weeks as plants grow. Water regularly to field capacity—each week when plants are small, but increase this to twice a week when plants are carrying a heavy crop.

Most tomatoes are grown on stakes about 2 m in length. Hammer stakes into the soil 5 cm from stem of plants after transplanting. Plants are pruned to 2 leaders (main stems) which are tied to the stakes. Break off laterals (which grow from leaf axils) with a sideways action when small or cut with a sharp knife when larger. The lateral to select for the second leader is the one immediately below the first flower truss. This lateral is more vigorous than others. Tie the leaders to the stake with garden twine just above a leaf stalk to stop it slipping down the stem. Ties should be about 30 cm apart. Make a figure-8 tie (see diagram) which allows the leader to increase in size. Take care that flower trusses are not squeezed between leader and stake. Carefully twist the leader so that each truss faces outwards. Diseases and pests can be a problem with tomatoes. Fruit fly (in warm climates) and tomato caterpillars are the worst pests and must be controlled. (See Chapter 17.)

For top quality fruit, pick when red ripe, although slightly coloured fruit ripens well indoors. Fruit keeps well in the refrigerator for 1–2 weeks (ripe) or 4–5 weeks (green). Fresh tomatoes cannot be quick frozen. Preserve them by bottling or cook them and then package for freezing.

There are more varieties of tomatoes than any other vegetable crop. This reflects their world-wide popularity and emphasizes the many different types available for selection and breeding. Fortunately, the number of home garden varieties is relatively small but readers will realise that many other varieties are listed by Departments of Agriculture and garden writers in each state.

Staking Varieties Grosse Lisse, released over forty years ago, is still the most popular garden variety. It is mid-season in maturity with medium to large globe-shaped fruits. An earlier maturing variety is Red Cloud. The bush is not as tall as Grosse Lisse but has excellent orange-red fruit. Rouge de Marmande is a very early variety which sets fruits at lower temperatures—20°C—than other varieties. It has Chinese-type fruit with distinct ribs. Yellow is another staking variety with bright yellow fruit. Apollo Hybrid, a recent introduction, is as early in maturity as Rouge de Marmande but is a more prolific cropper and an excellent home garden variety. As the name implies, Whopper is the ideal variety for gardeners who delight in describing how they grow large fruit. The flavoursome fruit is indeed very large, with a smooth skin and deep red colour. The plant is tall and vigorous.

Sweet 100 tomato

Non-Staking Varieties These varieties have wide, spreading bushes and do not need stakes for support. Because of their lateral spread they must be spaced 75–100 cm per plant each way. The most popular for home gardens are Egg Tomato or Early Crop. Both have deep red, egg-shaped fruits which are used fresh in salads and are excellent for bottling.

Dwarf and Mini-Tomatoes Patio Hybrid is a dwarf plant to 60 cm tall, but the fruit is medium to large size. A short stake alongside will help support plants. Tiny Tim is a true mini-tomato. Plants grow to 30–40 cm tall and bear masses of bite-size fruit. These varieties are ideal for growing in large pots or tubs, but don't forget to give them lots of water in hot weather.

TURNIPS

(*See Swedes and Turnips.*)

VINE CROPS

(*See Choko, Cucumbers, Marrows, Melons, Pumpkin and Squash.*)

WATERMELON

(*See Melons.*)

ZUCCHINI

(*See Marrows.*)

Vegetables

What to look for	Nutrition and kilojoules	Storage	Preparation	Method of cooking
BEANS Firm, long straight pods, crisp enough to snap. Good green colour.	Small mineral and vitamin content particularly vitamin C. Some fibre. 126 kilojoules per 100 grams.	Wash, drain and store in vented plastic bag in refrigerator. Use soon after harvest.	Wash, top and tail and remove strings. Slice diagonally or leave whole.	Steam or boil. May be lightly tossed in butter to glaze. Do not overcook.
BROCCOLI Compact flower heads with no sign of yellow. Leaves and stems should show no sign of ageing.	Good source of folic acid. Excellent source of vitamin A and vitamin C. Fair source of calcium and fibre. 147 kilojoules per 100 grams.	Keep dry. Handle as little as possible as flowers bruise. Store in vented plastic bag in refrigerator. Use within 1–2 days.	Wash. Steam flowers whole in bunches. Stems may be sliced and served as a separate vegetable.	Steam, boil or oven bake. May be served raw if finely sliced. Remember stems take longer to cook than flowers.
BRUSSELS SPROUTS Firm and compact with no limp leaves.	Excellent source of vitamin C. Good source of folic acid and fibre. 205 kilojoules per 100 grams.	Wrap in plastic. Store in refrigerator.	Wash. Trim stalk. Remove any poor quality leaves.	Steam or boil. May be deep fried in batter after initial cooking.
CABBAGE Firm heads. Outer leaves should be strongly coloured and not limp.	Good source of vitamin C. Some calcium and fibre. 109 kilojoules per 100 grams.	Trim lightly and remove outer leaves. Wrap in plastic and store in refrigerator. Use within a week of harvest.	Remove any poor quality leaves. Wash. Remove rib if desired. Slice finely or leave whole.	Drop a whole walnut into the water while cooking cabbage to minimise odour. May be boiled, steamed or stir fried.
CAPSICUM Well shaped, thick walled and firm, with a uniform glossy colour (deep red or bright green).	Very good source of vitamin C. Fair source of vitamin A. 109 kilojoules per 100 grams.	Store in plastic bag in refrigerator. Use within 5 days.	Simply wash and remove all seeds.	Delicious raw in salads. May be stuffed and baked or used in soups and casseroles.
CARROTS Firm, smooth and well formed. Deep orange to red in colour.	Outstanding source of vitamin A. Some fibre. 151 kilojoules per 100 grams.	Store in plastic bag in refrigerator.	Wash and scrape lightly. May be left whole, sliced or diced for cooking.	Steam, boil, braise or shred. Delicious raw or cooked.
CAULIFLOWER Should not have a ricey appearance or obvious flowers. Look for firm white compact heads without spots or bruises.	Very good source of vitamin C, fair source of folic acid and fibre. 109 kilojoules per 100 grams.	Remove all leaves as they absorb moisture from head. Store in plastic bag in refrigerator. Use before heads turn brown.	Wash and break into flowerets or leave whole.	Steam or boil and top with cheese sauce. Use raw in salads and soup.
CELERY Crisp, firm, well-coloured stalks with no blemishes or limp leaves.	Small mineral content, some fibre. 75 kilojoules per 100 grams.	Wash and store in plastic bag in refrigerator.	Remove leaves, wash stalks and cut to desired length. Remove any loose fibres.	Eat fresh, braise or stir fry. May be added for flavour to stews or soups.
CUCUMBER Should be green with no yellow colouring. Firm and fresh looking.	Low energy, high water content. 59 kilojoules per 100 grams.	Store in crisper in refrigerator. Use within a few weeks.	Wash and slice. Remove rind if desired.	Boil, steam or bake with filling. Most often eaten raw.
EGGPLANT Dark purple to purple black colour with glossy skin. Firm to touch.	High water content. Small amounts of most minerals and vitamins. Some fibre. 105 kilojoules per 100 grams.	Keep for about 7 days in refrigerator crisper.	Wipe over. Not necessary to remove skin. Discard stalk. Slice and leave sprinkled with salt 20 minutes to extract bitter juice. Rinse prior to cooking.	Bake, boil fry or mash.
LETTUCE Choose firm green heads with crisp, blemish-free leaves.	Some potassium, fibre and folic acid. 71 kilojoules per 100 grams.	Perishable. Store in plastic in refrigerator crisper and use as soon as possible.	Remove core, washunder running water, drain. Tear rather than cut leaves.	Usually eaten fresh. May be braised, stir fried or added to soup.

MUSHROOMS Look for firmness, white or creamy colour and unbroken shape. Avoid withered mushrooms.	Good source of niacin and riboflavin. Excellent source of potassium. 92 kilojoules per 100 grams.	Perishable. Store in paper bag in refrigerator. Use within 2–3 days.	Do not peel. Wipe over cap. Only remove stem if desired. Do not wash under running water.	Can be eaten raw, baked with a filling or sauteed in butter. Cook only lightly.
ONIONS Firm, with clear outer skin, no dark patches or signs of sprouting.	Small amount of vitamins and minerals. Rich in sugars. 147 kilojoules per 100 grams.	Store in cool, dry and dark area. May be stored in refrigerator.	Peel and cut in required style e.g. rings, quarters etc.	Sauté, boil, bake, cream or fry.
PEAS Pods should be bright green in colour. Very firm and full pods indicate over maturity.	Some protein and iron. Fair source of thiamin and folic acid. Good source of dietary fibre. 335 kilojoules per 100 grams.	Store in plastic bag in refrigerator. Use as soon as possible.	Remove shell and discard, unless using snowpeas.	Boil, steam, or braise with lettuce.
POTATOES Firm and unbroken skin with no green tinge. There should be no dark spots or green shoots.	Fairly good source of vitamin C. Good source of potassium and dietary fibre. Some protein. 335 kilojoules per 100 grams.	Store in cool, dry and dark area. Do not store in refrigerator.	Do not soak in water. Only peel if necessary.	Bake, boil, fry, steam or mash.
PUMPKIN Firm, bright and well coloured flesh.	Good source of vitamin A. Some fibre. 130 kilojoules per 100 grams.	Cool, dark storage until cut. Then remove seeds, wrap in plastic, and store in refrigerator.	Wipe over. Cut into suitably sized pieces, remove seeds, stringy pieces and skin if desired.	Bake, boil, steam or mash.
ROCK MELONS Smell is a good indication of flavour and ripeness. Avoid soft spots and look for a clean stem scar.	Excellent source vitamin C. Good source vitamin A. Fair source dietary fibre. Some iron. 105 kilojoules per 100 grams.	Ripen at room temperature for finer flavour. Wrap cut melon in plastic and store in refrigerator away from butter and milk.	Tends to flavour other foods when cut. Remove seeds and serve chilled. Slice as required.	Use in fruit salad, eat alone or with ice cream. Ideal as an entree, slice and serve with prosciutto, or fill with port.
SILVER BEET Glossy, bright green leaves that show no sign of limpness.	Excellent source of vitamin A. Good source of folic acid and vitamin C. Fair source of calcium and iron. Some fibre. 96 kilojoules per 100 grams.	Buy on day required. Store in plastic bag in refrigerator. Highly perishable.	Wash carefully. Tear rather than cut leaves. Stems may be served as a separate vegetable.	Eat raw in salad, steam or boil. Used as a wrapper for fillings.
SWEET CORN Husks fresh and green in colour. Kernels well filled, tender, milky, and pale yellow in colour.	Some protein and vitamin A. Good source of dietary fibre. 406 kilojoules per 100 grams.	Wrap in vented plastic bag and keep refrigerated.	Remove corn silk and outer leaves.	Boil, bake or steam.
TOMATOES Free of blemish. Firmly fleshed, and weight heavy in the hand.	Good source of vitamin C, some vitamin A. 88 kilojoules per 100 grams.	Only refrigerate when over ripe. Always remove from refrigerator 1 hour before eating to improve flavour.	Wash, dry and remove stalk. Remove skin only if necessary by plunging into boiling water.	Use fresh or stew, bake, sauté, stuff, or prepare as a sauce.
WATER MELON Large, well coloured bright fruit that is heavy in the hand. A yellowish underside is a good guide to ripeness.	Fair source of vitamin C. Some vitamin A. 113 kilojoules per 100 grams.	Store in cool place or in refrigerator. When cut, use promptly.	Wipe skin, serve chilled in slices or wedges. Use a melon baller for a quick dessert.	Great for picnics and in fruit salad, jams and pickles. Rind can be steamed and served with butter and nutmeg. Lovely as a refreshing drink.
ZUCCHINI Well shaped with firm glossy skin and good colour.	Low energy, high water content. 66 kilojoules per 100 grams.	Place in plastic bag in refrigerator.	Wash or wipe over. Use unpeeled, sliced or halved or cut in strips.	Boil, steam, bake or eat raw.

Table reproduced by courtesy of the NSW Department of Agriculture

SOWING GUIDE FOR VEGETABLES

This gives a summary for vegetable crops under the following headings:

1. Best months to sow in each climate zone.
2. Sowing method—seed bed (s) or sow direct (d)
3. Sowing depth (mm)

Vegetable	Tropical/Subtropical ● Subtropical only ▲ Tropical only ■												Temperate											
	J	F	M	A	M	J	J	A	S	O	N	D	J	F	M	A	M	J	J	A	S	O	N	D
Artichokes (Suckers)	●	●	●													●	●	●	●	●	●			
Asparagus (2 yr. crowns)					●	●	●											●	●					
Beans (dwarf)	▲	●	●	●	●	●	●	●	●	●	▲	▲	●	●							●	●	●	●
Beans (climbing)	▲	●	●	●	●	●	●	●	●	●	▲	▲	●								●	●	●	●
Beetroot		●	●	●	●	●	●	●	●	●			●	●	●				●	●	●	●	●	●
Broad Beans				●	●	●	▲									●	●	●	●					
Broccoli	▲	▲	●	●	●	●	●	●	▲	▲	▲		●	●	●	●	●							●
Brussels Sprouts	*Not Suitable*												●	●	●									●
Cabbage	▲	●	●	●	●	●	●	●	●	●	●	▲	●	●	●	●		●	●	●	●	●	●	●
Cape Gooseberry	●	●	●	■	■	■	●	●	●	●	●	●								●	●	●	●	●
Capsicum (Pepper)	●	●	●	■	■	■	●	●	●	●	●	●								●	●	●	●	●
Carrots		●	●	●	●	●	●	●	●	●	●		●	●	●				●	●	●	●	●	●
Cauliflower	●	●	●	●									●	●	●									●
Celery	●	●	●	●							▲	▲	●	●						●	●	●	●	●
Chicory			●	●	●	●	●	●	●				●	●	●					●	●	●	●	●
Chinese Cabbage	●	●	●	●	●	●	●	●	●	●	●	●	●	●	●	●			●	●	●	●	●	●
Choko (see Note 1)					■	■	●	●	●	●										●	●	●		
Cress	●	●	●	●	●	●	●	●	●	●	●	●	●	●	●	●	●	●	●	●	●	●	●	●
Cucumber (see Note 3)	●	●	●	■	■	■	●	●	●	●	●	●	●								●	●	●	●
Eggplant	●	●	●	■	■	■	■	■	●	●	●	●									●	●	●	●

*Time to picking is period from sowing to harvest for direct-sown crops, but for seedlings is period from transplanting to harvest.

†Select from winter or summer varieties.

4. Seedling emergence (days)
5. Sowing or planting distance between rows and plants within the row (cm)
6. Time of picking (weeks)
7. Quantity for family of four at each sowing—number of plants (p) or row length (m)

Cold												Sowing Method Seedbeds (S) Sow direct (D)	Sowing Depth (mm)	Seedlings Emerge (days)	Sow and thin or transplant to . . . *cm* apart		Time to Picking * (weeks)	Quantity for family of four at each sowing Plants (p) Row length (m)
J	F	M	A	M	J	J	A	S	O	N	D				Rows	Plants		
						•	•	•	•			D	150	—	100	100	20–28	3–5 p
					•	•						D	150–200	—	100	30–50	16–24	20–25 p
•									•	•	•	D	25	7–10	50–60	7–10	8–10	3–5 m
									•	•	•	D	25	7–10	100	10–15	10–12	1–3 m
•	•							•	•	•	•	D	12	10–14	20–30	7–10	10–12	2–3 m
		•	•			•	•					D	40	10–14	60–75	15–20	18–20	5–6 m
•	•									•	•	S or D	6	6–10	45–60	45–60	12–16	9–12 p
•	•								•	•	•	S or D	6	6–10	60–75	60–75	16–20	6–9 p
•	•	•					•	•	•	•	•	S or D	6	6–10	40–70	40–70	8–16	9–12 p
								•	•	•		S	6	14–28	100	100	20–28	2–3 p
								•	•	•		S or D	6	10–14	50–60	50–60	10–16	4–6 p
•	•							•	•	•	•	D	6	10–21	20–30	3–5	16–20	4–6 m
•										•	•	S or D	6	6–10	50–75	50–75	14–26	9–12 p
								•	•	•	•	S	6	14–21	30–40	30–40	20–22	16–20 p
•	•							•	•	•	•	D	12	10–14	20–30	3–5	16–20	1–2 m
•	•	•					•	•	•	•	•	D	6	6–10	30–40	30–40	8–10	6–9 p
Not Suitable												D	50–75	—	—	100	18–20	1–3 p
•	•	•	•	•	•	•	•	•	•	•	•	D	6	6–10	Sow seeds in pots or garden		4–6	(see Note 2)
									•	•	•	D	12	6–10	100	40–50	8–12	4–6 p
									•	•		S or D	6	10–14	60–75	60–75	14–16	4–6 p

Note 1—Usually grown on fence or trellis. Plant whole fruit which has started to sprout, with cotyledons at soil level.
Note 2—Make successive sowings as required.
Note 3—Early plants can be raised in punnets or pots.
Note 4—Many herbs are perennials and will grow for several years.

Vegetable	Tropical/Subtropical ● Subtropical only ▲ Tropical only ■												Temperate											
	J	F	M	A	M	J	J	A	S	O	N	D	J	F	M	A	M	J	J	A	S	O	N	D
Endive			●	●	●	●	●	●					●	●	●					●	●	●	●	●
Herbs (see Note 4)	●	●	●	●	●	●	●	●	●	●	●	●	●	●	●	●			●	●	●	●	●	●
Kohl Rabi	▲	●	●	●	●								●	●	●				●	●	●			
Leeks	▲	▲	▲	●	●	●							●	●	●	●					●	●	●	●
Lettuce†	●	●	●	●	●	●	●	●	●	●	●	●	●	●	●	●	●	●	●	●	●	●	●	●
Marrow (see Note 3)	●	●	●	■	■	■	●	●	●	●	●	●	●								●	●	●	●
Melons (see Note 3)	●	●	■	■	■	■	●	●	●	●	●	●									●	●	●	●
Mustard	●	●	●	●	●	●	●	●	●	●	●	●	●	●	●	●	●	●	●	●	●	●	●	●
Okra	●	●	■	■	■	■	■	●	●	●	●	●									●	●	●	●
Onions		●	●	●	●										●	●	●	●	●					
Onions (Spring)	●	●	●	●	●	●	●	●	●	●	●	●	●	●	●	●	●			●	●	●	●	●
Parsnip		▲	●	●	●	●	●	●	▲				●	●	●				●	●	●	●	●	●
Peas (dwarf)			●	●	●	●	●							●	●	●	●	●	●	●				
Peas (climbing)			●	●	●	●	●							●	●	●	●	●	●	●				
Potatoes (tubers)	▲	●	●	●	●	●	●	●	●				●	●					●	●	●			
Pumpkin (see Note 3)	●	●	■	■	■	■	●	●	●	●	●	●									●	●	●	●
Radish	●	●	●	●	●	●	●	●	●	●	●	●	●	●	●	●	●			●	●	●	●	●
Rhubarb (seed)	▲	●	●	●	●			▲	▲	▲	▲	▲								●	●	●	●	●
Rhubarb (crowns)	▲	▲	▲	▲		●	●	●	▲	▲	▲	▲	●	●				●	●	●	●	●	●	●
Rosella	■	■	■	■	■	■	■	●	●	●	●	●									●	●	●	
Salsify		▲	●	●	●	●	●	▲	▲				●	●	●				●	●	●	●	●	●
Shallots (bulbs)		■	●	●	●	●	●							●	●	●	●	●						
Silver Beet	●	●	●	●	●	●	●	●	●	●	●	●	●	●	●				●	●	●	●	●	●
Spinach	■	■	■	●	●	●	●				■	■		●	●	●	●	●						

*Time to pick:ng is period from sowing to harvest for direct-sown crops, but for seedlings is period from transplanting to harvest.

†Select from winter or summer varieties.

Cold												Sowing Method Seedbeds (S) Sow direct (D)	Sowing Depth (mm)	Seedlings Emerge (days)	Sow and thin or transplant to . . . *cm* apart		Time to Picking * (weeks)	Quantity for family of four at each sowing Plants (p) Row length (m)
J	F	M	A	M	J	J	A	S	O	N	D				Rows	Plants		
•	•							•	•	•	•	S or D	6	10–14	20–30	20–30	8–12	6–9 p
•	•	•					•	•	•	•	•	S or D	6	6–28	Sow seeds in pots or garden		12–20	(see Note 2)
•	•						•	•	•			D	6	6–10	30–40	10–15	8–10	1–2 m
•	•	•							•	•	•	S	6	10–14	15–20	15–20	12–20	40–50 p
•	•	•	•	•	•	•	•	•	•	•	•	S or D	6	6–10	20–30	20–30	8–12	9–12 p
									•	•	•	D	20	6–10	100	100	8–14	3–6 p
									•	•	•	D	20	6–10	150	100	14–16	2–3 p
•	•	•	•	•	•	•	•	•	•	•	•	D	6	6–10	Sow seeds in pots or garden		4–6	(see Note 2)
									•	•	•	S or D	6	10–14	100	50–60	16–20	4–5 p
			•	•	•	•	•					S or D	6	10–14	20–30	8–10	24–32	4–6 m
•	•	•	•				•	•	•	•	•	D	6	10–14	5–10	1–2	8–12	0.5–1 m
•	•						•	•	•	•	•	D	6	21–28	30–40	5–7	18–20	3–5 m
					•	•	•	•	•			D	25	7–10	40–50	3–5	12–16	3–5 m
					•	•	•	•	•			D	25	7–10	100	3–5	14–16	1–3 m
							•	•	•	•	•	D	100–150	—	60–75	30–40	16–20	50–60 p
									•	•	•	D	20	6–10	100	100	14–16	3–6 p
•	•	•	•					•	•	•	•	D	6	5–8	10–15	3–5	6–8	0.5–1 m
								•	•	•	•	S or D	12	10–21	40–50	40–50	16–20	12–15 p
•	•						•	•	•	•	•	D	80–100	—	40–50	40–50	8–12	12–15 p
Not Suitable												S	12	10–14	150	150	20–22	2–3 p
•	•						•	•	•	•	•	D	6	10–14	30–40	5–7	20–22	1–2 m
•	•	•	•	•								D	50–75	—	15–25	15–25	12–14	6–9 p
•	•						•	•	•	•	•	S or D	12	10–14	30–40	30–40	8–12	9–12 p
	•	•	•	•	•	•	•					D	12	14–21	30–40	30–40	8–10	12–15 p

Note 1—Usually grown on fence or trellis. Plant whole fruit which has started to sprout, with cotyledons at soil level.
Note 2—Make successive sowings as required.
Note 3—Early plants can be raised in punnets or pots.
Note 4—Many herbs are perennials and will grow for several years.

Vegetable	Tropical/Subtropical • Subtropical only ▲ Tropical only ▪												Temperate											
	J	F	M	A	M	J	J	A	S	O	N	D	J	F	M	A	M	J	J	A	S	O	N	D
Squash (see Note 3)	•	•	▪	▪	▪	▪	•	•	•	•	•	•									•	•	•	•
Swedes		•	•	•									•	•	•									
Sweet Corn	•	•	▪	▪	▪	▪	•	•	•	•	•	•	•								•	•	•	•
Sweet Potato (Shoots)	•	•	▪				•	•	•	•	•	•									•	•	•	
Tomato	•	•	•	•	•	•	•	•	•	•	•	•								•	•	•	•	•
Turnips		•	•	•	•								•	•	•	•								

*Time to picking is period from sowing to harvest for direct-sown crops, but for seedlings is period from transplanting to harvest.

†Select from winter or summer varieties.

GROWING VEGETABLES IN CONTAINERS

Many vegetables are suitable for growing in containers on balconies or patios if garden space is limited. If you follow the advice on 'Gardening in Containers' (Chapter 14) and select the correct varieties, there is every reason to expect success with container-grown vegetables.

Most vegetables can be grown in containers given the right conditions of soil, moisture, nutrients and aspect. Follow the cultural notes in this chapter and refer to the table above for the best times to sow.

The table below lists the most suitable kinds and varieties for this type of culture. All of them can be grown in a wide range of climates.

VEGETABLE	MOST SUITABLE VARIETIES	RECOMMENDED MIN. DEPTH OF CONTAINER
Cabbage	Earliball	25 cm
	Hybrid Sugarloaf	25 cm
Capsicum	All	40 cm
Carrot	Baby	25 cm
	Early Horn	25 cm
	Chantenay	25 cm
Cress	Curled Cress	10 cm
Cucumber	Pot Luck	40 cm
	Patio Pik	40 cm

Cold												Sowing Method Seedbeds (S) Sow direct (D)	Sowing Depth (mm)	Seedlings Emerge (days)	Sow and thin or transplant to . . . *cm* apart		Time to Picking * (weeks)	Quantity for family of four at each sowing Plants (p) Row length (m)
J	F	M	A	M	J	J	A	S	O	N	D				Rows	Plants		
									•	•	•	D	20	6–10	100	100	12–14	4–6 p
•	•						•	•				D	6	6–10	20–30	7–10	12–16	3–5 m
									•	•	•	D	25	6–10	50–60	20–30	12–16	20–24 p
Not Suitable												D	50–75	—	100	40–50	18–20	18–24 p
								•	•	•		S or D	6	10–14	50–60	50–60	12–20	12–15 p
•	•	•				•	•	•				D	6	6–10	20–30	7–10	10–12	3–5 m

Note 1—Usually grown on fence or trellis. Plant whole fruit which has started to sprout, with cotyledons at soil level.
Note 2—Make successive sowings as required.
Note 3—Early plants can be raised in punnets or pots.
Note 4—Many herbs are perennials and will grow for several years.

GROWING VEGETABLES IN CONTAINERS

VEGETABLE	MOST SUITABLE VARIETIES	RECOMMENDED MIN. DEPTH OF CONTAINER
Eggplant	All	40 cm
Herbs	Many types	20 cm
Lettuce	Cos	25 cm
	Mignonette	25 cm
	Buttercrunch	25 cm
Mustard	Finest White	10 cm
Onion	Spring or Shallot Bunching	20 cm
Pumpkin	Golden Nugget	30 cm
Radish	All	20 cm
Silver Beet	Fordhook Giant	25 cm
Tomato	Patio Hybrid	30 cm
	Tiny Tim	30 cm
	Small Fry	30 cm
Zucchini Marrow	Blackjack hybrid	40 cm
	Greyzini hybrid	40 cm

Chapter 9

THE FLOWER GARDEN

Although many flowering annuals, biennials and perennials need similar soil and cultivation, they have many and varied likes and dislikes. You need to know their different requirements to get the best results. Some flowers like full sunlight, while others prefer shade; some are sown in spring, others in autumn, while a further group can be sown almost year round. There are some which prefer fertile, moist soil, and others will grow in rather poor soil and drier conditions.

There are many hundreds of flowering plants in the gardening scene, but this chapter is confined to those of interest to most Australian home gardeners. Full details of growing them—soil preparation and cultivation—are not always given except when there is a special requirement to which attention should be paid. New gardeners should study the general information on soil and use of fertilizers in Chapters 3 and 4 and especially the methods for sowing seed and raising seedlings in Chapter 5.

ANNUALS

These plants complete their life span in one year. Most of them flower in 3–4 months from sowing. Flowering finishes as the seeds ripen and the plants die. Annuals are divided into two groups: summer flowering and winter/spring flowering. The former are sown in spring; the latter are sown in summer and autumn. In some flowers there are both annual and perennial types—alyssum, lupins and statice are good examples. Many so-called annuals are really perennials. In cool climates they may last for three or four years but in warmer districts their effective life is much shorter.

Annual flowers are always favourites for the home garden—no other plants give such a colourful display in such a

Gay annuals provide a colourful ribbon between fence and lawn

Asparagus is easy to grow, and produces crops for many years

Spring or salad onion

Sugarsnap is an outstanding edible podded pea. Cook them or use in salads

Purple King is a delicious bean. It turns bright green when cooked

Spring blossoms

(*Top left*) Rocks and low-growing plants complement well-established trees

(*Left*) Magnolia

(*Bottom left*) Tulips and hyacinths for spring colur

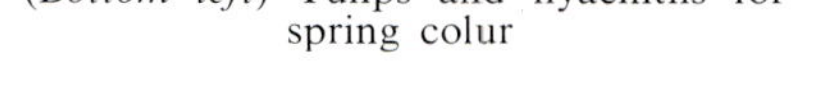

Cineraria flowers well in the garden or in pots. Ideal for shade

Portulaca gives best results when planted in a full sun aspect

Jubilee marigolds flower on vigorous, sturdy bushes

The Australian native strawflower (*Helichrysum*) is a popular everlasting for background or clump plantings

Beautiful Austral stock is a recent introduction. It was bred at Yates Research Farm near Sydney

Morning Glory can be grown as an attractive ground cover

Sunnygold Cosmos is a vigorous grower

Sweet peas contribute colour and fragrance to a garden or vase

Can Can is a mixture of delightful large-flowered pansies

short time and for so little trouble. They can be changed from season to season for varying effect and colour combination. They should not be confined to narrow borders only, but allowed to display their brilliance in large masses, clumps or drifts. They complement other garden subjects and plantings can be arranged so that flowers are in bloom (except in very cold districts) at almost any time of the year.

BIENNIALS

These plants usually live for two years. Like annuals, they flower the first year. At the end of this first flowering, the leaves die back or the plant becomes dormant. As the flowering period comes around again, plants break into fresh growth and another flowering follows. After this second flowering, plants die in the same way as annuals.

Cool and temperate climates suit most biennials, but there are a few exceptions, such as Canterbury Bells. Many others are grown successfully in warm climates if treated as annual plants. It is the cooler conditions between flowering which encourage the dormant period. Without a cold period the plant is less likely to make a good flush of growth in the second year. In cold climates, biennials are usually sown in spring. They flower in late spring and summer and then remain dormant through autumn and winter to restart growth for flowering again in warmer weather.

PERENNIALS

This term applies to plants which have an effective life of three years or more. Some, under favourable conditions—especially in cool climates—may remain in the garden permanently. But in this chapter we will concentrate on those perennials which flower the first season from seed. Most perennials bloom at the same time each year, but a few, such as carnations, have been bred and selected to bloom over a longer time—although there are, of course, flush flowering periods. As a rule, perennials which flower quickly from sowing seed, say, within six months, are those which should be freshly sown each year, especially in warm climates. This does not mean that some, for example, snapdragons will not stand cutting back at least once to give a good display at their second flowering.

Flowering bulbs, like anemone and ranunculus, are perennials because under good conditions they continue to flower each season without special attention. Dahlias and chrysanthemums, which are grown from tubers, are in the same category because the tubers carry over from year to year. But they are not usually classified as perennials in the same way as carnations and hollyhocks, where it is the plant itself which persists.

Carnations, chrysanthemums, dahlias and some other perennials may be propagated from seed or by vegetative methods —plant division, cuttings or layers. (See Chapter 6.) They are not always grown from seed because many named varieties do not breed true to type. Plants grown from seed will vary in type and colour so, if you grow a good strain from seed, you may discover a plant of particular quality which is worthwhile propagating vegetatively.

FLOWER GARDEN TERMS

There may be some confusion in the terms used to describe the purpose for which garden flowers are best suited. The following definitions may assist.

Marmalade daisy. For tough conditions

BEDDING PLANTS

Bedding or carpeting plants (usually annuals) are those best suited for planting in masses in a special bed of their own, although they can also be used for planting in rows, clumps or drifts.

BORDER PLANTS

The term 'border' usually applies to beds which are longer than they are wide. The bed may also contain a variety of plants (including shrubs). Border plants are those which lend themselves to planting with other plants or shrubs, either in rows or clumps. A herbaceous border—a term common in English and European gardening—consists of a perennial plant, like shasta daisy, which is cut down each winter to grow again the following spring. There are evergreen borders such as gazanias and violets too.

EDGING PLANTS

Edging plants are dwarf or prostrate plants (annuals, biennials or perennials) for dividing beds or borders from lawn, paths and paved areas. The term also applies to an outside line of dwarf plants bordering a flower bed.

DWARF, MEDIUM AND TALL FLOWERS

DWARF FLOWERS—30 cm or less

Ageratum, Alyssum, Aster (dwarf), Begonia (bedding), Bellis (English Daisy), Calendula, Candytuft, Carnation, Celosia (dwarf), Chrysanthemum (dwarf), Cineraria (dwarf), Cockscomb, Dianthus, Eschscholtzia, Forget-me-not, Fairy Pinks, Gazania, Globe Amaranth, Godetia (dwarf), Impatiens, Linaria, Livingstone Daisy, Lobelia, Lupin (dwarf), Marigold (dwarf), Mignonette, Nasturtium (dwarf), Nemesia (dwarf), Nemophila, Nigella, Ornamental Basil, Ornamental Chilli, Pansy, Phlox (dwarf), Polyanthus, Portulaca, Primula, Salvia (dwarf), Stock (dwarf), Sweet Pea (Bijou, dwarf), Torenia, Verbena, Viola, Virginian Stock, Zinnia (dwarf).

MEDIUM HEIGHT FLOWERS—30–60 cm

Acroclinium (Everlasting Daisy), Antirrhinum (Snapdragon), Aquilegia (Columbine), Aster, Aurora Daisy, Balsam, Boronia, Brachycome (Swan River Daisy), Calendula, Candytuft, Canterbury Bells, Carnation, Celosia, Centaurea (Corn-

Alyssum adds colour to edges, rockeries or wall gardens

flower), Chrysanthemum, Cineraria, Coleus, Dahlia (dwarf), Dianthus, Gaillardia, Geranium, Gerbera, Godetia, Gomphrena (Globe Amaranth), Gypsophila, Helichrysum (Strawflower), Honesty, Linaria, Lupin, Marigold (French), Molucella (Irish Green Bell), Nasturtium, Ornamental Chilli, Petunia, Phlox, Poppy (Iceland), Rudbeckia (Gloriosa Daisy), Salpiglossis, Salvia, Saponaria (Big Gyp), Schizanthus, Statice, Stock, Sweet William, Viscaria, Wallflower, Zinnia.

TALL FLOWERS—60 cm or over

Amaranthus, Aster, Cleome, Cosmos, Chrysanthemum, Dahlia, Delphinium, Hollyhock, Kochia, Larkspur, Lupin (tall), Marigold (tall), Salvia (tall), Scabiosa, Sunflower, Sweet Pea, Zinnia (tall).

FLOWERS FOR SHADE OR SEMI-SHADE

Ageratum, Alyssum, Begonia (bedding), Calendula, Aquilegia, Cineraria, Coleus, Cyclamen, Canterbury Bells, Forget-me-

The stately spikes of delphiniums have few equals as background plantings

Ageratum makes splendid displays of powder blue

not, Impatiens, Linaria, Lobelia, Nasturtium, Nigella, Pansy, Polyanthus, Primula, Schizanthus, Virginian Stock, Wallflower. (Viola and Pansy require half sunlight.)

SPRING FLOWERS

Acroclinium, Ageratum, Alyssum, Antirrhinum, Aquilegia, Candytuft, Canterbury Bells, Centaurea, Chrysanthemum (annual), Cineraria, Delphinium, Dianthus, Forget-me-not, Gaillardia, Godetia, Gypsophila, Helichrysum, Larkspur, Linaria, Lobelia, Lupin, Marigold (French), Mignonette, Nasturtium, Nemesia, Pansy, Polyanthus, Poppy, Primula, Saponaria, Scabiosa, Schizanthus, Statice, Stock, Sweet Pea, Sweet William, Viola, Wallflower.

SUMMER AND AUTUMN FLOWERS

Amaranthus, Antirrhinum, Aster, Balsam, Begonia (bedding), Carnation, Celosia, Chrysanthemum (perennial), Cockscomb, Dahlia, Dianthus, Eschscholtzia, Gaillardia, Gerbera, Gomphrena, Gypsophila, Marigold (African), Petunia, Phlox, Portulaca, Salpiglossis, Salvia, Sunflower, Torenia, Verbena, Viscaria, Zinnia.

HOW TO GROW INDIVIDUAL FLOWERS

ACROCLINIUM

(*See Everlasting Daisy.*)

AGERATUM

Also known as Floss Flower, this attractive blue annual will not tolerate frosts, and is usually grown for spring, summer and autumn display. All varieties make splendid border plants, and their soft blue flowers are excellent for garden display and indoor decoration. They will succeed on a variety of soils but respond to good conditions and added fertilizer. They are fairly drought-resistant but need regular watering in dry weather. They are at their best when grown in full sunlight, but will give fair results in semi-shade.

In warm climates, you can sow seed in almost all seasons, but in cold districts, spring and summer sowings are best. Seed can either be sown in seed beds and the seedlings transplanted, or sown direct in the garden bed. Cover seed with vermiculite or compost and keep moist until seedlings emerge. Seedlings are transplanted or thinned to a distance of 15–20 cm apart. Plants need very little care apart from normal cultivation and watering. Liquid feeds of Thrive, Aquasol or Zest at regular intervals will promote flowering. Cut back all spent blooms. Blue Mink is a popular dwarf variety and there are some good Fl hybrid kinds which are more even in size and which flower for a long time. Blue Blazer and Blue Tango are two of these. Also available is a packet of mixed blue and white dwarf varieties, very free flowering, growing to a height of about 20 cm.

ALYSSUM

Sweet Alice, as this plant is often called, is popular for edging and border work. It is ideal for rockeries and wall gardens as it flowers all the year in most climates. It does quite well in semi-shade but flowering is more prolific in open sunlight. It grows well in all types of soils, but thrives in good, friable soil with added fertilizer. Good drainage is essential as it resents damp conditions.

It temperate climates, you can sow seed at almost any time of the year but in cold districts it is best to sow during spring and autumn months. Seedlings can be raised in boxes or punnets for transplanting or seed can be sown direct in the garden in clumps or stations 7–10 cm apart and thinned if necessary. Cover seeds with vermiculite or compost and keep damp until seedlings emerge. Water plants regularly in dry weather, giving a good soaking, say, once or twice a week, rather than frequent sprinkling. Destroy weeds while they are small, otherwise the fine roots of the plants will be damaged when large weeds are pulled out. Give regular liquid feeds of soluble fertilizer as plants grow. This will promote flowering over a longer period. Cut back all spent flowers.

Varieties Carpet of Snow has masses of pure white flowers on dwarf bushes 10 cm tall. It is excellent for borders and edging, in rockeries and between bricks or stones in paths or paving. Royal Carpet has deep violet flowers on bushes the same height as Carpet of Snow, with which it combines for a beautiful colour combination. Wonderland is a rather smaller alyssum forming attractive mounds with rose-pink flowers. Cameo Mixture is a delightful blend of Carpet of Snow, Royal Carpet, Wonderland and other subtle colours which is ideal for mass planting. *Alyssum Saxatile* grows to about 15 cm with clusters of deep yellow flowers. It is spring blooming but plants last for 2–3 years. Ideal for borders, mass plantings and for adding colour to rockeries.

AMARANTHUS

This is a popular summer annual, widely grown for its brilliant foliage and its ability to stand very hot weather. Plants grow to 1–2 m. Amaranthus revels in hot sunny situations, but needs ample water during dry times. Prepare soil well a week or two before planting, with animal manure (if available) or compost, plus a mixed fertilizer such as Gro-Plus Complete at the rate of $\frac{1}{3}$ cup per square metre.

Sowing can be made in spring when the danger of frost is over and can be continued until early summer. It is best to sow a few seeds direct in the garden in clumps 40 cm apart and thin out to 1–2 seedlings. You can also raise seedlings and transplant to the same distance apart. Cover with vermiculite or compost and keep damp until seedlings emerge. If transplanting seedlings, discard any paler plants which are not well-coloured. Water regularly while plants are growing and keep down weed growth by shallow cultivation. As hot weather approaches a mulch of compost or grass clippings will protect shallow roots and conserve moisture. When the plants are 30 cm tall, give liquid feeds of soluble fertilizers and repeat this treatment every 10–14 days.

Varieties Flying Colours has a great variety of colours in the foliage—red, yellow, orange and green. Flaming Fountain has rather finer leaves in rich scarlet.

ANTIGONON

(*See Chapter 11.*)

ANTIRRHINUM

(*See Snapdragon*)

AQUILEGIA (Columbine)

This unusual and attractive flower has been improved by selection in recent years and the present-day strains are well worth a place in the garden. Aquilegia is a perennial best suited to cool climates. Plants prefer a sunny position, but they can also be grown in semi-shade. Sow seed in autumn, but in cool climates you can sow in early spring too.

Sow seed in seed beds or punnets and transplant seedlings when 5–7 cm tall, spacing 30–40 cm apart. Water regularly, especially in dry weather, and scatter a mulch of grass clippings or compost around each plant. Apply liquid feeds of Thrive or other soluble fertilizers as buds start to form. In cold districts, flowering in the first year is not as prolific as it is in subsequent years. In cool climates, plants will last for many years but in warmer climates it is best to start new seedlings at least every second year.

ARISTOLOCHIA

(*Dutchman's Pipe—See Chapter 11.*)

ASTER

Rich, bright flower colours have made asters a firm favourite with Australian gardeners. Although subject to aster wilt, good strains such as Giant Crego and

Aquilegias (Columbines) prefer a sunny position in the garden

King Aster will give good results if not planted in soil which has grown asters the previous year. Good drainage and a well-structured soil also help plants to resist the wilt problem.

Asters prefer a light sandy soil which is not heavily manured. They may not grow quite as well on heavy clay soils so these should be improved by the addition of compost for better crumb structure. In preparing the bed, apply a mixed fertilizer such as Gro-Plus Complete at the rate of $\frac{1}{3}$ cup per square metre. Careful soil preparation, well before the seedlings are transplanted, will pay dividends. Asters prefer an open sunny position.

There is little advantage to be gained from sowing seed too early in spring. Later sowings will bloom at the same time—about mid-summer. You can sow seed from spring (but wait until frosts are over) through until mid-summer. In mild districts, you can sow until late summer. Raise seedlings in seed beds, boxes or punnets as described in Chapter 5. After sowing cover the seeds lightly with vermiculite, compost, a light sandy soil or seed raising mixture.

When the seedlings are large enough, transplant them into the prepared bed, spacing them 20–30 cm apart each way. On heavy soils seedlings may have difficulty in establishing their root system. You can assist root development by adding some of the seed bed compost or a handful of sand in the hole where the seedlings are planted. In hot weather, spread a mulch of lawn clippings or compost over the surface of the bed to conserve moisture and keep the roots cool. This is a very important consideration in growing asters successfully. With normal attention plants should be flowering in four or five months from sowing the seed. Later sowings will bloom a little earlier. It is important to remember that asters are very susceptible to frost, so do not sow too late in summer because this will not give time for the plants to flower.

Varieties Giant Crego Mixed is a wilt-resistant strain with large double flowers made up of long curled and twisted petals. It contains all the flower colours for which asters are so well known and is a favourite for early flowering in summer gardens. King Aster is a vigorous strain with plants about the same height as Giant Crego. It flowers freely and has large, fully double blooms made up of attractive quilled petals with excellent lasting quality. Perfection Aster Mixed is an outstanding strain with large evenly packed flowers with beautifully textured petals. An excellent strain for cut flowers which last longer than other strains. Dwarf Colour Carpet is a colourful dwarf aster growing to 20 cm high. Ideal for borders and rock gardens. Bouquet grows to about 40 cm and produces a profusion of delightful, pastel-coloured flowers, pin-cushion shaped, 5 cm in diameter. Excellent colour mixture for cut flowers. Totem Pole is an entirely new strain producing sturdy plants to 60 cm tall. It has a bright colour range of very large, fully double flowers with curled petals similar to Giant Crego.

Colour Carpet dwarf asters make ideal border plants

AURORA DAISY (Arctotis)

This is a beautiful annual, flowering during winter, spring and summer. The large daisy-like flowers are produced freely on long, strong stems about 40 cm tall and come in a wide range of brilliant colours, including tangerine, rose, pinks, red, claret, lemon, orange and white. Some flowers show attractive two-tone effects while others are one colour.

Plants will grow on a wide range of soils and do very well on light sandy soils. Sow seed in autumn (for winter flowering) and also in spring. Seed may be sown direct in the garden bed or seedlings may be raised in boxes or punnets and transplanted when 5–7 cm tall. Space plants 30–40 cm apart. Water well during hot, dry weather and give liquid feeds of soluble fertilizers as flowering commences. A sunny position is best.

BABY BLUE EYES

(*See Nemophila*)

BALSAM

This colourful flower is an excellent plant for borders or massed display. It can also be grown in windox boxes, pots and troughs. Balsam likes the sun but will do well in semi-shade too. Prepare soil well and add organic matter and a pre-planting fertilizer.

Raise seedlings in boxes or punnets in early spring, but direct sowing is best delayed until all danger of frosts is past. Transplant or sow in clumps 30 cm apart each way. Plants eventually grow to a height of 50 cm and flower in about three months from sowing. Superb Double Mixed is the most widely grown variety.

BEGONIA

Few other flowers grow so well in heavily shaded places as the bedding or fibrous-rooted begonia. Begonias are well suited to grow in southerly and westerly aspects which receive no direct sunlight. While these plants thrive in shade, they do quite well in a sunny position, but need regular watering in dry weather. Prepare soil well with added compost plus a mixed fertilizer such as Gro-Plus Complete at $\frac{1}{3}$ cup per square metre. Spring and early summer are the best times to sow seed in most climates, but in warm climates with mild winters you can sow in autumn too.

The seed is almost as fine as flour, so seedlings need to be raised in seed boxes or punnets. Mark out rows 3–5 cm apart

Begonias grow well in sun or shade. Excellent for edging or massed display

with the point of a nail, or lay the edge of a ruler on the surface and press down gently. Sprinkle seed thinly along the rows and press it in. Cover with a light dusting of vermiculite, compost or seed raising mixture. Water gently with a fine spray so as not to disturb the surface. Alternatively, stand containers in shallow water until moisture seeps to the surface. Keep damp until seedlings emerge in 14–21 days. Prick out seedlings when small to a seed tray and grow them on until 5–7 cm high before transplanting to the garden bed. Space seedlings 20 cm apart for the dwarf bedding types. Water regularly until well established. Liquid feeds of soluble fertilizer at intervals of 10–14 days will promote more rapid growth, especially towards flowering time.

Varieties Strawberry Sundae is a delightful dwarf bedding begonia with dense compact growth which is covered in flowers in all but the frosty months of the year. Flower colours include dark and light reds, pinks of various shades and white.

Tuberous begonias are usually grown in pots or in shadehouses. They make beautiful pot plants and can be raised from seed or grown from bulbs. Double Exhibition Strain is available from seed and has double flowers in a very wide colour range. Bulbs are planted during winter and spring.

BELLIS (English Daisy)

Although perennials, these hardy daisies are best treated as annuals and fresh seed sown each autumn. They are very attractive and make a splendid edging plant with their fully double blooms carried on stems up to 15 cm high. They prefer open sunlight, but grow quite well in semi-shade. The plants are adapted to a wide range of soils but respond to good soil structure, so it is worthwhile adding organic matter and a pre-planting fertilizer when preparing the bed. In warm climates the seeds should be sown in autumn, but in very cold districts you can sow seed in spring too.

Raise seedlings in seed beds, trays or boxes as described in Chapter 5, covering the seed with vermiculite, compost or seed raising mixture. Transplant seedlings 10–15 cm apart each way. Give them ample water during dry weather and keep down weeds by shallow cultivation. This is important, as the plants are small and could easily be choked by weeds. When the first flower buds appear, give the plants liquid feeds of soluble fertilizers. This will promote flowering over a long period. Cut off all faded flowers regularly.

BIG GYP (Saponaria)

This plant is not related to Gypsophila, but the rose-coloured flowers are somewhat similar and are widely used for cut flowers with other blooms like gerberas and carnations. The flowers last well after cutting.

Big Gyp is adaptable to most soils but responds to soil improved with organic matter and an application of mixed fertilizer. Plants flower quickly from seed and in warm weather will produce flowers about two months after sowing. You can sow seed practically the whole year round, except in very cold or very hot weather. Many gardeners make small successive sowings, which provide a continuous supply of cut flowers. The seed germinates freely in the open garden where the plants are to flower. Sow thickly in rows about 12 mm deep spaced 20–30 cm apart and cover with vermiculite, compost or seed raising mixture. There is seldom any need to thin the seedlings and the plants will produce really good blooms even when grown closely together. When grown close together each plant forms a support for its neighbour too.

BORONIA

Boronia is one of our best-known wildflowers and the sweetly scented brown species is probably the most fragrant of all bush plants. This is the reason for its tremendous popularity.

The plants prefer light sandy soils but will grow well on heavier soils if sand and compost are added. They will grow in open sunlight but often do better in a lightly shaded spot. The soil should be well-drained, but capable of holding moisture too. Boronia seeds have a hard coat so it is best to soak them in hot water before sowing.

Sow seed in spring when the soil has become warm and cover with a light sprinkling of compost or leaf mould. Press down lightly with a piece of flat board to firm soil. Germination is very slow, and seedlings may take 6–10 weeks to emerge. Seed should not be allowed to dry out during this time. Seed can be sown in either punnets or small pots or direct in the garden where the plants are to flower. For direct sowing, the soil must be carefully prepared for both good drainage and moisture retention. If you have raised seedlings in punnets or pots, prick them out into individual pots when very small. When transferring them to the garden choose a cool day and water thoroughly after planting. Once established, give plants regular liquid feeds of soluble fertilizer at about half the usual strength. Established plants will resist frost but young plants need some protection during the first winter. After each flowering prune the plants back—they will last for several years under favourable conditions.

BRACHYCOME

(*See Swan River Daisy.*)

CALCEOLARIA

Calceolarias make very attractive pot plants for spring flowering. The plants are excellent for bush-houses and glasshouses (in cold climates) but are not really suitable for outdoor planting. They are rather particular in their soil and temperature requirements and care must be taken to grow them really well. In mild climates they can be grown in a bush-house with moderate warmth during the flowering period.

In temperate climates, sow the seed from mid-summer to early autumn, but in cold districts seed can be sown in both spring and summer and the plants grown successfully. Raise seedlings in seed boxes, trays or punnets as described in Chapter 5. Use a soil mixture made up of equal parts of finely sifted garden loam, sand and

Hardy, easy to grow calendulas. This is Honey Babe

vermiculite or compost. This mixture is friable and drains easily but will hold moisture as well. Sow the seed on the surface, barely covering it with a sprinkling of sand, vermiculite or sifted compost, but firm the soil down. Water gently with a very fine spray or by standing the box or punnet in a dish of water until moisture seeps to the surface.

Transfer young seedlings from the box or punnet into 7–10 cm pots and as they grow larger move them into larger pots, with a final planting into pots about 20 cm in diameter. When finally established and the plants growing strongly, give them weak liquid feeds of soluble fertilizer. Calceolarias resent excessive heat and need only moderate warmth at the flowering period. A wide range of colours is contained in Yates Superb Mixed.

CALENDULA (English Marigold)

Calendula is another flower which has become very popular because the modern strains are much superior to the older ones. This improvement has been possible due to the efforts of plant breeders who have developed and selected plants for vigour, flower size, colour and quality. Calendulas are hardy and easy to grow. In temperate climates, plants come into bloom during the winter months, provided sowings are made early enough. Calendulas grow best in fertile, well drained soil, but succeed on a variety of soil types. They prefer an open, sunny position but must be watered regularly during dry weather. In warm climates, sow seed in autumn or early winter for best results. In cold districts you can sow in autumn and also in spring.

Seed can be sown direct in rows, or a few seeds in clumps in the garden bed, or the seedlings raised in seed beds or boxes for transplanting. Whichever method you use, cover the seed to a depth of 12 mm with vermiculite, compost or seed raising mixture. Thin rows or clumps if required so that plants are spaced about 30 cm apart, or transplant seedlings at this distance. Transplant when seedlings are large enough to handle easily. It is better to have them on the large side rather than too small. Whether sowing direct or transplanting, always apply a pre-planting fertilizer to the bed when preparing the soil. This will improve the vigour of the plants and increase the size of the flowers. Remove spent blooms to prolong flowering, and keep down weeds by shallow cultivation. Rust is the most serious disease of calendulas and can be prevented by regular spraying with fungicide. (See Chapter 17.) Badly affected plants should be removed and burnt.

Varieties Campfire is an improved strain with semi-dwarf, vigorous, free-flowering plants. The large flowers are orange with a scarlet sheen. Pacific Beauty is a strain representing calendulas at their best. The colour range includes some lovely pastel shades. Honey Babe is a dwarf variety suitable for mass plantings and rockeries. The colours include yellow, gold, orange and shades between. Princess Mixed is a new semi-dwarf strain with fully double crested blooms which are ideal for cut flowers. Flowers are yellow, gold and orange.

CALIFORNIAN POPPY (Eschscholtzia)

Another addition to the list of sun-loving plants which thrive in warm summer months with brightly coloured flowers. Seed can be sown in spring and summer in both temperate and cool climates. You can sow seed direct in the garden or raise seedlings for transplanting. With either method, cover seed lightly with vermiculite or compost and keep damp until seedlings emerge. The seedlings are thinned or transplanted 30 cm apart. Keep down weeds by shallow cultivation and give liquid feeds of soluble fertilizers as plants commence to flower.

CANDYTUFT

There are two distinct types of candytuft. The most popular is called *umbellata*, and carries its flowers at the top of the stem. The second is called *coronaria* or Hyacinth-flowered Candytuft. It produces a mass of pure white florets, which resemble the hyacinth flower. Candytuft grows well in most soils, but they must be well drained. The plants prefer a sunny situation, sheltered from strong winds. Plants must not be crowded because they become too spindly and flowers are poor.

In temperate climates, sow seeds from autumn to early winter. In cold districts, seed can be sown in spring too. Prepare the bed well with organic matter added, plus a pre-planting fertilizer at $\frac{1}{3}$ cup per square metre. Seed can be sown direct in the garden in clumps 20–30 cm apart, or seedlings can be raised in seed boxes or punnets for transplanting at the same distance. The hyacinth-flowered varieties can be spaced rather close at 20 cm apart. Candytuft is an easy plant to grow. Keep down weeds by shallow cultivation and give liquid feeds of soluble fertilizer when plants commence to flower.

Varieties Fairy Mixed has massed heads in a wide range of pastel colours covering the compact plants. Hyacinth-flowered has a profusion of white flowers. Fairy Floss is similar to Fairy Mixed.

CANTERBURY BELLS

Canterbury Bells form a spire of large 'cup and saucer' flowers with the texture of fine porcelain. Colours range from dark blue to mauve, rose, soft pink and white. Excellent for garden display in clumps or drifts and for cut flowers in spring. Sow seed in autumn in seed boxes or punnets, just covering with vermiculite, compost or seed raising mixture firmed down gently. Keep soil moist until seedlings emerge—usually in 14–21 days. Transplant seedlings when 5–7 cm tall into the garden bed, spaced 30 cm apart each way. Give liquid feeds of soluble fertilizer when flower buds appear.

CARNATION

Like most garden flowers, this fragrant favourite had humble beginnings. It was introduced to England in a wild form about the sixteenth century but it was not until about 1900 that the Perpetual Flowering types were developed.

In Australia we are mainly interested in two types (either of which can be grown from seed)—the Chabaud (bedding carnation) and the Perpetual Flowering. Chabaud is the most widely known and is easily recognised by its deeply serrated or fringed petals and a strong clove-like scent. Perpetual Flowering types are the source from which most named varieties are obtained. But it should be understood that named varieties of carnation will not breed true to type from seed. If you require these named varieties, they must be started from cuttings or purchased as plants from nurseries.

Carnations are fragrant favourites for garden or house decoration

When starting carnations from seed, it is best to raise seedlings in seed beds, boxes or punnets using a good soil mixture as described in Chapter 5. Germination may be slow and rather erratic. For this reason the seed bed or containers must not be neglected for 10–14 days—perhaps longer—until seedlings emerge. Carnation seed can be sown direct but it is difficult to give the germinating seeds the same attention. Seed can be sown at almost any time of the year but spring or autumn are regarded as the best periods.

Whether carnations are started from seed, cuttings or nursery plants, the requirements for soil, cultivation and general care are the same. Carnations prefer an open position exposed to full sunlight and resent being crowded by other plants or shrubs. It is best if they are protected from westerly and southerly winds too. They will grow in a wide range of soils from heavy (but well structured) to light sandy soils. On heavy soils add organic matter and sand to improve both structure and texture. On sandy soils add organic matter in the form of animal manure, garden compost, peat moss or vermiculite to improve structure and increase water-holding capacity. Manure, compost, peat moss, vermiculite or any other materials for soil improvement should be dug into the soil some weeks beforehand. At the same time lime and a pre-planting fertilizer such as Gro-Plus Complete should be added. As a general guide, use a handful of lime per square metre on sandy soils but double this amount on heavier soils. The rate for mixed fertilizer is the same for both soil types—$\frac{1}{3}$ cup per square metre. Dig the bed to a depth of 10 cm to thoroughly mix the soil, moisture holding materials, lime and fertilizer. Leave for a few weeks and then gently break the surface soil a few days before planting, to destroy any weeds. The bed should contain sufficient nutrients to feed the plants through the main growing period. Just before flowering, scatter another ration of mixed fertilizer around the plants at the rate of 3 tablespoons per square metre. Alternatively, give liquid feeds of soluble fertilizer. Bad drainage is probably the greatest enemy of carnations and the plants are more likely to be attacked by root rot and collar rot diseases. All beds should be raised above the surface—about 10 cm for light soil and 15 cm for heavy soil.

Short stocky plants are the best as thin and straggly ones do not produce compact growth. Space plants 30–40 cm apart each way. When planting out, keep the lower leaves well out of the ground so that no more than about 12 mm of soil covers the top of the roots. Keep the soil surface well stirred—but not too deeply—to keep the plants free of weeds.

The greatest number of good blooms come from compact, sturdy plants. You can encourage sturdiness by nipping back the early flowering stems regularly. When the plants are about 15 cm tall, pinch them back to induce side shoots, which in turn are pinched back when they reach the same height. When plants have developed 8–12 shoots (depending on the growth habit of the variety), allow these to send up flower stems. When picking flowers, always break them off near the base of the plant. Each main stem should bear only one bloom if first class flowers are desired. This means the removal of all side buds growing from the leaf axils to leave the main bud only at the top of the stem. In some varieties, buds will have a tendency to burst at the calyx (the green collar beneath the flower). You can prevent this by fitting small rubber bands around them.

After about twelve months of flowering, plants may begin to look straggly. But if they still appear healthy and vigorous, cut them back hard to within 5 cm of the centre stem. With normal attention, these plants should produce a further flush of blooms as good as the previous flowering. In some soils—especially light soils—plants which have flowered abundantly may not be worth this treatment so it is best to renew them—either from seedlings or cuttings taken in autumn. Pests and diseases are not a serious problem in growing carnations. Aphids and thrips are the most common pests but can be controlled by spraying with insecticides. Rust and collar rot are the most serious fungus diseases. For control of both pests and diseases see Chapter 17.

Varieties Fragrance is a new dwarf strain which has similar flowers to the Chabaud type. This variety does not need stakes for support and cut flowers have good lasting quality. Gems of France is a beautiful mixture of Chabaud (fringed) and Enfant de Nice (plain edge) types in a good colour range. Lover's Bouquet is a true Chabaud type with large, fully double flowers with a typical clove fragrance.

CELOSIA

This annual beeding plant revels in a hot, sunny situation. Plants are susceptible to frost and seed must be sown in spring and early summer when the soil is warm. Plants grow well in almost any garden soil but prefer a soil to which organic matter plus a ration of mixed fertilizer has been added. Seed can be sown direct in the garden in rows or clumps or seedlings can be raised in seed beds, boxes, or punnets for transplanting. Thin seedlings or transplant to 20–30 cm apart. Celosia is an easy plant to grow. Water plants regularly—especially in hot weather—and give liquid feeds of soluble fertilizers as flowering commences. Destroy weeds by shallow cultivation to avoid damaging surface roots. A mulch of grass clippings around the plants discourages weeds and conserves moisture.

Varieties Forest Fire is a semi-dwarf variety to 90 cm tall with spectacular scarlet plumes and contrasting dark foliage. Golden Triumph grows to the same height with plumes in shimmering gold. Fairy Fountains is a new dwarf variety with plumes in scarlet, salmon, light pink, gold and yellow. An excellent variety for borders or mass bedding.

CENTAUREA

(*See Cornflower*)

CHRISTMAS BUSH

This popular native plant of New South Wales needs little introduction. The showy shrub or small tree grows to 3–6 m and produces masses of pink to scarlet bracts after the small white flowers have dropped in summer. Christmas bush can be raised from seed, although most people find it more convenient to buy advanced plants from nurseries. When sowing seed, choose a sandy soil or bush sand for the containers (small pots are probably the most suitable) and cover the seed with 12 mm of compost or sand and press firmly. Sprinkle with a further light layer of compost or leaf mould. Seeds are slow to germinate and the container must be kept damp (but not wet) until seedlings emerge. Transplant seedlings to their permanent position when 7–10 cm high. Christmas bush prefers a sandy soil and full sunlight. If late spring and early summer is dry, water generously for a good flowering.

CHRYSANTHEMUM

The most widely grown species of chrysanthemum is *Chrysanthemum indicum*. This is a perennial which can be grown from seed, although named varieties must be propagated by root divisions or cuttings taken in July or August. Shasta Daisy (*C. maximum*) is another perennial type. There are also annual chrysanthemums, of which Painted Daisy (*C. tricolor*) is the best known.

The perennial chrysanthemum (*C. indicum*) includes a very wide range of flower types—decorative single, semi-double and fully double. It is the source of all modern named and exhibition varieties. As previously stated, named varieties must be propagated vegetatively from root divisions or cuttings. Plants are usually available from nurseries in October/November. Seed of perennial chrysanthemums will produce plants with a variety of flower forms and colours—an interesting (and economical) way of growing them.

Chrysanthemums grow well in a variety of soils, but both heavy and light soils are improved with garden compost plus the addition of a mixed fertilizer at the rate of $\frac{1}{3}$ cup per square metre. It is a good idea to dig the soil over two or three times before planting, to bring it to an open, friable condition. Chrysanthemums dislike poorly drained soil. Raising the beds above the surrounding level is probably the easiest method. Plant them in a warm, sunny position, avoiding shady sites and those exposed to wind.

Spring is the best time for sowing seed. Raise the seedlings in seed beds, boxes or punnets as described in Chapter 5. Make up a good seed bed soil, sow the seed and cover with a sprinkling of vermiculite or compost about 6 mm deep. Keep damp until seedlings emerge in 10–14 days. Seedlings are ready to transplant when 7–10 cm high. For exhibition, space plants 100 cm apart each way, but for general garden display or cut flowers this distance can be reduced to 75 cm apart. Firm the soil well around the roots of each plant and water well. Continue regular watering until plants are established. Keep the plants growing strongly by giving liquid feeds of soluble fertilizer and cultivate the soil surface regularly to control weeds. Liquid feeds can be discontinued when the flower buds show colour.

When growing chrysanthemums, management practices will largely determine the habit of growth, the number of flower stems and the size and quality of the flowers. 'Stopping' consists of breaking out the growing tip of the plant to promote 3–4 lateral stems to develop. This is done when the plants are about 20 cm tall. Lateral stems can also be stopped when 5–10 cm tall to induce further laterals and a bushy plant. 'Disbudding' (the removal of surplus buds) influences the size and quality of the flowers. If you wish to have very large blooms, disbud to leave the centre or top bud. Flowering can also be delayed or extended by leaving the second or third bud. Small-flowered types of chrysanthemum are usually not disbudded. The main types of perennial chrysanthemums are: *Large Flowered Exhibition:* Giant, waratah-shaped blooms to 15 cm diameter with broad in-curving petals; *Decorative:* Similar flowers to the Exhibition type but rather smaller; *Spider or Quilled:* Petals are loose and finely rolled or quilled; *Anemone-centred:* Tightly packed centre giving a pincushion effect and surrounded by a row of outer petals often in contrasting colours; *Singles:* Daisy-like flowers with a distinct centre and 3–5 rows of petals; *Pompon or Button:* Clusters of small double, anemone-like flowers 3–5 cm in diameter; *Cascade:* Small single blooms on trailing stems giving a cascade effect. Suitable for pots, troughs and window boxes; *Charm:* Compact plants to 60 cm tall with masses of small daisy-like flowers for massed display.

All but the low-growing types of chrysanthemums (Anemone-centred, Pompone, Cascade and Charm) need staking by the time buds form. It is best to surround the plants with 2 or 3 light stakes and tie garden twine around them at intervals of about 30 cm.

Varieties The following varieties can be grown from seed. Autumn Brilliance is a perennial strain with a bright colour range of single and double flowers in late autumn. Goblin is a new dwarf annual variety with buttercup-yellow blooms above the attractive dark green foliage. Plants grow to 20 cm with a spread of 30 cm. It is an ideal variety for edging and in rockeries and window boxes or troughs. Sow in spring or autumn. Star daisy (*C. paludosum*) is another dwarf annual chrysanthemum with small, daisy-like white flowers with yellow centres. Plants grow to 30 cm and are covered in blooms which are excellent for cut flowers. Sow spring or autumn. Painted Daisy (*C. tricolour*), see under this heading.

CINERARIA

Cinerarias are among the garden flowers which prefer semi-shade. The plants can be grown in shady aspects in the open garden or in pots in a shadehouse. The pots can be moved indoors for decoration when flowering commences. In very warm climates, a shadehouse or bush-house is essential to grow them successfully.

Dwarf cinerarias and column stocks make a delightful combination

Modern varieties have a much wider colour range than older varieties and both large-flowered, tall types and compact, dwarf types are available.

Sow seed during late summer and autumn for flowering in late winter and early spring. The seeds are quite small and should be sown in seed boxes or punnets and lightly covered with vermiculite, compost or seed raising mixture and gently watered.

To grow strong seedlings it is a good idea to prick them out as soon as they are large enough to handle into 5–7 cm pots, or into larger boxes or trays, spacing them about 10 cm apart. If you wish to grow the plants on in pots, transfer them to 15 cm pots and finally to 20 cm pots for flowering. The soil for potting should be an open friable mixture which drains easily. Add a ration of mixed fertilizer to each batch of soil. As the plants grow, give regular liquid feeds of a soluble fertilizer, especially towards flowering time.

The main insect pests of cinerarias are aphids and leafminer. Spray the plants regularly with insecticide to control them. (See Chapter 17.)

Varieties The new variety of tall cineraria, 'Tapestry' is the 1983 Flower of the Year. Gives a brilliant show in both sun and shade. Colours range through crimson, scarlet, pink and many shades of blue, with some flowers being zoned with white. The best dwarf strain is Multiflora Mixed with smaller flowers on compact plants which are excellent for pot culture.

CLEOME

(*See Spider Flower.*)

CLIANTHUS

(*See Sturt's Desert Pea.*)

COBAEA

(*See Chapter 11.*)

COCKSCOMB

Cockscomb is an annual bedding plant very similar to celosia but the flowers are tight velvety crests instead of feathery plumes. See Celosia for cultivation notes. The best variety is Dwarf Jewel Box.

COLEUS

Coleus has brilliant, coloured foliage and is an ideal plant for growing in pots in a shadehouse or in shaded parts of the garden. Although they prefer semi-shade, the plants need a warm sheltered position and are very susceptible to frost.

Sow seed in boxes or punnets in spring or early summer using the soil mixture described in Chapter 5. Seed is quite small so cover lightly with vermiculite, compost or seed raising mixture. Transplant seedlings when 5 cm high and space them 30 cm apart in the garden. If growing the plants in pots select seedlings with the best colouring—this can be determined when the plants are quite small. Potting soil should be open and friable with organic matter and a ration of mixed fertilizer added. The stems of seedlings are soft and tender so handle them carefully to prevent damage. When the plants are established, water regularly so that the soil is always damp but not wet. Pinch back main shoots to encourage lateral branching. Give liquid feeds of soluble fertilizer every 7–10 days to promote large leaves and intensify colours. Pinch out any flower buds to prolong leaf growth.

COLUMBINE

(*See Aquilegia.*)

CORNFLOWER (Centaurea)

Centaurea cyanus is the well-known blue cornflower. Seed mixtures also contain other colours—rose, maroon, lavender and white. Cornflowers grow well in temperate climates but are ideally suited to cool climates. They need good drainage and a well-structured and fertilized soil. Plant them in a position which receives morning sun.

Sow seed from early autumn until early winter in prepared seed boxes or punnets. Cover the seed with vermiculite, compost or seed raising mixture and water gently. Transplant seedlings to the garden and space them 40–50 cm each way. Cultivate regularly to control weeds and give soluble fertilizers every 10–14 days when plants show flower buds. Double Mixed is the best variety. Plants grow to 60 cm tall with a wide range of flower colours.

COSMOS

These tall, colourful plants make an excellent background for low-growing annuals and give plenty of cut flowers for indoor decoration. Cosmos are easy plants to grow and need little attention. They will probably be more widely grown when home gardeners have realised the tremendous improvement which has taken place —in both flower form and colour range —in these plants over recent years. The flowers, with their fine colours and good lasting quality, make cosmos well worth growing in any garden.

Cosmos will grow on a wide range of soils but respond best to well-drained, friable soils with a mixed fertilizer such as Gro-Plus Complete added. The plants need a sunny position and because of their height should be sheltered from strong winds. You can start sowing seed in spring and continue through to summer. Seed sown in January in most districts will produce plants which will flower before cold weather or frost arrives.

It is best to raise seedlings in seed beds, boxes or punnets for transplanting into the garden, but as the seeds are quite large you can sow a few seeds direct in clumps spaced 40–50 cm apart in the garden and thin to the strongest 1–2 seedlings. When transplanting seedlings space them at the same distance apart. Keep the soil between the plants well cultivated and in hot weather spread a mulch of grass clippings to keep the roots cool and moist. Staking plants is rarely needed because when they are grown close together they support each other.

Varieties Mammoth Single Mixed has large rose and crimson flowers with a sprinkling of white. Plants grow to 120 cm tall. Sunny Gold is a vigorous grower with a profusion of 5 cm diameter yellow double flowers and an attractive foliage. Plants grow to about 1 m. Highly recommended.

CYCLAMEN

The delicate flowers in white, pinks, mauves and reds make cyclamen one of the most popular plants for winter and early spring flowering. They can be grown in an open friable soil in a semi-shaded position in the garden but are more often grown in pots. They do well indoors but need a well-lit, well-ventilated but draught-free spot—preferably with an hour or two of sunlight each morning.

The plants form a corm or bulb and can be carried on from year to year but corms more than two or three years old are best replaced by new plants grown from seed. Sow seed from late summer to early autumn in a seed box or punnets using a soil mixture made up of garden loam, vermiculite or peat moss and coarse sand as outlined in Chapter 5. Stand containers in water until moisture seeps to the surface, allow to drain for 24 hours or so and press the seeds into the dark damp soil mixture to a depth of 6 mm, spacing them 2–3 cm apart. Seedlings emerge over a period of 4–6 weeks. Prick them out when small into individual 5–7 cm pots using a similar soil mix, and transfer them to larger pots as they grow. Give liquid feeds of Thrive or Aquasol (at half strength) every 3–4 weeks. Water regularly by standing the pots in water until soil is wetted and then allow to drain. Do not over-water—plants will benefit from short spells of dryness.

Buds form the following autumn and respond to regular liquid feeds until flowering ceases in late spring. Then, as leaves turn yellow, reduce the water supply and tip pots on side for final drying. The corm or bulb which has formed during the flowering period can remain in the pot or can be taken out and stored in damp sand or peat moss before repotting in late summer. Water sparingly until new growth starts.

DAHLIA

Named in honour of Dr Andreas Dahl, a Swedish botanist, this popular flower is a native of Mexico and found its way to the Botanic Gardens, Madrid, in 1789. At that time, only three kinds of flowers were known—a double purple, a single rose colour and a single red. Tubers from each of these were sent to Kew Gardens, London, but they did not survive. Some years later, another lot brought from France was grown successfully. Today there is a tremendous array of dahlias, both in colour range and flower form. They have become one of the most popular plants for late summer and autumn flowering. Dahlias are easy to grow and given reasonable attention will produce a mass of blooms for many weeks.

You can grow dahlias from seed or from tubers of named varieties which are usually available in late spring. There is

Dahlia is an outstanding colour maker for summer gardens

a lot of interest in growing them from seed because there is always the possibility of obtaining an unusual variation in colour or flower form from the plants you grow. Plants will flower within 3–4 months after planting out seedlings and will continue flowering for many weeks. During this time, the plants are forming small tubers. Tubers of the best plants can be saved for replanting next season. Named varieties of dahlias will not grow true to type from seed.

Dahlias prefer a sunny position which is sheltered on the southern and western sides. They need a well-drained soil with an open friable structure. In sandy soils add organic matter in the form of compost, leaf mould or peat moss. Animal manure is also suitable but too much of it will produce too much leaf growth at the expense of flowers. If the bed for dahlias has been in constant cultivation, you will need to dig this over before planting and add well-rotted organic matter at the same time. Do not use lawn clippings which have not been decomposed. Where soil has been idle for a season or two, dig it roughly in winter and again just before planting in spring. Very deep digging is not necessary. Spade depth is ample in most soils. A ration of lime is recommended if the soil is acid. Apply no more than $\frac{2}{3}$ cup per square metre on sandy soils and $1\frac{1}{2}$ cups per square metre on heavy soils. A light scattering of a mixed fertilizer will be beneficial.

Sow seeds in spring or early summer—plants from early sowings will flower in January and those from late sowings in autumn; autumn blooms will appear fresher and not be bleached by hot sun. Use seed boxes, trays or punnets, covering the seeds with about 12 mm of vermiculite, compost or seed raising mixture. Seedlings may take 14–28 days to emerge so keep the containers damp for this length of time. Transplant seedlings when 5–7 cm tall. Growing plants should be supported by a stake to protect them against wind damage. After planting out, water regularly to make sure that the soil is never dry. Slugs and snails should be controlled immediately young shoots appear. When the plants are about 30 cm tall, apply a ration of mixed fertilizer, giving about 50 g to each plant. Spread the fertilizer in a circle around the plant, rake it in and water thoroughly.

When the buds appear, give liquid feeds of soluble fertilizers every 10–14 days to promote large blooms and prolong flowering. It is important to remove all spent flowers regularly to encourage further bud formation. Flowers for indoor decoration should be picked in the cool of the evening and the stem ends dipped in boiling water for about 30 seconds. Take care that the rising steam does not scald the blooms by tilting the container slightly or covering them with a double thickness of newspaper wrapped around the stems.

If you wish to propagate dahlias from tubers, these should not be lifted until the plants have completely died down—usually early winter. Dig them with a garden fork, taking care not to damage the crown and tubers attached to it. The lifted crowns can be stored in the shade under trees or in a corner of the garden shed during winter. As warm weather starts in spring, tubers will form 'eyes' or shoots. Cut between the shoots to leave some stem tissue surrounding the shoot and with one or two tubers attached. Set the tubers so that the root end is covered by about 10 cm of soil but the shoot is at soil level.

Varieties Like chrysanthemums, there is a wide range of types of dahlias, summarised as: *Decorative:* Large, heavy, double flowers 15–25 cm in diameter produced on long stems. Well grown plants are 2 m or more tall; *Hybrid Cactus:* Flowers with narrow, curled petals but smaller than Decorative types. Plants grow to 1.5 m tall; *Charm:* Rather smaller flowers than Cactus dahlias. Usually in pale pastel colours. Very free-flowering and plants grow to 1.2 m; *Collarette:* Flowers have a distinct centre with space between petals. Plants grow to 60–75 cm tall; *Nymphaea or Water Lily:* Daisy-like flowers with space between petals—usually in pastel shades with one colour fading into another. Plants 60–75 cm tall; *Paeony-flowered:* Small semi-double flowers in soft colours; *Pompon:* Small, tightly packed flowers in a wide colour range.

All tall-growing dahlias need some support. Place three stakes about 15 cm apart around the base of the plant. Spread these outward so that they are about 30 cm apart at the top. Encircle with twine at 40 cm height and again at 80 cm. Then join the centre of each tie to form a second triangle and encourage the branches to grow naturally through the crossing sections. As with chrysanthemums, stopping (pinching out the growing point) and disbudding are often practised in the tall-growing dahlias, but this is rarely necessary with dwarf bedding types.

The large-flowered, tall-growing types are the Decorative, Hybrid Cactus and Charm Dahlias. These are grown from tubers available from nurseries as named varieties. Both types may reach a height of 150 cm or more so need to be planted about 1 m apart. The smaller dahlias usually grow to 60–75 cm tall and can be spaced about 50 cm apart. Many smaller types can be grown from seed. The following mixtures are readily available: Hi Dolly (Unwin's Dwarf) is an early-flowering type with a lavish assortment of interesting and bright colours. The bushy plants grow to 60 cm and flower over a long time, especially if the blooms are picked regularly. Flowers are both double and semi-double. An ideal mixture for mass displays and borders. Pompon Mixed has a wide colour range of golfball-size flowers, some of which

Pompon dahlias

are bicoloured. Ideal for cut flowers for decorative arrangement. Redskin is a new variety with attractive purplish or metallic coloured foliage which contrasts beautifully with the bright flower colours in this mixture. Plants grow to 60 cm tall and flowers, about the same size as Hi Dolly, are both double and semi-double.

DELPHINIUM

Delphiniums have few equals for tall, stately spikes of flowers in rich shades of colour. Delphiniums are best treated as annuals in warm temperate climates but in cold districts the plants will last for several years, providing the summer months are mild. Delphiniums prefer a well drained, fertile soil with plenty of organic matter and a ration of mixed fertilizer added. Prepare the bed well to have the soil in friable dark damp condition for transplanting the seedlings. In temperate climates, you can sow seed in autumn, winter and early spring, but in cold districts restrict sowing to autumn and spring only. Once established, delphiniums are very hardy and tolerate frost well, but in very cold districts seedlings from autumn sowings should be protected during the first winter.

Sow seed in a good soil mixture in seed boxes, trays or punnets. Germination is often slow and seedlings may take 21–28 days to emerge. Transplant seedlings when large enough to handle conveniently, spacing them about 50 cm apart each way. As the plants grow, keep the soil well cultivated to control weeds. During summer months a mulch of compost or grass clippings will prevent evaporation and keep the roots cool. Delphiniums are gross feeders and will respond dramatically to dressings of liquid fertilizer at regular 2–3 weekly intervals.

Varieties Pacific Giants is a magnificent, tall hybrid variety. The giant flower spikes are closely packed with satin-textured blooms in glorious colours which range from white through shades of pink and lavender to pale blue, mid blue, royal blue and purple.

DIANTHUS

Dianthus is a close relative of carnation and its sowing times and cultivation requirements are the same (see carnation in this chapter). The plants are perhaps more adaptable to a wider variety of soils and they tolerate very dry conditions. They make excellent rockery and edging plants and produce large quantities of fragrant flowers in reds, mauves, pinks and white. Dianthus is usually treated as an annual, but if the plants are cut back after the first flowering and given liquid feeds of soluble fertilizer they will flower again in the second year.

Varieties Dianthus Pastel Carpet is an excellent dwarf grower with good-sized flowers in striking combinations of red, pink and white shades. Very uniform and ideal for borders, beds, pots and troughs. Mini Glow is a delightful, dwarf variety growing to 10 cm high, excellent for rockeries and low borders. Fairy Pinks or Pinks (*Dianthus plumarius*) are grown in the same way. The plants persist from year to year and are useful as perennial borders or rockery plants. Flowers are single or semi-double. Sweet William (*Dianthus barbatus*) is another close relative.

DUTCHMAN'S PIPE

(*See Chapter 11.*)

ENGLISH DAISY

(*See Bellis.*)

ESCHSCHOLTZIA

(*See Californian Poppy.*)

EVERLASTING DAISY

Everlastings are always handy for indoor decoration because they can be dried and kept without water for many weeks. Everlasting daisies will grow on a wide range of soils but need good drainage and an open sunny position. A ration of mixed fertilizer such as Gro-Plus Complete will promote vigorous growth and increase the size of flowers. Everlasting daisies can be sown in both autumn and spring.

It is best to sow seeds direct into the garden where the plants are to flower. Mark out shallow rows 12 mm deep, sprinkle seed sparsely along them and cover with compost or vermiculite. Thin seedlings to 20–30 cm apart. Alternatively, sow a few seeds in clumps at this distance and cover as before. Keep the soil moist until the seedlings are well established.

Everlasting daisies (Acroclinium) need good drainage and a sunny position

The plants usually need very little attention apart from regular watering and shallow cultivation to destroy weeds. Take care not to cultivate too closely to the plants. When buds start to form apply a side dressing of mixed fertilizer and water in. Alternatively give liquid feeds of soluble fertilizer every 10–14 days. If flowers are required for indoor decoration, cut them in full bloom, tie in bunches and hang head downwards for a few weeks until the stems are dry.

FAIRY PINKS

(*See Dianthus.*)

FORGET-ME-NOT

The English forget-me-not is still a very popular flower for edging, borders and rockeries. Although the flowers are traditionally blue with yellow centres there are also varieties in pink and white. The plants thrive in moist, semi-shaded situations with morning sun for a few hours each day. An open, friable soil is needed to grow them to perfection. Sow seed in late summer or early autumn in seed boxes, trays or punnets. Cover the seed thinly with compost or vermiculite as described in Chapter 5. Seed may be slow to germinate and seedlings may take 21–28 days to emerge. Keep the containers damp but not wet for this period. Transplant seedlings when large enough to handle and space them 20–30 cm apart. Blue Bird is a dwarf free-flowering variety with bright blue flowers. It is best treated as an annual but very often will seed naturally and new seedlings will appear each year.

GAILLARDIA

There are both annual and perennial species of this plant and each has its own special merits. The annual types flower quickly from seed but the perennials have a longer flowering period and can be cut back several times. You can sow seed of the annual types either in autumn or spring in most districts but the perennial is best sown in autumn. Seed can be sown in seed boxes or punnets and the seedlings transplanted, spacing them 30 cm apart each way. Alternatively, sow a few seeds direct in clumps at the same distance apart and thin to the strongest 2–3 seedlings.

Varieties Lorenziana Double Mixed (*G. lorenziana*) is the annual species with pompone-like flowers in shades of yellow, gold and crimson. Plants grow quickly from seed to a height of 40 cm and flower throughout summer. Hiawatha is the perennial (*G. grandiflora*) and has larger single flowers in shades of gold, orange and scarlet, many with contrasting colours in the centre.

GAZANIA

This dwarf perennial plant with slender, leathery leaves and daisy-like flowers in shades of cream, yellow, pink and mahogany is a native of South Africa. It revels in full sunlight and does well in dry situations. Excellent for banks, rockeries and low borders. Established plants can be divided after flowering in spring and summer. Sow seeds in boxes or punnets in spring or summer and transplant seedlings when 5 cm tall, spacing them 20 cm apart. Give liquid feeds of soluble fertilizers when buds appear in spring. The best variety is Sunshine, with large flowers in a wide colour range.

GERANIUM

Geraniums are ideal perennial plants for pots, window boxes, or massed flower display in beds and borders, especially in sunny situations. Plants of named varieties can be purchased from nurseries or started from cuttings, but geraniums can also be grown from seed. Sow seeds in boxes or punnets in spring or early summer—the seeds need a temperature of 25°C to germinate. Prepare an open, friable soil mixture for the containers. Scatter seeds on surface, cover lightly and press down firmly. Water gently or stand container in water until moisture seeps to the surface. Keep the soil moist until seedlings emerge. Transplant seedlings when large enough to handle into pots filled with a similar soil mixture and grow on until well established for transplanting to tubs or the open garden. When planting for massed colour in the garden space plants 40–50 cm apart. The best variety for raising from seed is Gaytime, which contains a large range of colours for spring and summer flowering.

Gazania, a native of South Africa, flourishes in dry, sunny positions

Gerberas are unsurpassed for hot sunny conditions

Geraniums, or more correctly Pelargoniums, are very adaptable perennials. Provided the plants are in full sunlight, they will thrive in both warm and cool climates. They withstand light frosts. The soil in which they grow must be well drained but not overly rich. Too much nitrogen or animal manure tends to promote excessive leaf growth at the expense of flowers.

Geraniums may become straggly as they grow so regular pruning is needed to keep the plants compact and to encourage more flowers. Light pruning can be done after each flush of flowers during the warmer months, followed by a general pruning and 'clean up' in autumn when flowering is over. Vigorous stems should be cut back by one to two thirds, making the cut just above a node or joint with a bud facing outwards. At the same time, remove dead or diseased wood and any inward-growing or crossing stems. When new growth starts, 'tip pruning' or 'pinching out' may be necessary to encourage further branching.

Geraniums are very easily propagated by cuttings, which are best taken in autumn. Both tip cuttings and stem cuttings develop roots easily. Take the cuttings with two to four nodes or joints and make the cut just below the lowest one. With tip cuttings, remove all but the top leaves. With stem cuttings, remove lower leaves and retain the top leaf only, which may be cut in half to reduce loss of moisture by transpiration. Cuttings may be planted directly in the garden providing the soil is crumbly, but it is more reliable to place them 4–5 cm deep in pots filled with coarse sand or a mixture of coarse sand (3 parts) and peat moss or vermiculite (1 part). Most cuttings develop roots within three or four weeks from planting.

GERBERA (South African Daisy)

Gerberas are naturally adapted to hot, sunny conditions and for this reason they have become one of the most popular flowers in Queensland and other tropical or subtropical parts of Australia. They do well in temperate climates too, providing they are grown in a sunny position sheltered from strong winds. The dainty flowers come in a wide range of colours and are excellent for indoor decoration, especially when combined with light fern or Gypsophila. Gerberas are adapted to light soils but will grow well on heavier soils provided they have been improved by the addition of organic matter and sand. The soil must be open and well drained and it is best to grow them on beds raised at least 20 cm above the surrounding level. On acid soils, gerberas respond to an application of lime and a mixed fertilizer added when the bed is being prepared.

Sow seed in spring or early summer when the weather is warm. Prepare an open, friable soil mixture to use in seed boxes, trays or punnets as described in Chapter 5. A mixture with rather more sand (say 1 part sand to 1 part of other materials) is recommended. Scatter the seed on the surface, cover with compost or vermiculite to a depth of 6 mm and press firmly so the seeds make contact with the dark damp soil. Seed containers must be kept in a warm spot until seedlings emerge—usually about 14–21 days. The soil in the containers can be kept warmer if they are kept indoors and covered with glass or a clear plastic bag. The cover should be removed when seedlings appear and the container transferred to a warm, sheltered but shady spot in the garden.

It is best to prick out the seedlings when they are quite small. Transfer them into a larger container spaced 5 cm apart or into individual 5–7 cm pots. They can then be grown to a good size before planting in the open garden, spaced 40–50 cm apart each way. When planting it is important to keep the crowns well above the surface, especially on heavy soils. Soil washing into the centre of the crowns often promotes rotting. Gerberas can be grown in the same position for two or three years. Old plants can be lifted, divided and replanted. This is best done in late summer or early autumn in temperate areas before cold weather sets in. In tropical climates you can divide the plants at almost any time of the year. The best variety for seed is Yates Giant Mixed which contains a wide range of vivid colours.

GLOBE AMARANTH (Gomphrena)

This attractive summer-flowering annual grows to a height of 30 cm and has round, clover-like, 'everlasting' flowers in rich purple. It is an adaptable plant in regard to soil but responds to good soil preparation plus added fertilizer. It is excellent for bedding, low borders or edging in a sunny position.

Sow seeds in spring or early summer in seed boxes or punnets and transplant seedlings to the garden, spaced 30 cm apart. Alternatively, sow a few seeds direct in clumps at the same distance and thin to 2–3 seedlings at each position. Keep the plants well watered in dry times and mulch with compost or grass clippings to conserve moisture and deter weeds. Give liquid feeds of soluble fertilizer when buds appear. Little Buddy, with masses of purple, globe-shaped blooms, grows to 20–30 cm tall and is the most popular variety. Flowers may be used fresh for indoor decoration or may be bunched and hung, head downwards, in a cool place to dry for 'everlastings'.

GLOXINIA

These exquisite pot plants are ideal for growing in glasshouse, shadehouse or fernery and flower well indoors in a position which suits them. The velvety, bell-shaped flowers have rich colours overlaid with mottling or spotted effects. Deep red, rose, violet or purple are the predominant colours overlaid with white, pink, mauve or lavender-blue markings.

Sow seed in winter or early spring—under glasshouse conditions, autumn sowing is possible. Seed is extremely small and rather slow to germinate, so extra care and attention is needed. Sow seed in seed trays, punnets or pots using an open, friable, rather sandy soil mixture as outlined in Chapter 5. Enclose containers in a plastic bag until seedlings emerge. When seedlings are large enough to handle, prick them out into small individual pots and transfer to larger pots as they grow. Plants reach a height of 30 cm and flower in mid to late summer. Give liquid feeds of soluble fertilizer when buds appear. Small bulbs are formed during the flowering period. These can be replanted the following spring.

GODETIA

Godetias are attractive annuals which have been aptly called 'Farewell to Spring', because they flower late in spring, helping to bridge the gap between spring and summer flowers. The dwarf type, to 60 cm tall, has masses of single blooms on top of the plant and is more widely grown than the tall azalea-flowered type with double flowers.

The plants prefer a sunny aspect sheltered from strong winds. They will grow on a wide range of soils but tend to make too much foliage in fertile soils. Prepare the bed with a mixed fertilizer such as Gro-Plus Complete at $\frac{1}{3}$ cup per square metre, but avoid fertilizers high in nitrogen and side dressings of water-soluble fertilizers. In most climates, sow seeds in autumn to early winter; spring sowings can be made in cold districts. You can raise seedlings in boxes or punnets for transplanting when 5–7 cm high. Space seedlings 30 cm apart each way. Alternatively, sow a few seeds direct in clumps spaced at the same distance and thin to 2–3 seedlings at each position. The plants tolerate dry conditions well but respond to occasional watering as they grow. Cultivate or hand weed around the plants but take care not to disturb the rather shallow root system. Dwarf Mixed is the leading variety of Godetia.

GOURDS

(*Ornamental—See Chapter 11.*)

GYPSOPHILA

Gypsophila, with its dainty white flowers, is an extremely useful plant for flower arrangements. It is also attractive grown in odd clumps in the garden. The plants need a well prepared soil with organic matter and mixed fertilizer added. If soil is acid an application of lime is recommended. Select a sunny but sheltered position. Seeds can be sown at almost any time of the year except in the very coldest or hottest months. By making small successive sowings you can have a continuous supply of flowers for indoor decoration. Sow seeds direct in the open garden, either in rows 20–30 cm apart or sow a few seeds in clumps at the same distance. Cover with vermiculite or compost and keep damp until seedlings emerge. Thinning the seedlings is rarely needed as plants flower well when closely spaced. This way plants support each other too.

HELIANTHUS

(*See Sunflower.*)

HELICHRYSUM

(*See Strawflower.*)

Helichrysum subilifolium is an Australian native with yellow flowers borne on long stems 40–50 cm above a compact rosette of leaves. Suitable for planting here and there in the garden or in clumps. Ideal for brightening up the odd corner. For cultural directions see Strawflower to which it is related.

HOLLYHOCK

One of the tallest flowers in the garden, hollyhocks are an old English favourite. The plants grow to 2–3 m or more with magnificent spikes of large, closely-packed flowers. They prefer full sunlight and a well-sheltered position. In exposed situations, plants need staking or a support such as a trellis to which to tie them. Hollyhocks are adapted to both light and heavy soils but will respond best to fertile soil with organic matter and a mixed fertilizer added.

The best time to sow is late summer and autumn. In most temperate climates, when frosts are not severe, annual types will flower in spring. Perennial types often do not flower well until the second year. Sow the seed in boxes or punnets and cover with vermiculite or compost to a depth of 6 mm. Seedlings emerge in 14–21 days and are ready for transplanting in 6–8 weeks, spacing the seedlings 30–40 cm apart. Do not crowd hollyhocks with other plants and leave space around them for cultivation. Small bedding plants which grow no more than 30–40 cm are suitable companions. When hollyhocks are 30 cm tall, give them a side dressing of mixed fertilizer and another application when buds appear. Hollyhocks are very attractive to snails and slugs so protect the plants with a regular scattering of Baysol or Defender. Plants are also susceptible to rust fungus, especially in humid weather, so spray regularly with fungicide to control. (See Chapter 17.)

Varieties Summer Carnival is an early-flowering annual hollyhock growing to 3 m in height with magnificent spikes of double flowers in a wonderful colour range. Double Mixed is a perennial type but similar in size and flower colours.

HONESTY (Lunaria)

Honesty is a biennial plant but is usually grown as an annual. Plants grow to 60 cm with attractive but rather insignificant lavender or purple flowers in spring. The decorative seed pods, which are prized for dried flower arrangements, mature in mid-summer. Plants do best in a cool, partly shaded position in the garden.

Sow seed in autumn in most districts; spring sowings can be made in cold climates. Raise seedlings in boxes or punnets using an open, friable soil mixture as described in Chapter 5. Transplant seedlings when large enough to handle, spacing them 40–50 cm apart. Give liquid feeds of soluble fertilizers towards flowering time. To preserve the pods for indoor decoration, cut the stems when pods are ripe and allow them to dry. When completely dry peel the outside of the pod by flicking between forefinger and thumb to reveal the silvery, transparent lining.

IMPATIENS

Impatiens, also known as Sultan's Balsam or Busy Lizzie, is a close relative of balsam, but the plants are less compact and bear single flowers in shades of pink, salmon and deep rose. Impatiens is an excellent plant for moist, shady areas and does well when grown in pots, troughs, window boxes and hanging baskets.

Seed can be sown in autumn or spring and is best raised in seed boxes, trays or punnets with a light covering of vermiculite or compost. When seedlings have grown their second leaf, prick them out into small individual pots. At a height of 5–7 cm, transfer the seedlings to larger pots or plant in the open garden spaced 30–40 cm apart. Water regularly in dry weather and give liquid feeds of soluble fertilizer as flowering commences. Baby Mixed, with a good range of flower colour, is the most popular variety.

IPOMOEA

(Morning Glory—See Chapter 11.)

IRISH GREEN BELLFLOWER

Irish Green Bellflower is also known as Molucella or Molucca Balm. It is an attractive bedding or accent plant with closely-packed, pale green, bell-shaped flowers (bracts) born on stems 60 cm long. The stems are much in demand for modern flower arrangements. Sow seed, in autumn or spring, direct in the garden, because seedlings do not transplant easily. Sow a few seeds in clumps spaced 20–30 cm apart and cover with vermiculite or compost. Thin each clump to the strongest 1–2 seedlings. The plants are very adaptable to light and heavy soils and need little attention apart from regular watering.

Summer cypress (Kochia) changes colour from light green to burgundy in autumn

KANGAROO PAW

There are several species of these attractive and unusual native plants. This floral emblem of Western Australia (*Anigozanthus manglesii*) is a herbaceous plant growing to one metre or more tall with velvety, claw-shaped blooms in red, green and black. Like many Australian native plants, kangaroo paw prefers light sandy soils with good drainage. Add bush sand if the soil is heavy. Plants do best in full sunlight. Sow seed in boxes, punnets or pots using an open, rather sandy soil mixture. Seed germination is slow and erratic so keep the containers moist, but not wet, until seedlings emerge. Transplant them to their permanent position when 10 cm tall spacing them 30–40 cm apart.

KOCHIA

Kochia is also known as Mock Cypress or Summer Cypress because of its symmetrical shape and dense, soft-green summer foliage turning to shades of russet and variegated form of this useful plant. These attractive annuals can be used as individual speciments here and there in the garden, as background plants to other annuals or even as a temporary hedge. The plants prefer a warm, sunny situation and do best on well prepared soil with a mixed fertilizer added. Sow seed in spring or early summer in boxes, pots or punnets with a very light covering of vermiculite or compost. Keep moist until seedlings emerge. Transplant seedlings when 5–8 cm tall, spacing them at least 60 cm apart so plants can display their symmetrical shape. Plant about 40 cm apart for a hedge effect. Water regularly in dry weather.

LARKSPUR

Larkspurs are tall, spring-flowering annuals which are ideal for accent planting or as background plants to low-growing bedding or border plants. The plants grow to 60–75 cm and the flower spikes, in pink, rosy-red, light blue and dark blue, are excellent cut flowers. Larkspurs are adaptable to a wide range of soils but respond to well-drained fertile soils with added organic matter plus a mixed fertilizer. If soil is acid, an application of lime is recommended. Soil should be prepared well beforehand and in 'dark damp' condition for sowing the seeds direct in position. Plants prefer full sunlight and protection from strong winds. In exposed situations, they may need staking.

In most temperate climates sow seed in autumn and early winter, but in cold districts spring sowing is successful. Seeds germinate best at a temperature of about 15°C, so that early sowing in late summer or early autumn may not be as successful as later sowing when soil is cooler. Sow a few seeds direct in clumps or stations spaced 20–30 cm apart each way. Cover seeds with vermiculite or compost about 3 mm deep and keep moist until seedlings emerge in 14–21 days. Thin seedlings to 2–3 to each position. This close spacing gives a good mass of colour and helps the tall plants to support each other. Control weeds by regular, shallow cultivation and give liquid feeds of soluble fertilizer when buds appear. Rainbow Mixed, the most widely grown variety, has an excellent range of colour from pale pink to deep blue.

LINARIA

Linaria is an adaptable and colourful little annual for flowering in winter and spring. Plants grow 30–40 cm tall with spikes of flowers like tiny snapdragons in delicate pastel shades. Like larkspurs, linarias are adaptable to a wide range of soils but respond to good soil structure, added fertilizer and an application of lime if soil is acid. Prepare the bed well beforehand to have it in friable, dark damp condition for direct sowing.

Sow seed from early autumn to early winter in temperate climates but spring sowings are successful in cold districts too. Like larkspur, seeds germinate well in cool soil. Sow seeds direct in shallow rows quite thickly. Thinning is only needed if the seedlings are overcrowded. Alternatively, sow a few seeds in clumps about 10–15 cm apart—usually all seedlings can be retained to give a dense mass of colour. Plants can be smothered by weeds in the early stages of growth. Shallow cultivation or hand weeding is needed from the seedling stage onwards to avoid weeds becoming too large. Water plants regularly, especially in dry weather, and give liquid feeds of soluble fertilizers every 10–14 days when the plants are well established. After the first flowering, cut back plants to promote a further flush of blooms. Fairy Bouquet is the leading dwarf variety with flowers in shades of cream, yellow, gold, apricot, pink and mauve. It is excellent as a low border or for clump growing in rockeries, and flowers very quickly from sowing.

LIVINGSTONE DAISY
(Mesembryanthemum)

Livingstone daisies are among the brightest of flowers for late winter and spring. The dwarf plants are 15 cm tall and covered with tightly-packed flowers in yellow, pink, cerise and purple. They are ideal plants for carpeting, edging and rockeries. The plants are very adaptable to light or heavy soil, withstand dry conditions well but need good drainage. They must be grown in full sunlight as in shade the flowers remain closed, even on cloudy days.

In temperate climates, seed can be sown from early autumn to early winter but in cold districts early spring as well. Sow seed in boxes or punnets for transplanting 10–15 cm apart or sow a few seeds direct in clumps at this distance and cover with compost or vermiculite. Seeds may take 14–21 days to germinate so keep moist until seedlings emerge. Thin each clump to 2–3 seedlings. Water regularly until plants are established and then only if weather is dry. Providing a pre-planting fertilizer has been added to the garden bed before sowing or transplanting, additional fertilizer is rarely necessary.

LOBELIA

Lobelia, another dwarf, spring flowering plant, is excellent for massed colour effects, for edging, rockeries and window boxes. Few plants have flowers in the richest and deepest blue of lobelia, which blends dramatically with white alyssum, yellow violas or dwarf marigolds. Plants respond to a friable, fertile soil with mixed fertilizer added. A sunny aspect, especially morning sun with protection from strong winds, is preferred.

Sow seed in autumn in boxes or punnets. Seed is small so prepare a friable soil mixture which holds moisture well. Sow seed on surface and press lightly and firmly into the soil and lightly cover with compost or vermiculite. Water gently or stand the container in water until moisture seeps

Livingstone daisies revel in warm, sunny positions

to the surface. Transplant seedlings when quite small (2–3 cm), spacing them 10 cm apart. Water regularly and give liquid feeds of soluble fertilizer when buds appear.

Varieties Crystal Palace grows to 15 cm tall with bronze-green foliage and rich, dark blue flowers. String of Pearls grows to about the same height but has flowers in shades of pink, mauve and rose-purple as well as crisp white and clear sky blue. Basket Lobelia is an interesting variety with trailing stems up to 30 cm in length. The leaves are bright green and the flowers, rich blue with a white eye, are borne in great profusion. Ideal for hanging pots and baskets. Use four seedlings per 30 cm container.

LUPIN

Lupins belong to the legume group of plants and are able to add nitrogen to the soil because of the nitrogen-fixing bacteria contained in their root nodules. (See Chapter 4.) New Zealand Blue lupins are frequently used as a green manure crop to improve the soil. Most garden varieties of lupins are winter-growing annuals which flower in spring and early summer. They are very adaptable to climate and soil and grow well in warm and temperate districts. The Russel Lupin is a perennial type which is best suited to cold climates such as the eastern highlands of New South Wales, southern Victoria and Tasmania.

Lupins do not need very fertile soil.

Lupins

Too much fertilizer—especially those high in nitrogen—tends to favour an abundance of foliage at the expense of flowers. Prepare the soil well beforehand so it will be in friable dark damp condition for direct sowing in autumn. On most soils, an application of lime (about $\frac{2}{3}$ cup per square metre for light soil, $1\frac{1}{2}$ cups per square metre for heavy soil) is recommended. Also add a ration of mixed fertilizer high in phosphorus, such as Gro-Plus Complete, at $\frac{1}{3}$ cup per square metre. Select a well drained, sunny position to grow lupins.

The seeds are quite large (although size varies with different species and varieties) so are easy to handle for direct sowing into the garden. If the soil is in dark damp condition at sowing, seeds will germinate quickly and easily with little need for extra watering until seedlings emerge. Too much moisture in the early stages of germination may do more harm than good. (See Chapter 5.) Dust seed with fungicide before sowing. It is best to sow a few seeds in clumps, the spacing depending on the size to which the plants will grow. For dwarf varieties allow 20 cm between plants, for tall, large-seeded varieties (and Hartwigii) allow 30–40 cm, and for Russel Lupins allow 50 cm. Seed of Russel Lupins is more expensive, so it may pay to sow individual seeds in small pots and transfer seedlings to their permanent positions when 7–10 cm tall.

For dwarf types, thinning is rarely needed, but for larger varieties thin each clump to 1–2 seedlings. Once established, lupins need very little attention apart from regular watering in dry weather. When buds appear, weak liquid feeds of soluble fertilizers will promote larger blooms and prolong flowering. For cut flowers, stand stems in boiling water for 15–20 seconds; take care steam does not damage lower blossoms. Pick spent flowers regularly.

Varieties Pixie is a dwarf variety growing to 15–20 cm, and is ideal for mass bedding or borders. Flower colours in pale blue to dark blue, light to medium pink, mauve and white. Hartwigii lupins are semi-tall plants with hairy foliage and attractive flower spikes in light to dark blue, mauve, pink, yellow and white. Russel Mixed (*L. polyphyllus*) are perennial lupins which grow 1–2 m tall with magnificent flower spikes in yellow, gold, apricot, salmon pink, maroon, purple and shades of blue. These herbaceous perennials will shoot each year from the crown in cold districts and usually last for four or five years.

MARIGOLD

Marigolds are summer-flowering annuals and are best sown in spring or early summer because they prefer the warm weather and are susceptible to frost. African marigolds are rather taller with large flowers while French marigolds are shorter, more compact—some are dwarf types—and have smaller flowers. Some French marigolds can also be sown in autumn, providing there are no frosts. Marigolds prefer friable fertile soils with added organic matter and fertilizer but they tolerate poor soils too. They need a warm sunny aspect which is well sheltered from wind.

Commence sowing seed in spring after all danger of frost is over; sowing can be continued until mid-summer for late summer and autumn flowering. Generally speaking, a frost-free period of five months is needed from sowing. You can sow seed in boxes or punnets for pricking out into larger containers when the seedlings are about 1 cm tall and are ready for transplanting when 7–10 cm high. As seeds germinate quickly and seedlings are quite vigorous, you can sow seeds direct in rows or a few seeds in clumps too. The distance apart for transplanting or sowing direct depends on the height and spread of the plants. Generally allow a distance of 40 cm apart each way for tall varieties, 30 cm apart for shorter varieties and 20 cm apart for dwarf types. Marigolds are shallow-rooted plants so need regular watering in dry weather. A mulch of grass clippings, compost or leaf mould will prevent moisture loss, keep the roots cool and discourage weeds. Give liquid feeds of soluble fertilizer as buds appear and remove spent blooms regularly to prolong flowering.

Varieties

AFRICAN MARIGOLDS: Jubilee has vigorous, sturdy bushes to 75 cm tall with huge, tightly-ruffled flowers in lemon, gold and orange. Golden Girl is another vigorous hybrid to 60 cm tall with large, double flowers in rich gold. Jezebel is similar to Golden Girl with an unbelievable profusion of large double blooms from soft, creamy primrose to deep glowing orange. These varieties are extremely showy and produce masses of flowers suitable for cutting.

FRENCH MARIGOLDS: Honeycomb grows to about 30 cm and is covered in flowers in shades of rich brown and yellow. An excellent bedding variety. Flamenco is a new dwarf variety for bedding and low borders. The compact bushes grow to 30 cm, with masses of double blooms in lemon, yellow, gold and mahogany. Petite Yellow is a very dwarf type, growing to 15–20 cm high. It is very free-flowering with double, clear yellow blooms for many months. Petite Orange is the orange counterpart of Petite Yellow with the same free-flowering ability. Freckle Face is a pretty mixture of dwarf types. All three are excellent for mass bedding, low borders or as rockery plants. Rockery Mixed is a new mixture of dwarf marigolds to 25 cm high. Plants have attractive, fern-like foliage and small, single flowers in shades of yellow, gold, orange and red.

Jezebel marigold provides a profusion of bloom ranging in colour from primrose to to deep orange

MARMALADE DAISY (Rudbeckia)

Marmalade daisy is an attractive bedding or border plant to 40 cm tall with masses of flowers in summer. The blooms are gold-yellow with a purple-black centre cone. The plants are best treated as annuals but in cold districts they can be grown as herbaceous perennials. They are adapted to a wide range of soil and climate but prefer a fertile, friable soil and full sunlight. Sow seed in spring or early summer in seed boxes or punnets for transplanting, or direct in the garden in clumps spaced about 30 cm apart. Give liquid feeds when flower buds appear.

MIGNONETTE

This dwarf spring flowering annual is an 'old world' favourite, more for its spicy aroma than for its small, orange-yellow flower spikes. It is a good subject for low borders, edging and rockeries and for growing in containers such as troughs or pots. The plants respond to friable, fertile soils to which organic matter and mixed fertilizer has been added. On acid soils, an application of lime is recommended.

Seedlings of mignonette do not transplant well so it is best to sow seed direct in the garden bed. You can sow seed in autumn, winter and spring in warm temperate climates but in cold districts autumn and spring only. It is best to scatter seed thinly in shallow rows or a few seeds in clumps where the plants are to flower. Cover the seeds with vermiculite, compost or seed raising mixture. Seed germination is often erratic so keep the soil moist but not wet for about 14 days. When seedlings emerge, thin them if overcrowded. Established plants need little attention apart from occasional watering. Give liquid feeds of soluble fertilizer every 10–14 days, especially when buds appear. Remove spent flower spikes to prolong flowering.

MINA

(*See Chapter 11.*)

MORNING GLORY

(*Ipomoea—See Chapter 11.*)

MOLUCELLA

(*See Irish Green Bellflower.*)

NASTURTIUM

Nasturtiums are very adaptable, colourful annuals. Old varieties were rather straggly plants, but plant breeders have made great improvements in the modern varieties now available. Not only are plants more compact but the flowers are produced well above the foliage to give a brighter colour display. Nasturtiums will grow on a wide range of soils, but they do best on moderately fertile soils on which the plants produce less foliage and flower more prolifically. They prefer open sunlight and rather dry conditions but will make quite a good showing in partial shade. Nasturtiums make excellent bedding plants and are also good for growing in rockeries, troughs, tubs or large hanging baskets. Plants flower in 10–12 weeks from sowing.

In warm temperate climates, you can sow seed from spring to early autumn but in cold districts make sowings in spring only. Prepare the bed for direct sowing to have the soil in dark damp condition, but avoid using compost or animal manure, both of which provide conditions favouring excessive leaf growth. A mixed fertilizer such as Gro-Plus Complete which is high in phosphorus can be scattered over the soil at $\frac{1}{3}$ cup per square metre and raked in before sowing. Seeds are large, easy to handle and germinate in 14–21 days. Sow a few seeds in clumps spaced 20–30 cm apart. Thinning is rarely needed and this gives a greater mass of colour. Water moderately until plants are well established, but then keep them on the dry side to encourage flowering. Do not give liquid feeds of soluble fertilizers as these will encourage leaf growth.

Varieties Jewel Mixed has compact plants and contains a mixture of the choicest colours available—primrose, gold, orange, mahogany and red. Cherry Rose is one of the brightest varieties with semi-double cherry rose flowers which contrast dramatically with the foliage. Whirly Bird is a recent introduction which is distinct from other varieties because the flowers are without a distinct spur. The plants are dwarf and very compact and carry the flowers, in a bright range of colours, well above the foliage. Alaska is another exciting new variety. The compact, bushy plants have marbled or variegated leaves in green and white. Flowers are single and range in colour from yellow through orange to red. All the new varieties grow to 20–30 cm. Roulette is a new variety carrying a heavy crop of flowers above the foliage. Ideal for baskets.

NEMESIA

Nemesia is one of the brightest and most colourful bedding and border plants for late winter and spring flowering. Again, the trend has been to develop dwarf *compacta* types in preference to the tall *strumosa* types. Nemesias are adapted to light or heavy soil but prefer friable, fertile soil to which organic matter and mixed fertilizer have been added. Prepare the soil well beforehand so that it is in good condition for direct sowing. Direct sowing is the best method, although seedlings can be raised and transplanted if preferred. The plants need a warm, sunny aspect with good drainage.

In warm climates, sow seed in autumn or early winter but in cold districts sowings can be made in both autumn and spring. For the dwarf varieties, sow seed direct in rows or better still, a few seeds in clumps spaced 15–20 cm apart. Cover the rows or positions with a light sprinkling of vermiculite, compost or seed raising mixture. Seedlings usually emerge within 10–14 days and thinning is rarely needed. Alternatively, seedlings can be raised in boxes or punnets, pricked out into larger containers if crowded and transplanted when 5–7 cm high. If transplanting seed-

For colourful bedding plants use nemesia in clumps or for massed display

lings, harden them off by withholding water for a few days and then give a good watering the night before you transplant. Space seedlings 15 apart each way. When the plants are established they need little attention. For bushier plants you can pinch out the leading stems. Water regularly and give liquid feeds of soluble fertilizers every 2–3 weeks, especially towards flowering time. Plants will flower in 14–16 weeks from seed sowing.

Varieties Carnival Mixture is the leading dwarf variety. It has large flowers in cream, yellow, gold, orange, scarlet and red on strong, bushy plants 20–30 cm tall. Blue Gem is also a dwarf plant growing to 20 cm with smaller flowers in a clear sky-blue. It is one of the best blue-flowered annuals for spring display and contrasts well with Carnival as a foreground border or in clump plantings.

NEMOPHILA (Baby Blue-Eyes)

Nemophila is another charming dwarf annual for late winter and spring flowering. Plants grow to 20–30 cm with fern-like foliage and small, sky-blue, saucer-shaped flowers. It is an excellent fill-in plant for clump plantings in garden beds or in rockeries. Like nemesia, the plants prefer a friable, fertile, well drained soil in a warm, sunny position.

Seedlings do not transplant well and it is best to sow seeds direct in the garden. In most districts, autumn sowings are best but in cold districts, spring sowings can be made too. Sow seeds thinly in rows or a few seeds in groups or clumps spaced 15 cm apart, and cover lightly with vermiculite or compost. Thin out seedlings if overcrowded. Water regularly and give liquid feeds of soluble fertilizer when buds appear.

ORNAMENTAL BASIL

This attractive annual is also known as Dark Opal Basil because of its purple-bronze foliage. Plants grow to 30–40 cm tall and leaves have the same spicy aroma as the herb Sweet Basil. The spikes of small, lavender-white flowers are not spectacular but the plants are recommended for their foliage and aroma. They are ideal for troughs, pots or rockeries or spotted here and there among summer bedding plants. Sow seeds in spring or early summer in boxes or punnets for transplanting seedlings 20–30 cm apart or sow direct in the garden in clumps at the same distance. Plants grow well in sun or semi-shade and need little attention apart from regular watering and weeding. Remove flower spikes as they appear to keep a neat and tidy appearance.

ORNAMENTAL CHILLI

Ornamental Chilli is a variety of capsicum or pepper, the plant we grow in the vegetable garden. The fruits are much smaller but very attractive and change colour with maturity from green or purple to yellow, orange and scarlet. In mild climates, plants will over-winter to grow again the following spring but they are usually grown as annuals. They are warm season plants and frost susceptible, so the best sowing time is spring to early summer. They need a warm, sunny position with shelter from strong winds. They prefer a fertile soil to which organic matter and a mixed fertilizer has been added during preparation. Ornamental chillis are excellent for individual specimens here and there in the garden or for container growing in tubs or large pots. Plants grow to about 25 cm.

A new dwarf bush variety grows to about 30 cm and produces a multitude of very colourful large fruit well displayed above the foliage. In well-grown plants foliage is almost covered by the fruit. Unlike other ornamental chillies, fruit on this variety is edible and is therefore safer to have in the garden when there are young children present.

You can sow a few seeds direct in clumps spaced 60 cm apart but it is best to raise seedlings in boxes or punnets and transplant them when 7–10 cm high, spacing them at the same distance. Regular watering, together with mulching, is needed in dry weather. Do not give extra fertilizer until the small white flowers appear. Then give liquid feeds of soluble fertilizer every 2–3 weeks to encourage and prolong flowering and fruiting. The fruits, or berries as they are often called, are extremely hot, so take care to keep young children away from them.

PAINTED DAISY .
(Chrysanthemum tricolor)

This annual chrysanthemum is a good subject for clump planting here and there in the garden or as a background plant. The plants grow to 60–75 cm with large white flowers zoned with yellow, red and purple. They flower in spring and early summer from seed sown in autumn. In cold districts, seed can be sown in spring too. They do best on well-drained, fertile soils with organic matter and fertilizer added. They need a warm, sunny position sheltered from strong winds.

Sow seed in boxes or punnets, covering the seed with vermiculite or compost to a depth of 6 mm. Transplant seedlings when 7–10 cm high, spacing them 40–50 cm each way. Water regularly and give liquid feeds of soluble fertilizers when buds appear. Plants flower within 12–14 weeks from transplanting. Remove spent blooms to prolong flowering.

PANSY

The pansy was originally known as Heartsease or Wild Pansy and is closely related to viola, violetta and violet, all of which belong to the genus *Viola.* Pansies come in a wide colour range but are distinguished by velvety black or dark-coloured blotches or markings. These spring-flowering, biennial plants are treated as annuals and are excellent for carpet bedding or low borders. They need a well-drained, friable, fertile soil but the tender roots should not be in contact with concentrated fertilizer. It is best to prepare the bed well beforehand by adding compost or animal manure together with a ration of mixed fertilizer ($\frac{1}{3}$ cup per square metre) thoroughly incorporated into the top 10–15 cm of soil. Pansies prefer sunlight for most of the day but shade from hot afternoon sun is an advantage.

The multicoloured painted daisy flowers profusely in spring and early summer

Seed can be sown from midsummer to early winter in most districts but in cold areas, autumn and early spring are recommended sowing times. Sow seed in boxes or punnets using an open, friable soil mixture as described in Chapter 5. Sow the seed thinly in shallow rows covering lightly with vermiculite or compost. Germination may be slow (21–28 days) so containers must be kept moist, but not wet, for this length of time. Prick out the seedlings when quite small, spacing them 5 cm apart in seedling trays filled with a similar soil mixture. Grow them on for a few weeks in partial shade until the seedlings are larger and sturdy enough for planting out. Transplant the seedlings into the well prepared dark damp soil of the garden bed, spacing them 20 cm apart for bedding and 30 cm apart for large exhibition blooms.

Water plants regularly but do not overwater. Keep weeds under control by careful shallow cultivation, or with a mulch of grass clippings or compost tucked around the plants. This will help to conserve moisture, keep the soil crumbly and encourage a thick mat of surface roots. Once the plants are well established, give weak liquid feeds of soluble fertilizer every 2–3 weeks. If the first flowers are small, remove some of the buds to increase the size of those which remain. Always pick spent blooms regularly to encourage new buds and prolong flowering. The main pests are aphids which cluster under the foliage. (See Chapter 17 for control.)

Varieties Can Can is a specially formulated mixture of the largest pansies available. It contains blotched and fancy types in a wide range of colours. Super Jumbo

(*Above*) Hollyhocks are ideal for stately backgrounds

(*Top right*) Snapdragons (*Antirrhinum*) are ideal cut flowers

The dried seed pods of Honesty (*Lunaria*) are most useful in floral arrangements

Sweet William has a wide variety of colours and markings

(*Right*) The ever-popular anemone adds brilliance and colour to spring gardens

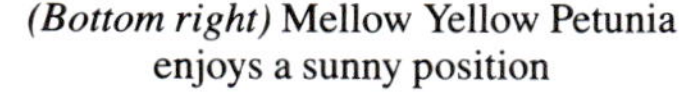

(Bottom right) Mellow Yellow Petunia enjoys a sunny position

Coleus gives variety in colour and form to shady spots

is a very early and free-flowering strain. Special breeding techniques with the Jumbo Giant strain have produced extra large blooms in a wide range of beautiful pastel shades and strong colours. Roggli or Swiss Giants are popular strains in Australia. The glorious flowers are sometimes ruffled and borne on long stems from sturdy plants. The colour range includes red, brown, blue and yellow with striking blotches. Superb Giants is a new blend of beautiful blooms with a full colour range and different markings. It contains the best of the Jumbo Giants, Swiss Giants and Engelmann's strains. Ullswater Blue is one of the prettiest pansies of the Roggli type. The flowers are rich blue with darker markings. An excellent variety for bedding, borders or indoor decoration.

PASSIONFRUIT

(*See Chapter 16.*)

PETUNIA

Petunias are one of the most colourful annuals for the summer garden. Like many other flowering plants, petunias have been improved tremendously by plant breeders in recent years. The *multiflora* (bedding type) of petunia are sturdier and more compact with flowers in strong, clear colours. The *grandiflora* types have magnificent single, double or frilled blooms and the plants are stronger and sturdier than old varieties. All petunias are sun-loving plants and will tolerate dry conditions once they are established. Plants should be sheltered from strong winds, especially the taller, large-flowered types. They grow well on a wide range of soil textures from light to heavy but respond to fertilizer applied when the bed is prepared. Too much fertilizer and water produces soft sappy plants which flower poorly. Petunias are excellent for mass bedding and borders but can also be grown (especially the *grandiflora* types) in tubs, troughs and large pots on a sunny terrace or patio. There are semi-trailing types too for window boxes and hanging baskets for a cascade effect.

Seed can be sown in spring after all danger of frost is past and continued through to midsummer in most districts. In cold areas with a shorter growing season, sow seed in early summer. In warm climates, such as Brisbane and further north, sowings can continue to late summer or autumn. Petunia seed is very small and extra care should be taken in raising the seedlings in boxes or punnets. Prepare an open, friable soil mixture as described in Chapter 5. Scatter the seed on the surface and cover very lightly with vermiculite, compost or seed raising mixture. Press firmly and water very gently. Alternatively stand the container in water until moisture seeps to the surface. Petunia seed needs fairly warm (25°C) conditions to germinate so it is best to keep the containers indoors until the seedlings show through—then move them outside immediately, but to a sheltered position. Small containers can be covered with glass or plastic bags. This helps to maintain an even temperature and prevents drying out. Prick out the small seedlings into seedling trays, spacing them 3–5 cm apart, and grow them on until 5 cm high for transplanting to the garden. Transplant the small-flowered bedding types at a spacing of 25–30 cm each way but allow 30–40 cm for the large-flowered varieties. Always scatter snail baits around the newly planted seedlings to protect them from slugs and snails. When established, petunias need little attention apart from occasional watering if the soil becomes dry. Plants in tubs, pots or hanging baskets dry out very quickly and need more regular watering. Give weak liquid feeds of soluble fertilizer when buds appear.

After the first flush of flowers, plants can be cut back and given a liquid feed for a second flush of flowers in late summer or early autumn.

Varieties

SMALL-FLOWERED *multiflora* BEDDING PETUNIA (30–40 cm tall): Dazzler is a hybrid, small-flowered type with compact vigorous plants which are very resistant to wet weather. Colours are brilliant and varied—a superb petunia for garden display. Rose of Heaven is an excellent variety for bedding or borders. The bushy, compact plants carry masses of single flowers in a brilliant, rose colour.

LARGE-FLOWERED *grandiflora* PETUNIA (40–60 cm tall): Colour Parade was one of the first Japanese F1 hybrids and is the most popular strain. Single flowers are a gay mixture of colours including carmine, salmon, bright red, dark blue, clear white and several different shades of pink. Elegant Cascade is a specially blended F1 hybrid mixture of large-flowered, single petunias in a wide colour range. It is ideally suited for tubs, large pots, window boxes and hanging baskets. Giant Victorious is one of the most prized flowers for the keen gardener. The double, ball-like flowers are up to 10 cm in diameter. Colours include white, cream, rose, pink, mauve, scarlet and purple. Titan is a compact *grandiflora* mixture with a brilliant range of colours. Mellow Yellow is a new variety with blooms of creamy yellow with deeper markings. It produces a profusion of medium-sized flowers on a compact bush. Petticoat petunias are Picotee types of outstanding colour and frilled edges to the bloom. Available in red and white and purple and white.

PHLOX

Phlox is one of the brightest summer flowers for mass bedding or borders. Dwarf varieties, the most widely grown, are 20 cm in height but tall varieties reach 40 cm or more. The range of flower colour is magnificent, many with contrasting white centres and others star-shaped with pointed petals. The plants prefer full sunlight but perform well in any situation which has sun for part of the day. They respond to friable, well-drained soil with added organic matter plus a mixed fertilizer. Prepare the soil well for direct sowing in spring or early summer. In warm northern areas you can sow in late summer and autumn too, providing there are no frosts.

Colour Parade, a Japanese hybrid petunia, is one of the most popular strains

Seeds can be sown in boxes or punnets if preferred but direct sowing is best because seeds germinate easily. Phlox seeds are ideal for sowing in clumps spaced 10–15 cm apart each way. Scatter a few seeds in each clump or station and cover with vermiculite, compost or seed raising mixture to a depth of 3 mm. Alternatively, seeds can be sown thinly in rows at the same distance apart. With both methods, thinning is rarely needed and this close spacing of plants gives a denser mass of colour. When seedlings have emerged (14–21 days) keep them well watered until plants are established. Then water regularly but do not over-water. Phlox will tolerate fairly dry conditions. Give liquid feeds of soluble fertilizers when buds appear. When flowering commences, avoid overhead watering as flowers last better when dry. Watering around the base of the plants and a mulch of grass clippings or compost helps to keep soil moist and protects the shallow roots. Remove spent flowers to promote new buds and prolong flowering.

Varieties Drummondii Dwarf is the most widely grown variety and is ideal for carpeting, low borders and in rockeries. The plants are compact and grow to 20 cm with a very wide colour range which includes pink, lavender, salmon, scarlet, crimson, blue and white with some flowers with light-coloured centres. Bright Eyes Mixture is a rather taller strain to 30–40 cm. All flowers have a white or contrasting centre to a wide range of strong colours. Twinkle Mixture is a very dwarf, compact variety which flowers freely. The lovely star-shaped blooms are interesting and attractive in a good colour range. Derwent Dwarf Mixed is a free-flowering, compact strain with bushy plants covered with very large flowers in beautiful colours.

PINKS
(*See Dianthus.*)

PIN-CUSHION FLOWER
(*See Scabiosa.*)

POLYANTHUS
These spring flowering, primrose-like plants are herbaceous perennials but are often grown as annuals. The modern strain—Pacific Giants—has clusters of large florets on tall, strong stems in a range of colours which include apricot, gold, pink, scarlet, red, blue and white. Polyanthus grows well in a cool, sheltered, partially shaded position or shadehouse. When grown in shaded situations, the plants may last for two or three years and can be divided when they are dormant. The plants can also be grown in open beds in full sun and make ideal pot plants. If grown in the open, move the plants to shade after flowering and take plenty of soil with the roots. Polyanthus prefer a friable, well-structured soil and respond to liquid feeds of soluble fertilizer.

Sow seeds in late summer and early autumn in boxes, punnets or pots. Seeds are small and germination may be slow and erratic. Prepare a good seed raising soil as outlined in Chapter 5. Scatter the seed along shallow rows in the boxes or on the surface of the soil mix in punnets or pots. Cover very lightly (3 mm) with a moisture-holding material such as vermiculite, compost or seed raising mixture. Press down gently but firmly with a board and keep the surface moist until seedlings emerge in 3–4 weeks. As germination is erratic, the seedlings may not all be ready for transplanting at the one time. Transplant seedlings when large enough to handle. If growing the plants in pots, use the same mixture as for raising seed. Do not let the pots dry out, and give liquid feeds, especially towards flowering. Pacific Giants is the leading variety with a superb colour range.

POOR MAN'S ORCHID (Schizanthus)
These spring flowering annuals with orchid-like flowers in shades of pink to violet are very adaptable and should be more widely grown. They thrive in semi-shaded positions but can be grown in the open garden provided they are shaded during the hottest part of the day. They are ideal for growing in pots or hanging baskets under trees or in ferneries and shadehouses. When grown in pots or baskets, the soil mixture should be open and friable with equal parts of garden soil, sand and moisture holding material such as vermiculite, compost or peat moss. Add a mixed fertilizer such as Gro-Plus Complete at the rate of a heaped tablespoon for each bucket of mixture.

In temperate climates, sow seed from late summer to early winter, but in cold districts, late summer and autumn only. It is best to sow seed in boxes or punnets as it is quite small and germination may take 14–21 days. Cover the seed very lightly, firm down with a piece of flat board or the bottom of a punnet and keep the surface damp until seedlings emerge. Transplant seedlings to the open garden when 5 cm tall (30 cm apart each way) or transfer to small pots for growing on to larger pots or hanging baskets. To promote bushy growth of the fern-like foliage, pinch back the leading stems regularly. Keep the plants well watered at all times and when established, give liquid feeds of soluble fertilizer every 10–14 days to promote flowering.

POPPY (Iceland)
Iceland poppies are probably the most popular flowers for late winter and spring. They are magnificent bedding plants, growing to 60 cm, and the cut flowers are unsurpassed for indoor decoration. Modern strains of Iceland poppies are a vast improvement on old varieties, both in flower size and colour range. Colours include lemon, yellow, gold, orange, pink, salmon, red shades and white. If spent blooms are picked regularly, the plants will flower for many months. Iceland poppies need plenty of sunshine, good drainage and a friable, fertile soil. They revel in 'good going', so it pays to prepare the bed well beforehand with plenty of compost or animal manure, together with a pre-planting ration of mixed fertilizer. Sunlight, especially morning sun, is needed to 'pop' the buds so select a warm, sunny (but wind-sheltered) bed to grow them. In most temperate climates like Sydney, you can start sowing in late summer and continue through to autumn. Early sow-

ings will produce plants to flower in winter. In cold districts such as Melbourne, summer to early autumn sowings are recommended. In warm northern areas sowings can be made later, from early autumn to winter.

Iceland poppy seed is very small and the seedlings are small and delicate. It is best to sow seeds thinly in boxes or punnets as described in Chapter 5 and cover with a very light scattering of compost, vermiculite or seed raising mixture. Keep the surface moist until seedlings emerge in 10–14 days. Prick out the seedlings when quite small, spacing them 3–5 cm apart in other boxes or seedling trays. Grow them on to a good size, hardening them off to more sunlight as they grow. Transplant seedlings 20–30 cm apart each way into the dark damp garden bed. Plant out on a cool day and keep the crown of the plant slightly above the surface. Planting seedlings too deep may cause the crown to rot. Keep the seedlings well watered with a gentle spray until established. With a spacing of 20–30 cm between plants, the foliage will eventually cover the soil but mulching with grass clippings or compost will conserve moisture, keep down weeds, and avoid possible damage to surface roots when cultivating. Root damage weakens the plants and may result in twisting of flower stems. Give liquid feeds when buds appear and then at 2-week intervals. Pinch out early buds until the plants have formed good clumps. Remove spent blooms to prolong flowering. For indoor decoration, pick flowers early in the morning in full bud or bud-opening stage. Dip the stems into boiling water for 30 seconds before arranging the blooms.

Varieties Artist's Glory is the most popular Iceland poppy for general garden display. It is a specially formulated mixture containing strong-stemmed flowers in shades of lemon, yellow, gold, apricot, rose pink, salmon and white and many with distinctive picotee edges. This strain flowers over a long period. Sunglow is a beautiful strain with predominant flower colours of gold, orange and red plus a range of in-between pastel shades. The plants are vigorous and free-flowering. Springsong has the largest and most brilliantly coloured flowers of all Iceland poppies. It is an F2 strain with a wide colour range and many picotee and two-toned flowers.

Iceland poppies are magnificent, long-flowering bedding plants

PORTULACA

Also known as Pig Face or Sun Plant, this summer flowering annual is excellent for low borders, edging, banks and rockeries. The plants are 15–20 cm tall and covered in bright daisy-like flowers in lemon, mauve, pink, salmon, crimson and white. Portulaca will grow on a variety of soils but responds to added fertilizer which should be incorporated in the soil when the bed is prepared. The plants do best in full sunlight, need good drainage and will tolerate dry conditions.

Sow seed in spring (after soil has become warm) or early summer. Seeds can be raised in boxes or punnets if preferred and seedlings transplant easily. But it is best to sow seeds direct in the garden, either in rows or patterns or in clumps spaced 10 cm apart. Cover the seed with about 6 mm of vermiculite, compost or seed raising mixture and keep damp until seedlings emerge in 10–14 days. With direct sowing there is usually no need to thin the seedlings unless they are very crowded. If thinning is necessary, seedlings —if lifted carefully—can be replanted.

The plants are prostrate and creep over the ground, protecting their roots and smothering weeds, but a light mulch of grass clippings or compost will assist them to become established. Water the plants in dry weather but do not over-water as they prefer rather dry conditions. Give liquid feeds of soluble fertilizers when buds appear about 6 weeks after sowing. The most widely grown variety is Double Mixed (Sunnybank) which contains a bright colour range.

PRIMULA

Primula malacoides is really a perennial but is always grown as a spring-flowering annual. It has come a long way from the old-type primula with small, mauve flowers on long stems. The modern primula is more compact and the flowers are much larger with a range of colour which includes mauve, pink, carmine, purple, ruby red and white. Most of this improvement in flower size and colour range has been carried out in Australia. Primulas have traditionally been regarded as shade-loving plants, but most varieties available in Australia grow equally well in full sunlight. They are ideal plants for bedding or borders but are also attractive when grown in troughs, pots and window boxes. They prefer friable, fertile soil with added organic matter and mixed fertilizer as a pre-planting dressing. On acid soils they benefit from an application of lime (1 to 2 cups per square metre according to soil type, see Chapter 4) during preparation of the garden bed.

Portulaca is ideal for a colourful summer carpet

In most climates sow seed from mid-summer to early autumn, but in warm areas like Queensland sowings can be made later. Like Iceland poppy, the seed of primula is small and the seedlings delicate. Prepare an open, friable soil mixture as described in Chapter 5 for filling seed boxes or punnets. Sow seeds in very shallow rows or scatter on the soil surface and press firmly with a board. Seed should be barely covered with vermiculite, compost or seed raising mixture. Seedlings may take 21–28 days to emerge, so keep surface moist for that period. Prick out seedlings when small into other boxes or seedling trays, spacing them 3–5 cm apart and grow them on until large enough to transplant. Space seedlings in the garden 15–20 cm apart each way. Choose a cool, cloudy day for transplanting and keep the new plants well watered until they are established. Give regular liquid feeds of soluble fertilizer when buds appear. If growing primula in pots use an open, friable potting soil made up of equal parts of garden loam, sand and compost or leaf mould and add $\frac{1}{4}$ cup of lime and a heaped tablespoon of mixed fertilizer to each bucket of mix. Proprietary potting mixtures are also suitable.

Varieties

Primula malacoides: Carmine Glow is a vigorous variety with compact, sturdy plants 20–25 cm in height. It has been selected over many years for its large, carmine-rose flowers which hold well in open sunlight. Gilham's White is very similar in growth form but the flowers are pure white. It does well in open sunlight or shade. Royalty has flowers in an attractive shade of pink and prefers semi-shade. Lollipops is a specially formulated

mixture of the dwarf annual primulas. Colours include carmine-rose, lavender-pink, ruby red and white. An excellent mixture for bedding or borders in open sun or shade.
Primula obconica: This is an evergreen perennial best suited to shady situations or for growing in pots or baskets in ferneries or shadehouses. The plants grow to 15–20 cm with large flowers in shades of rose, mauve, lavender-blue, crimson and white. The plants are excellent for indoor pot plants when in flower. Seed sown in late summer will produce plants to flower in spring. It is best to replant or repot in early winter.

RUDBECKIA
(*See Marmalade Daisy.*)

SALPIGLOSSIS

Salpiglossis is a tall bedding or background annual for summer flowering. The trumpet-shaped flowers are 5 cm long in shades of gold, bronze, red and violet. The plants require a friable, fertile soil with organic matter and mixed fertilizer added. Prepare the soil well beforehand for direct sowing in spring or early summer. Plants need a warm sunny aspect.

As seedlings do not transplant well, it is best to sow seeds direct in rows spaced 15 cm apart and thin seedlings to the same distance. Alternatively, sow a few seeds in clumps or stations and thin each position to 1–2 seedlings. Cover seeds to a depth of 6 mm with vermiculite or compost and keep damp until seedlings emerge in 14–21 days. Once plants are established, water regularly, especially in dry weather, and give liquid feeds of soluble fertilizer every 2–3 weeks. Plants will flower 12–14 weeks after sowing. The most widely grown variety is Emperor Mixed, with a good range of flower colour.

Lollipops primula—a multi-coloured mixture of the best dwarf types

SAPONARIA
(*See Big Gyp.*)

SALVIA (Flowering Sage)

Salvias are summer flowering perennials but are best grown as annuals. In warm temperate climates, plants will continue flowering into autumn or winter and may be worth growing on into the second year. Plants grow 30–60 cm in height and are excellent for mass bedding or borders. Modern varieties, in traditional scarlet and a number of other colours, are more compact and bushier than older strains. Salvias prefer a well-drained, friable soil to which organic matter and mixed fertilizer has been added during soil preparation. Plants do best in full sunlight and need protection from strong winds.

Sow seeds in spring or early summer when the weather is warm because the seed is difficult to germinate in cold soil. You can raise seedlings in boxes or punnets for transplanting or sow seeds direct in the garden. Space seedlings or sow a few seeds in clumps 30–40 cm apart for tall varieties, but 20–30 cm apart for dwarf, compact plants. When plants are established, they can be pinched back when 10 cm tall to encourage lateral shoots for bushy plants. Plants prefer moist but not soggy conditions so do not over-water. Occasional liquid feeds of soluble fertilizer will keep them growing strongly.

Varieties Splendens Bonfire is the traditional salvia with fiery scarlet blooms on long spikes. The modern strain of this variety is not as tall as older strains and rarely grows higher than 60 cm. Dwarf Scarlet has compact, semi-dwarf plants to 30 cm tall. Flowers are equal in size and colour to Bonfire.

Salpiglossis likes a warm, sunny aspect

Blue Mist (*Salvia farinacea*) is a tall plant to 60 cm and is excellent for background effect or when grown in clumps here and there in the garden. The deep Wedgwood-blue flowers are borne on long slender spikes. This variety can be cut back in warm weather for further flowering and in warm districts the clumps can over-winter for another flowering in late spring.

SCABIOSA

Scabiosa or Pin-cushion Flower is a spring flowering annual which has been improved in flower size and colour range over recent years. The plants are suitable for bedding or background and grow to about 60 cm. They require a sunny aspect sheltered from strong winds. Prepare the bed well beforehand, adding a mixed fertilizer at $\frac{1}{3}$ cup per square metre, and a ration of lime at 1–2 cups per square metre, depending on soil texture. The soil must be well drained but plants need ample water in dry weather.

Sow seed in autumn to early winter in temperate districts but in autumn and spring where colder. Seed is best sown in boxes or punnets and covered with 6 mm of vermiculite or compost. Keep soil moist until seedlings emerge in 14–21 days. Transplant seedlings when 5–7 cm high, spacing them about 40 cm apart. Give liquid feeds of soluble fertilizer when flower buds appear about 12 weeks from transplanting. Flower colours include shades of pink, mauve, purple and white.

SCHIZANTHUS

(*See Poor Man's Orchid.*)

SNAPDRAGON (Antirrhinum)

A severe rust fungus has, perhaps, led to a decline in popularity of snapdragons for the home garden in recent years. However, many of the new varieties are more resistant to this disease and more effective fungicide sprays should encourage gardeners to grow these magnificent flowers. Snapdragons are perennial plants but are best treated as annuals and sown from seed each year, although they can be cut back after the first flowering to give a second flush of blooms. The plants need well drained friable soils to which organic matter has been added liberally. Also apply a pre-planting fertilizer and a ration of lime when the bed is being prepared. Keep the soil in good condition by cultivating when it is dark damp until ready for transplanting. Snapdragons prefer full sunlight but do quite well if they are in sun for only part of the day.

Seed can be sown at almost any time of the year in temperate climates, but autumn

Snapdragons (Antirrhinum). Tetraploid strains produce ideal cut flower spikes on strong stems

is the best period for a spring display. In cold districts, you can sow seeds in spring or early summer too. As seeds are small, they should be sown in seed boxes or punnets using a friable, moisture-holding soil mixture as described in Chapter 5. Scatter the seed in shallow rows or broadcast on the surface and press firmly with a piece of board. Cover the seed very lightly with compost, vermiculite or seed raising mixture and keep damp until seedlings emerge—usually in 10–14 days. Seedlings are ready for transplanting in about 6 weeks from sowing when they are 3–5 cm high. Space seedlings of tall varieties about 40 cm apart each way, but dwarf varieties can be planted at 30 cm or even closer. If plants tend to produce buds too soon, nip them back to encourage lateral growth and 8–10 flower spikes. Give liquid feeds of soluble fertilizer at this stage. When cutting flowers for indoor decoration, or removing spent blooms, cut the stalks back to 5–7 cm from the crown to encourage a second crop of flowers on long stems. Continue to give liquid feeds while flowering continues.

Varieties Tetra Mixed is a tetraploid strain which grows to 60 cm with large, ruffled flowers in shades of yellow, gold, rose, lilac, tango, deep red and white. It has shown some resistance to rust fungus and is an excellent variety for cut flowers. Semi-Dwarf Mixed grows to 40–50 cm and is best suited for bedding and borders. The large flowers on strong, straight stems are excellent for cutting and come in lovely shades of pink and rose. Little Darling is a compact growing mound-shaped bush with attractive blooms in a beautiful range of clear colours. The bush grows to about 30 cm, flowers are prominently carried. An outstanding variety. Tom Thumb is a dwarf variety about 20 cm high. It has a bright colour range and is excellent for low borders and rockeries.

SPIDER FLOWER

Spider Flower (Cleome) is an unusual, shrubby annual growing to a height of 1–2 metres. It is an excellent background plant or for planting here and there in the garden. The flowers are pink, lilac, mauve or white with long, spidery stamens followed by decorative seed pods. The plants are very adaptable and flower well providing they are exposed to sunlight for part of the day. They do well on light or heavy soils which are well-drained and have had a mixed fertilizer added during preparation. Sow seeds in spring or early summer or again in early autumn. It is best to sow a few seeds direct in the garden in clumps or stations spaced 50–60 cm apart, and thin to the strongest seedlings. Water regularly and cultivate between plants. Give liquid feeds when buds appear.

STAR DAISY

(*See Chrysanthemum.*)

STATICE

Statice or Sea Lavender has attractive 'everlasting' flowers for garden display or for cutting and drying. The plants reach 60 cm and are suitable for accent in the garden bed. They need a moderately fertile soil and a well drained, warm, sunny aspect. Statice is a perennial plant but it is usually grown as a spring-flowering annual. Cut back for a second flowering.

In temperate climates, sow seed from autumn to early spring, but in cold districts avoid sowing during the colder months. It is best to sow seed direct into a well-prepared bed to which a mixed fertilizer has been added. Sow a few seeds in clumps or stations spaced 30–40 cm each way and thin to the strongest seedling. The seeds are large and may have some pieces of the dried petals attached. It does not

Statice is an attractive everlasting flower for cutting and drying

matter if these show above the soil. Cover the seeds with about 12 mm of compost or vermiculite. Seedlings often germinate slowly and erratically and may take 28 days or more to emerge, so keep the soil damp for this length of time. When the plants are established they need little attention. Providing a pre-planting fertilizer has been added before sowing, the plants usually flower well, but give them a side dressing of mixed fertilizer towards flowering time if they appear backward. The flowers are borne on long stems in clusters and range in colour from white through yellow to rose and blue. For everlastings, cut the flowers when mature, tie them in bunches and hang them head downwards in a cool, dry place. If well dried, the flowers will last for a long time without losing colour. The most widely grown variety is *sinuata* hybrid, with an excellent colour range.

STOCK

Stocks are one of the most popular spring flowering annuals. They are magnificent for garden display in beds and borders, excellent for indoor decoration and worth growing for their fragrance alone. Stocks have come along way from the sixteenth century when Peter Mathioli, an Italian botanist, first classified them in a group of plants which included his own name (*Mathiola incana*). At that time, flower colour was limited to purple only but today, as a result of plant breeding and selection, there is a tremendous colour range from white through cream, yellow, pink, lilac, mauve, to red, port wine and purple. Dwarf and tall varieties are now available too, but possibly the greatest advance has been the development of strains with a very high percentage of double flowers.

Oddly enough, stocks belong to the same botanical family (*Cruciferae*) as many of our important cool season vegetables—broccoli, cabbage, cauliflower, and turnips—so many of their requirements for climate, soil and fertilizer are very similar. Stocks, like the cabbage group, are really biennials but are grown as annuals. The plants prefer friable, fertile soils with plenty of organic matter (in the form of compost, animal manure or leaf mould) added. Also apply a mixed fertilizer such as Gro-Plus Complete at $\frac{1}{3}$ cup per square metre and a liberal ration of lime, except on alkaline soils. All these should be added to the soil beforehand and dug in to a depth of 15–20 cm so that the soil is in first-class condition for either transplanting seedlings or direct sowing. Drainage is also important because stocks will not tolerate wet feet. Raising the beds 15–20 cm above the surrounding surface will usually give adequate drainage. Generally the plants prefer rather dry conditions but should be given sufficient water to keep them growing steadily during the season. Stocks prefer a sunny position; tall varieties should be sheltered from strong winds.

In temperate and cold climates, seeds can be sown from midsummer to autumn but in warm northern districts autumn sowings are best. Seeds can be sown in seed boxes or punnets for transplanting and this method may be preferable for early sowings. Good results are possible by sowing direct, especially when moisture-holding materials, such as vermiculite, compost or seed raising mixture are used to cover the seeds. Stock seeds germinate quickly (10–14 days) and seedlings grow quite rapidly, so direct sowing avoids transplanting shock. When sowing direct, scatter a few seeds in clumps or stations and keep moist until seedlings emerge. Thinning the seedlings in each clump is only necessary if they are too crowded. It does not matter if a few seedlings are close together, providing there is sufficient space between each clump. Whether you transplant seedlings or sow direct, the planting distance will depend on the variety (dwarf or tall). As a guide, dwarf and column varieties are spaced 20–30 cm apart, but allow 30–40 cm for taller branching types.

When the plants are well established, cultivate regularly between individual plants or clumps to control weeds and water them when necessary, but do not overdo it. Give liquid feeds of soluble fertilizer as the plants grow, especially towards flowering time. An exception is the column type, which may develop lateral branches and new buds from the centre of the florets when it is fed too generously. Another point of interest is that plants which produce single flowers

The beautiful new Austral stock produces a profusion of spring blooms

can often be discarded in the seedling stage. Seedlings which are taller, thin-leafed and more vigorous usually produce single flowers, so many gardeners discard them when transplanting or thinning out. The method is not infallible, but has been used with some success. Most double strains available today will produce from 60–85 per cent double-flowered plants, so direct sowing without thinning will usually give good results. When plants are flowering well remove all spent blooms to promote new buds to form and prolong the display.

Varieties Austral is a beautiful new strain bred and selected at Yates Research Farm. Austral won the Flower of the Year award in 1976. Plants are vigorous, growing to 50 cm with a brilliant colour range including white, cream, apricot, various shades of pink and lavender, red and purple with other attractive bicolours. A short, closely packed central spike is soon followed by many lateral spikes—an excellent garden variety. Giant Perfection (Giant Imperial) is the popular, semi-bush type stock. The central flower spike is followed by lateral spikes. Individual flowers are large with a high percentage of doubles. Colours include white, cream, buff, many shades of pink, lavender, lilac, purple and brilliant red. Hi-Double (Trysomic) is a heavily branching strain from America. The plants are tall but bushy and produce 80–85 per cent double flowers in a full range of colours. Giant Column is a tall strain reaching 50–75 cm. It forms a strong flower spike packed with large double florets in a beautiful colour range. Ideal for clump plantings in the garden. Because they are non-branching, space between plants can be reduced to 20–30 cm. Dwarf

Double is an attractive midget stock growing to 25 cm tall with tightly-packed flower spikes of double florets. It has the perfume and full colour range of the taller, branching types. It is excellent for borders and combines well with low bedding plants like alyssum, nemesia, pansy and viola. It is also suitable in exposed situations where taller strains may be damaged by wind.

STRAW FLOWER (Helichrysum)

Straw Flower, an Australian native, is a popular 'everlasting' with paper-textured flowers in white, gold, mauve, pink and red. The plants grow to 75 cm or more and are ideal for background or for planting here and there in the garden. They will do well on most garden soils with base fertilizers added, but need a warm, sunny position. They also tolerate hard conditions better than most garden annuals but respond to regular watering in dry weather. Sow seeds either in autumn or spring in boxes or punnets, covering the seed lightly with moisture-holding material. Transplant seedlings when 5–7 cm high, spacing them 30 cm apart. Cut flowers when half open, tie in bunches and hang head downwards in a cool place for drying. Dried flowers will last for many months.

STURT'S DESERT PEA (Clianthus)

Sturt's Desert Pea, or Glory Pea, is a prostrate, trailing plant which is found in the dry inland areas of Australia. The large pea-like flowers are borne in clusters. The standard petal and keel are a brilliant red with a large black-purple blotch at the base of the standard. In their natural state, these plants thrive on quite poor, rather sandy soils which are alkaline in reaction. Once established, they tolerate extremely hot and dry conditions but resent any disturbance of the root system. So the plants need a well-drained soil in a warm, sunny position. The soil must not be fertilized heavily, but it is recommended that lime and a little superphosphate should be added. The plants do quite well in a window box or in troughs or pots.

Although the plants are biennials, they are best treated as annuals from seed sown in late spring or early summer. Sow seeds in punnets or pots for transplanting later, but take care to lift as much soil as possible with the roots. Seeds can also be sown direct, spacing the clumps about 40 cm apart. The seed coat is very hard and takes up water very slowly. You can assist germination by nicking the seed coat with a sharp knife, but take care not to damage the eye (*hilum*) of the seed. If the seed is sown in punnets or pots, cover with glass or clear plastic to give extra warmth until seedlings emerge. Keep seedlings watered until well established. After that, little attention is needed apart from careful hand weeding so that the roots are not disturbed.

SUNFLOWER (Helianthus)

Although sunflowers are the 'giants' of the summer flowering annuals, modern varieties are much shorter than older varieties, which grew to 3 m or more. Flower form and colour range have been improved tremendously in recent years. As their name implies, plants must be grown in full sunlight in a sheltered position to avoid wind damage. They are adapted to both light and heavy soils but do best when the soil has been improved by adding organic matter and a mixed fertilizer at $\frac{1}{3}$ cup per square metre. The soil must be well drained.

Sow seed in spring or early summer when weather and soil are warm. Although seed can be sown in boxes or punnets and transplanted when small, it is best to sow seeds direct in the garden by scattering a few seeds in clumps 50–60 cm apart each way and thin to 1–2 seedlings. The plants need little attention when established, apart from watering in dry weather, and cultivation to control weeds. Flowers for indoor decoration should not be too old before cutting. Remove spent blooms to prolong flowering.

Varieties Sungold is a quick-growing, free-flowering strain to 100 cm tall. The fluffy double flowers are a deep gold and may reach 15–20 cm in diameter. Bronze Shades is a rather taller strain (150 cm) with single flowers which are excellent for indoor decoration. Flowers are in shades of bronze and terracotta, many tipped with yellow and pink.

SWAN RIVER DAISY (Brachycome)

A native of Western Australia, this delightful little annual grows to 20–30 cm tall with dainty single flowers in blue, mauve or white. It is useful for low borders or as an accent plant. Its rather spreading growth makes it ideal for rockeries or container growing. Brachycome will grow on quite poor soil and prefers light, well-drained soil in a sunny position.

Sow seed in spring or early summer either in seed boxes or punnets for transplanting seedlings 20 cm apart, or sow a few seeds direct in clumps at the same distance. The seed is quite small so cover very lightly with vermiculite or compost. When the plants are established, they need little attention as they withstand hot, dry conditions. Do not over-water.

SWEET PEA

Sweet peas are one of the most popular spring-flowering annuals for garden display and for cut flowers. Most varieties grow 2–3 m tall, so need support in the form of a trellis, tripod or wire mesh cylinders. There is also a dwarf strain (Bijou). The dwarf plants grow to 60 cm tall and are ideal for borders, rockeries, window boxes and pots. Sweet peas have been improved tremendously in flower form, size and colour range since they were first introduced to England as a rather insignificant little purple flower from Sicily in the seventeenth century. In 1901 the Spencer Summer Flowering strain emerged as a sport in the gardens of the Earl of Spencer. Ten years later, as the result of another mutation, Mr Arthur Yates, founder of our Australasian firms, produced Yarrawa Spencer which was the first early-flowering variety. In due course other colours were added until a complete range of shades and colourings became available. In today's modern strains colours now range from pure white, cream through numerous shades of pink, lavender and mauve to light and dark reds, blue and purple. The *gigantea* strain is a further improvement with large, closely spaced blooms of beautiful texture and clear, crisp colours.

Sweet peas need ample sunlight and usually do poorly when shaded. If growing on a trellis the rows should run north-south so that the vines receive as much sun as possible. Good drainage is essential and it is best to raise the bed 15–20 cm above the surrounding surface so that water is shed quickly after heavy rain. Soil preparation is most important too. Except on naturally alkaline soils, lime should be added to the soil at about 1 cup per square metre on light soils and 2 cups per square metre on heavier soils. Then spread a generous layer of compost, animal manure or well-rotted grass clippings. Now add a mixed fertilizer such as Gro-Plus Complete at $\frac{1}{3}$ cup per square metre or at a heaped tablespoon per metre along the row where seed is to be sown. The organic matter will improve the structure of both light and heavy soils, but as these materials are very often low in phosphorus and potassium the addition of fertilizer will make good any deficiency. On the other hand, avoid using fertilizers which are high in nitrogen. Fork or spade the lime, organic matter and fertilizer into the surface soil (10–15 cm) so that the materials are mixed well. Then dig the bed over to spade depth, loosening rather than turning the soil so the topsoil stays on the surface. Always remember to dig or cultivate the soil when dark damp—especially on heavy soil—to preserve a good crumb structure. Give the bed a gentle but thorough watering and leave it to settle for a week or two. Then rake the surface to destroy any weed seedlings and to bring the soil to a crumbly condition for sowing the seeds direct.

In most temperate climates, seed can be sown from midsummer to late autumn, but March or April are usually the best months. In cold districts, spring sowings can be made too. As the germinating seeds are susceptible to damping off (especially in cold soil) it is advisable to dust the seed with fungicide before sowing. (See Chapter 5.) It is often a good idea to build the wire netting trellis before sowing. This will avoid disturbing or treading on the bed after the seeds are sown. Mark out shallow drills 2–3 cm deep and press the seeds into the soil, spacing them 5–7 cm apart. If the soil is loose and crumbly, cover the seed and lightly tamp down with the back of the rake. On very heavy soils, cover the rows with vermiculite or compost. If the soil is dark damp at sowing, additional watering is usually unnecessary until the seedlings emerge in 10–14 days. In very dry weather or on very sandy soil, extra watering may be needed to keep the

Decorative dahlias

surface damp.

Over-watering is one of the main reasons for poor germination. For the same reason, soaking seeds in water before sowing may also do more harm than good (see Chapter 5.) If you wish to pre-germinate seed, spread them out on wet blotting paper or wet towelling or mix them with moist vermiculite or seed raising mixture in a saucer or dish. This way they absorb water quickly but also get plenty of air which is essential to germination. You can plant out the swollen seeds in a few days but handle them very carefully if they have started to germinate. Another point to remember is that some seeds may be smaller than others and may look pinched or shrivelled. These seeds are often the darker colours—red, mauves and blues—and if you discard them you may not have a full colour range. Always sow all seeds in the packet or, if you have too many seeds, sow an average sample.

When the plants are 15–20 cm tall place some twigs along the row to help them to reach the netting. If the plants are very spindly it may be necessary to cut them back slightly to promote sturdier growth. Laterals will appear when the plants are 20–30 cm tall. For general garden display allow these to grow on, but for exhibition blooms cut some out. Water the plants regularly, especially in dry weather, and give liquid feeds of soluble fertilizer every 10–14 days when buds appear. Remove spent blooms to prolong flowering. Sweet peas are usually not unduly troubled by pests and diseases. For control measures see Chapter 17.

Varieties Sunshine Mixture has a splendid colour range of early-flowering *gigantea* sweet peas. Seventh Heaven is a similar mixture of early-flowering *gigantea* types but contains a wider colour range. Royal Favourite is a rather later-flowering strain. The plants are extremely vigorous and bear large ruffled blooms on long, sturdy stems. The flowers are heavily perfumed and very suitable for exhibition purposes. Colourcade is the most widely grown and popular of all the strains of sweet peas. This mixture of early-flowering *gigantea* strains contains a complete range of all sweet pea colours and shades. Treasure Chest contains six separate and named colours of the early-flowering *gigantea* type. Colours are selected by popular vote each year. Yates Prize Packet is a connoisseur's mixture. Each packet contains a special blend of the best colours in the early-flowering *gigantea* strain. Recommended for the show exhibitor. Bijou is an outstanding dwarf sweet pea. Plants grow to 60 cm tall and flowers are beautifully perfumed in a full colour range. The flowers are as large as those of tall varieties and borne on stems 20 cm long. An excellent variety for borders, rockeries, window boxes and pots, but remember that plants need full sunlight. A new and exciting variety, Pixie Princess, is excellent for borders, rockeries, window boxes and pots. Grows to about half the height of Bijou and is available in a wide range of colours. Late flowering. Tiffany is a new early flowering vigorous sweet pea featuring very large flowers on strong stems. Many blooms are true doubles, deeply frilled and bicoloured.

SWEET WILLIAM

Sweet William (*Dianthus barbatus*) is a biennial or short-lived perennial growing to 40 cm tall. It is usually treated as an annual; seeds can be sown from midsummer to autumn in most climates, but in both autumn and spring where colder.

The cultivation of this free-flowering, attractive plant for beds, borders or rockeries is very similar to dianthus and carnation, to which it is closely related (see Carnation and Dianthus).

Varieties Double Mixed has a large range of colours and a variety of markings in shades of pink, mauve and red. It flowers in late spring. Red Express is an early-flowering variety with large, pillar-box red flowers on long stems. Excellent for cut flowers.

TORENIA

Torenia is a summer flowering annual with predominantly blue flowers like small snapdragons. The plants grow to 20–30 cm and are suitable for low borders, edging, rockeries and window boxes. For borders or edging, the blue flowers combine beautifully with a background of taller annuals in yellow, pink or red. Torenia is very adaptable to soil type but it needs a well-drained sunny situation. Apply a mixed fertilizer at $\frac{1}{3}$ cup per square metre when preparing the bed.

Sow seeds in spring or early summer in seed boxes or punnets for transplanting seedlings when 5 cm high; space them 15–20 cm apart. Alternatively, sow a few seeds direct in clumps or stations at the same distance and thin each clump to 2–3 seedlings. Water regularly in dry weather and give liquid feeds of soluble fertilizer when buds are forming. Little Gem is the most widely-grown variety.

VERBENA

Verbena is a trailing perennial which may persist well for 2–3 years, but it is usually grown as an annual. The plants grow to 30 cm tall and continue to flower for many months. Flower colours include pink, mauve, red, purple—many with a white eye. Verbena is very adaptable to soil and grows well in most garden soils, but needs good drainage. It prefers full sunlight but will tolerate some shade.

Seed can be sown in all the warm months from spring to autumn. You can raise seedlings in boxes or punnets for transplanting or you can sow a few seeds direct in clumps spaced 25–30 cm apart and thin each clump to 2–3 seedlings. Seed is slow to germinate, seedlings taking 21–28 days to emerge, so take care that the soil is kept damp but not wet for this length of time. Once established the plants need very little attention. They tolerate quite dry conditions and grow well without extra fertilizer. Cultivate between plants to destroy weeds in the early stages but the bushy plants will soon cover the ground to form a dense mat. Cut back plants after flowering to promote a second flush of blooms. Gaiety Mixture is an excellent variety with a brilliant range of flower colours, excellent for planting in odd corners of the garden and in rockeries.

VIOLA

Violas are close relatives of pansies and are grown in exactly the same way. Violas can be planted rather closer together than pansies. The plants will continue flowering for a very long period, especially if the spent blooms are picked regularly. For general notes on soil, sowing and cultivation, refer to Pansy in this chapter.

Varieties Space Crystals has large, velvety flowers and is excellent for beds or borders. The flowers are self-coloured without markings and include yellow, apricot, rose, red, and shades of blue and white. Bambini is a dwarf variety producing a prolific number of small blooms in a wide variety of colours. An excellent type for garden beds and growing in containers. Johnny Jump Up is a delightful, small-flowered viola in purple with streaked yellow centres. Excellent for borders, rockeries and containers, it flowers prolifically from early spring to summer.

VIRGINIAN STOCK

Virginian Stock is a dainty little spring flowering annual with tiny flowers in white, cream, lavender and pink. The plants grow to 20 cm tall and are excellent for low borders, edging and odd corners. They prefer a warm, sunny aspect but tolerate part-shade. They are very adaptable little plants and grow well on most soils, but add a mixed fertilizer when preparing the bed. In temperate climates, sow seed in autumn, but in cold districts in both autumn and spring. Seedlings can be raised in boxes or punnets for transplanting but it is best to sow direct—thinly in rows or a few seeds in clumps spaced 15 cm apart. Thinning is rarely necessary.

Plants flower very quickly from seed. Give liquid feeds when buds start to show.

VISCARIA

Viscaria gives masses of dainty summer flowers in shades of pink, mauve, lavender and blue. The plants have thin, branching stems and grow to 30–40 cm. They need a warm, sunny position but are adaptable to both light and heavy soil to which a mixed fertilizer has been added during preparation. Sow seed in spring or early summer direct in the garden, scattering a few seeds in clumps spaced 15 cm apart. Cover with vermiculite or compost. Seedlings rarely need thinning providing there is space around each clump. Water regularly in dry weather and mulch around plants to conserve moisture and discourage weeds.

WALLFLOWER

Wallflower is another old-world spring flowering annual which has been improved in flower form and colour range, but is still as fragrant as ever. The plants grow to 60 cm and are excellent for bedding or borders or planted here and there in clumps or drifts in the garden. The plants prefer a warm, sunny aspect sheltered from strong winds. Like stocks (wallflowers belong to the same family) the plants respond to friable, fertile, well drained soil. Prepare the bed well beforehand, adding liberal quantities of compost or animal manure. On most soils, except those which are alkaline, add a ration of lime at 1–2 cups per square metre depending on soil type. Then apply a mixed fertilizer such as Gro-Plus Complete at $\frac{1}{3}$ cup per square metre. Fork or spade these additives into the soil to a depth of 15–20 cm and leave in the rough state for 4–6 weeks. Cultivate again about a week before transplanting so that the soil is in a crumbly condition.

In most districts sow seed from midsummer to autumn. Seed sown before April will produce plants which flower in late winter and early spring. Prepare an open, friable, seed raising mixing for boxes or punnets as described in Chapter 5. Sow the seeds in shallow rows and cover lightly with vermiculite or compost. Water gently and keep damp until seedlings emerge in 10–14 days. Prick out seedlings when small into other containers (filled with the same soil mix) spacing them 3–5 cm apart. Grow them on until 5–7 cm high and then transplant them to the garden bed, spacing them 20–30 cm apart. Keep the plants growing strongly by regular watering and destroy weeds by surface cultivation but take care not to damage the shallow roots. When buds appear, give liquid feeds of soluble fertilizers at intervals of 10–14 days. Russet Shades is widely grown and produces large fragrant blooms in rich yellow, brown, mahogany, deep ruby red, on vigorous plants growing to 60 cm tall. The double-flowered Winter Delight is very fragrant and has double blooms in many colours. Early blooming and ideal for cutting–recommended.

WARATAH

The waratah is a stately native shrub growing to 2–3 m. The large, cone-shaped flowers in vivid red are the floral emblem for New South Wales. In its natural state it grows on the rather poor sandy soils of the coastal bushland. Waratahs can be grown from seed sown in spring or early summer in a mixture of bush sand and leaf mould in boxes or punnets. Sow seeds 12 mm deep and keep moist for 21–28 days when seedlings appear. Transplant seedlings when 5 cm high or transfer to larger pots filled with a similar soil mixture before finally planting out. The soil needs to be well drained and plants resent the heavy use of fertilizers. Once plants are established, fertilizers containing blood and bone such as Gro-Plus for Camellias and Azaleas or slow-release fertilizers are suitable in moderation. After flowering in spring or early summer, cut the plants back, leaving a few leaves at the base of the flower stem.

ZINNIA

Zinnias are one of the most popular summer flowering annuals. Modern varieties, both tall and dwarf, have a wide range of flower forms and a tremendous range of colours. Zinnias are excellent for mass bedding but many of the dwarf varieties make suitable border plants. Zinnias must be grown in full sunlight for best results and need shelter from strong winds. Soil should be well drained and improved with compost or animal manure, a ration of mixed fertilizer and lime if the soil is too acid. Prepare the bed well in advance so that the soil is in a crumbly condition for transplanting or direct sowing.

In temperate districts seed can be sown in early spring to late summer but in cold districts late spring or early summer sowings are best. Seed germination is often disappointing if seeds are sown in cold soil too early in the season, especially when sown direct in the garden. Germination is much improved if sowing is delayed until the soil warms up to 20°C or more. The seeds are quite large and can be handled easily, but it is advisable to dust them with fungicide to protect against damping off. Early sowings can be made in seed boxes or punnets using an open, friable seed raising soil. Extra warmth can be provided by covering the containers with glass or clear plastic until the seedlings emerge (usually 7–10 days). Grow seedlings on until 5–7 cm high before transplanting. When sowing direct, scatter a few seeds in clumps or stations and cover lightly with vermiculite, compost or seed raising mixture. Thinning each clump is rarely needed unless the seedlings are overcrowded.

Whether transplanting seedlings or sowing direct, the distance between plants or clumps will depend on the variety and the size to which it will grow. As a general guide, space tall types 40 cm apart each way; semi-dwarf types 30 cm apart and dwarf types 20 cm apart. Always pinch back the centre shoot of zinnias to promote the growth of laterals and a bushy plant. Once established, the plants need little attention apart from regular watering (especially in dry periods) and cultivation to destroy weeds. In very hot weather, a good mulch of grass clippings or compost will conserve moisture and keep the roots cool. Give liquid feeds of soluble fertilizer when the plants are half-grown. To control mildew, a fungus which attacks the plants in late summer and autumn, see Chapter 17.

Varieties Gold Medal (dahlia-flowered) is a blend of the best colours in this strain. Large double blooms on strong stems. Plants grow to 100 cm in height. Showman (State Fair) is a mixture resembling the dahlia-flowered type. Huge double flowers in a bright range of colours are borne on strong stems. The bushy plants have some resistance to mildew and grow to 75–90 cm tall. Envy is an unusual dahlia-flowered zinnia with flowers in soft lime to emerald-green. Plants grow 60–90 cm in height. Happy Talk (Semi-dwarf Coquette) has large, ruffled, double flowers with attractive quilled petals. They resemble cactus dahlias. The mixture contains bright colours and pastel shades. Plants grow to 75 cm tall. Lilliput (Pom Pom) has ball-shaped, double flowers 3–5 cm in diameter in shades of yellow, pink, red and pastel colours. Plants grow to 60 cm in height. Pulcino is a new variety with a wide range of colours on dwarf bushes growing to about 50 cm in height. 8 cm diameter flowers are borne profusely on the bush which is hardy and compact. Sprite resembles the pom-pom type. Most flowers are double in a bright colour range. Plants grow to 60 cm. Persian Carpet (*Zinnia haageana*) has colourful gaillardia-like flowers in shades of lemon, orange, lavender, crimson and maroon, with white or gold markings on the tips of the petals. Plants grow to 60 cm and are excellent for massed bedding or clump planting in rockery pockets with other annuals. Little Star (*Zinnia linearis*) has masses of small, star-shaped yellow flowers over a long period. Plants are dwarf (to 30 cm tall) and are excellent for borders or rockeries. Thumbelina is a remarkable dwarf zinnia with medium-sized flowers in shades of yellow, orange, pink, scarlet, red and white. Plants are very bushy, growing to 20 cm, and are excellent for borders, edging, rockeries and window boxes. Mini-Pink is similar in size and bushy habit to Thumbelina and is smothered in double pink flowers.

FLOWER SEED MIXTURES

A number of mixtures of flower seeds for spring or autumn sowing are available for mass display, borders or special situations. These mixtures provide a wide variety of flower form and colour range.

Bambi This annual flower seed mixture makes a colourful border. It is very popular with children (and adults too). Sow seed from spring through to early autumn.

Bouquet of Blue This mixture contains

Perennial Phlox have flowers of pink, mauve, salmon and white

A rockery is the perfect setting for spring flowers

annual and some perennial blue flowers of medium height. It includes ageratum, centaurea (cornflower), delphinium, forget-me-not, larkspur, lupin, nigella, stock and other interesting blue flowers.

Rockery Mixed A blend of attractive and useful plants for rockery planting. The mixture contains about twelve different kinds, most of which are dwarf types but vary in colour and flowering season.

Spring Bouquet This cheerful mixture of spring flowering annuals contains carnation, cornflower, everlasting daisy, painted daisy, poppy, snapdragon, stock and wallflower. Sow seed in autumn.

Summer Fiesta A colourful mixture of summer flowering annuals for spring sowing. It includes aster, balsam, celosia, cosmos, marigold, petunia, phlox and zinnia.

Wild Flowers of the World A colourful mixture of wild flowers gathered from many parts of the world.

SOWING GUIDE FOR FLOWERS

Flower	BEST MONTHS TO SOW FOR CLIMATE																																				How to Sow: Seedbed (S) Direct (D)	Sowing Depth (mm)	Seedlings Emerge (Days)	Transplant or Sow Direct and thin to (. . . *cm* apart)	Approx. time to flowering (weeks)
	Tropical/Subtropical • Subtropical only ▲ Tropical only ■												Temperate												Cold																
	J	F	M	A	M	J	J	A	S	O	N	D	J	F	M	A	M	J	J	A	S	O	N	D	J	F	M	A	M	J	J	A	S	O	N	D					
Acroclinium (see Everlasting Daisy)																																									
Ageratum	•	•	•	•	•	•	•	•	•	•	•	•	•	•	•	•	•			•	•	•	•	•	•	•							•	•	•	•	S or D	6	14–21	15–20	14
Alyssum		▲	•	•	•	•	•	▲					•	•	•	•	•			•	•	•	•	•	•	•	•	•					•	•	•	•	S or D	3	10–14	7–10	8
Amaranthus						•	•	•	•	•										•	•	•	•	•									•	•	•		D	6	14–21	40	14
Antirrhinum (see Snapdragon)																																									
Aquilegia (Columbine)			•	•	•	•								•	•	•	•									•	•	•					•				S	3	21–28	30–40	28
Arctotis (Aurora Daisy)			•	•	•	•	•	•	•					•	•	•	•									•	•	•					•	•			S or D	6	18–21	30–40	16
Aster	•				■	■	•	•	•	•	•	•								•	•	•	•	•									•	•	•	•	S	6	10–14	20–30	14
Aurora Daisy (see Arctotis)																																									
Baby Blue Eyes (see Nemophila)																																									
Balsam	•	•				■	■	•	•	•	•	•								•	•	•	•									•	•	•	•	•	S or D	6	10–14	30	8
Begonia, bedding						•	•	•	•	•										•	•	•	•										•	•	•		S	1	14–21	20	16
Begonia, Tuberous					•	•	•	•										•	•	•	•											•	•	•			S	1	14–21	pots	28
Bellis (English Daisy)			•	•	•	•								•	•	•	•									•	•	•					•	•			S	3	10–14	10–15	12
Big Gyp (Saponaria)			•	•	•	•	•	•	•	•	•		•	•	•	•	•			•	•	•	•	•	•	•	•	•	•			•	•	•	•	•	D	6	14–21	20–30	8

Boronia	• • • •	• • •	• • •	S or D	3	45–75	40	56
Brachycome (see Swan River Daisy)								
Calceolaria	• • • •	• • • •	• • • • •	S	12	14–21	pots	20
Calendula	• • • •	• • • • •	• • • • •	S or D	12	10–14	30	10
Californian Poppy (Eschscholtzia)	• • • • • •	• • • • •	• • • •	S or D	3	10–14	30	8
Candytuft	• • • • ▲	• • • • •	• • • • •	S or D	6	14–21	20–30	12
Canterbury Bells	• • • • •	• • • • •	• • •	S	3	14–21	30	14
Carnation	• • • • • • •	• • • • • •	• • • • • • •	S	6	10–14	30–40	28
Celosia	• • ■ • • • • • •	• • • • •	• • • •	S or D	6	10–14	20–30	12
Centaurea (see Cornflower)								
Chrysanthemum	• • • • •	• • • •	• • • •	S	6	14–21	75–100	24
Cineraria	• • • •	• • • •	• • • •	S	3	10–14	30–40	20
Cleome (see Spider Flower)								
Clianthus (see Sturt's Desert Pea)								
Cockscomb	• • ■ • • • • ■ ■	• • • • •	• • • •	S or D	6	10–14	30	12
Coleus	• • • • • • • • •	• • • • •	• • • •	S	6	14–21	30 or pots	10
Columbine (see Aquilegia)								
Cornflower (Centaurea)	• • • • •	• • • •	• • • • •	S	3	14–21	40–50	14

SOWING GUIDE FOR FLOWERS

Flower	BEST MONTHS TO SOW FOR CLIMATE: Tropical/Subtropical ● Subtropical only ▲ Tropical only ■												Temperate												Cold												How to Sow: Seedbed (S) Direct (D)	Sowing Depth (mm)	Seedlings Emerge (Days)	Transplant or Sow Direct and thin to (. . . *cm* apart)	Approx. time to flowering (weeks)
	J	F	M	A	M	J	J	A	S	O	N	D	J	F	M	A	M	J	J	A	S	O	N	D	J	F	M	A	M	J	J	A	S	O	N	D					
Cosmos						●	●	●	●	●										●	●	●	●	●									●	●	●	●	S or D	6	14–21	40–50	12
Cyclamen	●	●	●	●	●								●	●	●	●	●							●	●	●	●	●								●	S	3	28–42	Pots	64
Dahlia (seed)						●	●	●	●	●										●	●	●	●	●									●	●	●	●	S	12	14–28	50–100	16
Delphinium			●	●	●									●	●	●	●	●	●	●						●	●	●					●	●			S	3	21–28	50	20
Dianthus			●	●	●	●	●	●	●	●				●	●	●	●	●	●	●	●	●	●			●	●	●					●	●			D	3	10–14	15–30	20
English Daisy (see Bellis)																																									
Eschscholtzia (see Californian Poppy)																																									
Everlasting Daisy (Acroclinium)			●	●	●	●	●	●	●					●	●	●	●			●	●	●				●	●	●					●	●			D	12	21–28	20–30	14
Fairy Pinks (see Dianthus)																																									
Forget-Me-Not			●	●	●								●	●	●	●	●								●	●	●	●								●	S	3	21–28	20–30	12
Gaillardia		■	●	●	●	●	●	●	●	●				●	●	●	●			●	●	●				●	●	●					●	●			S or D	6	14–21	30	16
Gazania						●	●	●	●	●										●	●	●	●	●									●	●	●	●	S	6	14–21	20	12
Geranium (seed)						●	●	●	●	●										●	●	●	●	●									●	●	●	●	S	3	14–28	40–50	16
Gerbera (South African Daisy)		■	■	■	■	●	●	●	●	●										●	●	●	●	●									●	●	●	●	S	6	14–21	40–50	30–50
Globe Amaranth (Gomphrena)	●	●	●			■	■	●	●	●	●	●								●	●	●	●	●									●	●	●	●	S or D	6	14–21	30	12

Gloxinia	• • • • •	• • • •	• • •	S	3	21–28	Pots	30
Godetia	• • • •	• • • • •	• • • • •	S or D	6	10–14	30	12
Gypsophila	• • • •	• • • • • • • • • •	• • • • • • • • • •	D	6	10–14	20–30	10
Helianthus (see Sunflower)								
Helichrysum (see Strawflower)								
Hollyhock	• • • •	• • • •	• • • •	S	6	14–21	30–40	28
Honesty (Lunaria)	• • • • • • •	• • • • • • •	• • • • •	S	3	14–21	40–50	12
Impatiens	• • • • • • • • • • • •	• • • • • • •	• • • • •	S	3	14–21	30–40	12
Irish Green Bellflower (Molucella)	• • • • • • •	• • • • • • •	• • • • •	S	6	14–21	20–30	12
Kangaroo Paw	• • •	• • •	• • •	S	6	30–90	30–40	16
Kochia	■ ■ ■ • • •	• • • • •	• • • •	S	3	10–14	40–60	–
Larkspur	• • •	• • • • •	• • • • •	D	3	14–21	20–30	20
Linaria	• • • • •	• • • • • •	• • • • •	D	6	10–14	10–15	10
Livingston Daisy (Mesembryanthemum)	• • • •	• • • •	• • • • •	S or D	3	14–21	10–15	20
Lobelia	• • • •	• • • •	• • • • •	S	3	10–14	10	14
Love-in-the-Mist (see Nigella)								
Lunaria (see Honesty)								
Lupin	• • • • • • •	• • • • • •	• • • • • •	D	12	10–14	20–50	16–32

SOWING GUIDE FOR FLOWERS

Flower	BEST MONTHS TO SOW FOR CLIMATE: Tropical/Subtropical ● Subtropical only ▲ Tropical only ■												Temperate												Cold												How to Sow: Seedbed (S) Direct (D)	Sowing Depth (mm)	Seedlings Emerge (Days)	Transplant or Sow Direct and thin to (. . . *cm* apart)	Approx. time to flowering (weeks)
	J	F	M	A	M	J	J	A	S	O	N	D	J	F	M	A	M	J	J	A	S	O	N	D	J	F	M	A	M	J	J	A	S	O	N	D					
Marigold, African	●	●	●	●	■	■	■	●	●	●	●	●								●	●	●	●	●								●	●	●	●		S or D	6	10–14	20–40	12
Marigold, French	●	●	●	●	●	●	●	●	●	●	●	●	●	●	●	●	●								●	●	●	●	●				●	●			S or D	6	10–14	20–40	12
Marmalade Daisy (see Rudbeckia)																																									
Mignonette		●	●	●	●	●								●	●	●	●	●	●	●	●					●	●	●					●	●			D	6	10–14	15–20	12
Molucella (see Irish Green Bell)																																									
Nasturtium		●	●	●	●	●	●	●	●				●	●						●	●	●	●	●									●	●	●	●	D	12	14–21	20–30	10
Nemesia			●	●	●	●								●	●	●	●									●	●	●					●	●			S or D	6	10–14	15–20	14
Nemophila (Baby Blue Eyes)		●	●	●										●	●	●	●									●	●	●					●	●			D	3	10–14	15	12
Nigella (Love-in-the-Mist)		●	●	●	●	●								●	●	●	●									●	●	●					●	●			S or D	3	21–28	20–30	14
Ornamental Basil			■	■	●	●	●	●	●											●	●	●	●	●									●	●	●	●	S or D	6	10–14	20–30	—
Ornamental Chilli			■	■	●	●	●	●	●											●	●	●	●	●									●	●	●	●	S or D	6	10–14	60	20
Painted Daisy		●	●	●	●									●	●	●	●	●								●	●	●					●	●			S	6	14–21	40–50	14
Pansy		●	●	●	●								●	●	●	●	●								●	●	●	●					●	●			S	6	21–28	20–30	16
Petunia			■	■	●	●	●	●	●											●	●	●	●	●									●	●	●	●	S	3	10–14	25–40	12
Phlox			■	■	●	●	●	●	●											●	●	●	●	●									●	●	●	●	S or D	3	14–21	10–15	10

Pinks (see Dianthus)								
Pin-Cushion Flower (see Scabiosa)								
Polyanthus	• • • •	• • • •	• • •	S	3	21–28	15–20	24
Poor Man's Orchid (Schizanthus)	• • • •	• • • • •	• • • •	S	3	14–21	30	14
Poppy, Iceland	• • • •	• • • • •	• • • •	S	3	10–14	20–30	24
Portulaca	• • • • •	• • • • •	• • • •	S or D	6	10–14	10	6
Primula	• • •	• • • • •	• • • •	S	3	21–28	15–20	24
Rudbeckia (Marmalade Daisy)	• • • • •	• • • • •	• • •	S or D	6	10–14	30	14
Salpiglossis	■ ■ ■ ■ ■ ■ ■ • • • • •	• • • • •	• • •	D	6	14–21	15	12
Salvia	• • • • • • • • • • • •	• • • • •	• • • •	S or D	3	14–21	20–40	12
Saponaria (see Big Gyp)								
Scabiosa (Pin-Cushion Flower)	• • • • •	• • • •	• • • • •	S or D	6	14–21	40	14
Schizanthus (see Poor Man's Orchid)								
Snapdragon (Antirrhinum)	• • • • • • • •	• • • • • • •	• • • • • • •	S	3	10–14	30–40	16
Spider Flower (Cleome)	• • • • • •	• • • • • • •	• • • • • •	D	3	14–21	50–60	12
Star Daisy (see Painted Daisy)								
Statice	• • • • •	• • • • • • • •	• • • • •	D	6	14–28	30–40	20
Stock	• • •	• • • • •	• • • •	S or D	3	10–14	20–40	20

SOWING GUIDE FOR FLOWERS

Flower	BEST MONTHS TO SOW FOR CLIMATE																																				How to Sow: Seedbed (S) Direct (D)	Sowing Depth (mm)	Seedlings Emerge (Days)	Transplant or Sow Direct and thin to (. . . *cm* apart)	Approx. time to flowering (weeks)
	Tropical/Subtropical ● Subtropical only ▲ Tropical only ■												Temperate												Cold																
	J	F	M	A	M	J	J	A	S	O	N	D	J	F	M	A	M	J	J	A	S	O	N	D	J	F	M	A	M	J	J	A	S	O	N	D					
Strawflower (Helichrysum)		●	●	●	●	●	●							●	●	●	●			●	●	●				●	●	●					●	●	●		S	3	10–14	30	16
Sturts Desert Pea (Clianthus)					●	●	●	●												●	●	●	●	●									●	●	●	●	S or D	6	14–28	40	24
Sunflower (Helianthus)	●	●	●	●	●	■	■	●	●	●	●	●								●	●	●	●	●									●	●	●	●	S or D	6	10–14	50–60	12
Swan River Daisy (Brachycombe)					●	●	●	●												●	●	●	●	●									●	●	●	●	S or D	3	14–21	20	16
Sweet Pea		■	●	●	●								●	●	●	●	●								●	●	●	●					●	●			D	25	10–14	5–7	14
Sweet William (see Dianthus)																																									
Torenia	●	●	●	■	■	■	■	●	●	●	●	●								●	●	●	●	●									●	●	●	●	S or D	6	10–14	15–20	16
Verbena		●	●	●	●	●	●	●	●				●	●						●	●	●	●	●	●								●	●	●	●	S or D	6	21–28	25–30	10
Viola		●	●	●	●								●	●	●	●	●								●	●	●	●					●	●			S	6	21–28	20	16
Virginian Stock		●	●	●										●	●	●	●									●	●	●					●	●			S or D	6	10–14	15	14
Viscaria					●	●	●	●												●	●	●	●	●									●	●	●	●	D	6	14–21	15	12
Wallflower		●	●	●	●								●	●	●	●	●								●	●	●	●									S	6	10–14	20–30	24
Waratah					●	●	●	●												●	●	●	●	●									●	●	●	●	S or D	12	21–28	120	50–100
Zinnia	●	●	■	■	■	■	●	●	●	●	●	●								●	●	●	●	●									●	●	●	●	S or D	6	7–10	20–40	12

Chapter 10

FLOWERING BULBS

There are many kinds of flowering bulbs or bulbous plants which are easy to grow in the home garden. In this chapter, we include the true bulbs and also those plants which are started from corms, rhizomes and tubers. To a botanist the differences are important but for the average home gardener they are all sufficiently similar to be regarded as bulbs. However, it may be useful to describe briefly the characteristics of each kind of bulb.

TRUE BULBS

True bulbs have an onion-like structure and consist of layers of fleshy 'scale leaves' which are closely folded on each other. The fleshy scales enclosing the flower shoot are storage tissues filled with plant foods such as protein, starch and sugar, all of which are formed during the previous season's growth. For this reason, it is important to leave plants of true bulbs to die down naturally each year to provide as much nourishment as possible for the next season. It also explains why true bulbs can be grown successfully in bowls of fibre which contain very little in the way of nutrients. When grown in soil (in pots or the open garden) true bulbs will product daughter-bulbs or bulbils which in time will become large enough to flower. Good examples of true bulbs are daffodil, jonquil, hyacinth, tulip and lilium.

CORMS

Corms do not have fleshy scales but consist of a shortened, swollen stem of solid storage tissue. The leaves arise in the axils of the scale-like remains of leaves of previous season's growth. A new corm is formed on top of the old corm, which shrivels and dies. Small daughter corms called cormels may also develop. Examples of corms include anemone, crocus, freesia, gladiolus, ixia and ranunculus.

RHIZOMES

Rhizomes are underground stems, usually thick and swollen, containing storage tissue. They develop roots, leaves and flowering stems from the nodes or joints. Good examples are flag or bearded iris, lily-of-the-valley and Solomon's seal.

TUBERS

Tubers can be either swollen stems or swollen roots for storage. New shoots arise from axillary buds on stem tubers or from buds on the short piece of stem on root tubers. Good examples of 'bulbs' of this kind are cyclamen, arum lily, and tuberous begonia.

BULBS IN THE GARDEN

Flowering bulbs are very adaptable and can be planted to give attractive and different garden effects. Some bulbs—especially anemone and ranunculus—are suitable for massed beds or borders. Lachenalias are good for low borders too. But most bulbs are best planted in bold clumps in the garden or rockery.

Because of different flowering times, many kinds can be planted together to give a continuous and varied colour display. A garden planted with a variety of bulbs is assured of some colour from early spring to summer or early autumn—jonquils and lachenalias in early spring followed by daffodils, anemones and ranunculi. Late spring brings hyacinths, freesias, sparaxis, tritonias and watsonias with calla lilies, hippeastrums, gladioli and many others in summer.

Some spring flowering bulbs can be planted under trees and on lawns or grassy slopes to become a permanent feature of the garden scene. In cold climates, bulbs such as bluebell, crocus, daffodil, jonquil, grape hyacinth, snowflakes, sternbergia and tulip become naturalized in these situations. In temperate districts, daffodils, jonquils, freesias, grape hyacinth and snowflake are best for lawn planting.

GENERAL BULB CULTURE

In the garden, a well-drained, sandy loam which is not overly rich is the best for bulbs and bulbous plants. But you can improve heavy soils by adding coarse sand and well-rotted organic matter to make them more friable. (See Chapter 3.) A light dressing of mixed fertilizer such as Gro-Plus Complete should be incorporated in the soil during preparation and prior to planting. Direct contact between fertilizer (or fresh manure) and the bulbs should be avoided. Most bulbous plants will respond to feeds of liquid fertilizers when flower buds appear.

Some bulbs will grow in full sunlight, others tolerate some shade. The best situation is given in the notes for each kind of bulb. Depth of planting varies with the size of the bulb but a good general rule is to plant at a depth equal to twice the width of the bulb. Planting time, together with approximate planting depth and plant spacing, is given in the notes and also summarized in the Bulb Planting Guide at the end of the chapter.

BULBS IN OUTDOOR CONTAINERS

Many bulbs grow to perfection in tubs, pots or troughs which can be moved about the garden, terrace or balcony. Smaller containers can be brought indoors when the plants flower. Daffodils, jonquils, hyacinths (including grape hyacinths), bluebells, lachenalias, freesias and tulips can be grown this way. Containers, which must have drainage holes, should be at least 15 cm deep to allow for good root growth. Cover the drainage holes with a piece of broken pot, coke or charcoal. Use a friable, free-draining soil to which a little mixed fertilizer has been added, or a proprietary potting mixture, and fill the pots to the depth required. Set the bulbs in place—about half the distance apart for normal planting in the garden—and fill in remainder of soil to within 2–3 cm of the top to leave room for watering. Keep the containers in a cool shady place until leaves emerge and then move into a sunny but sheltered position. Keep the soil moist at all times but do not over-water. Give liquid feeds of soluble fertilizer if plants appear backward and especially when buds appear.

BULBS IN FIBRE FOR INDOORS

This fascinating way of growing bulbs is ideal for the flat or home unit dweller. Bulb fibre is available from garden stores or nurseries to be used in pots or bowls of your own choice. No drainage is needed but bowls should be at least 7 cm deep. Hyacinths, daffodils and tulips are favourites for this method. It is best to plant only one variety of daffodil or one colour of hyacinth in each bowl. This way all plants in the bowl will flower together. Make sure the bulbs you buy are firm and free of mould. If not quite ready for planting, store the bulbs in a cool, dry cupboard. To promote earlier flowering on tall strong stems, place bulbs of hyacinths and tulips in the crisper tray of the refrigerator (7–10°C) for 3–4 weeks before planting.

Plant the bulbs in the moistened fibre so that their tips are just below the surface when the remainder of the fibre is packed around. Spacing is closer for bulbs in fibre, so allow 2–3 cm between bulbs. Water the bowl well and tip on its side to drain away any excess. Keep the bowl in a cool, dark spot and the fibre moist but not wet. Shoots emerge in 6–8 weeks, although if bulbs have been chilled they may start a week or two earlier. When shoots are clearly visible, move the bowls to a well-lit, airy room. When grown near a window, bowls must be turned occasionally to keep the plants erect. Keep the fibre damp but not wet while the plants are growing. If over-watered, tip the bowl on its side to drain. Bulbs grown in fibre should be planted out in the garden next season.

SPRING FLOWERING BULBS

ALSTROEMERIA (Peruvian Lily)

A tuberous plant adaptable to soil and situation. Large flowers in red, pink and yellow (some spotted or streaked in contrasting colours) are borne in clusters on stems 60–75 cm tall. Clumps are best left undisturbed for a few years. Plant in autumn to early winter 15 cm deep and 30 cm apart.

ANEMONE (Windflower)

Excellent spring flowers in reds, pinks and blues for mass bedding and borders. They contrast well with ranunculus which flower at the same time. Flowers of the St Brigid strain are semi-double or double while the poppy-like flowers of *coronaria* are single. Both are ideal for cutting. Anemones prefer full sunlight and a well prepared friable soil with a ration of mixed fertilizer added. The plants will carry over for several seasons but the best display is from new corms each year. Plant the corms in autumn 3 cm deep and 15 cm apart, making sure the flat part of the corm is uppermost. Some gardeners prefer to start the corms in seedling trays and later transplant.

BABIANA

Babiana has violet-blue or mauve flowers rather similar to freesias and produced on 20 cm spikes. The plants need full sunlight and like other bulbs native to South Africa (freesia, ixia, sparaxis and tritonia) will tolerate rather dry conditions. Plant from late summer to autumn 5 cm deep and 7 cm apart. After the plants have died down the bulbs can be lifted and stored for replanting next season, but the plants can be left for several years if preferred.

BLUEBELLS (Scilla)

Also known as Wood Hyacinth, these dainty lavender-blue, bell-shaped flowers are ideal for clump planting in the garden or in a rockery. The plants are also useful for grassy banks and under trees where they become naturalized. They grow well in full sun or semi-shade. The Spanish bluebell is rather larger and taller (30 cm) than the true English bluebell. Plants can be left undisturbed for a number of years. Plant bulbs in early to mid-autumn 7 cm deep and 10 cm apart.

BRODIAEA (Firecrackers)

The small, tubular flowers in rosy-red with green tips are borne on stems 30–40 cm tall. The plants prefer semi-shade and a well drained soil and are suitable for borders, clump planting or in rockeries. Plant bulbs in autumn 7 cm deep and 15 cm apart.

Hyacinths grow well in containers

CHIONODOXA (Glory of the Snow)

This charming little bulbous plant from the mountains of Crete is only suitable for cold climates. The clumps grow to 15 cm tall with blue and white star-shaped flowers in early spring. The plants need full sunlight and good drainage and should be left undisturbed once they are established. Plant bulbs in autumn 7 cm deep and 10 cm apart.

CLIVEA (Kaffir Lily)

An evergreen, bulbous plant growing to 60 cm tall with clusters of large, trumpet-shaped, orange flowers followed by attractive red berries. It does best in warm temperate climates and prefers semi-shade. It is excellent for growing under trees and makes a good specimen for large pots or tubs. Clumps are divided in autumn or early winter and the bulbs planted just below the surface and 30 cm apart.

CROCUS

One of the earliest spring flowering bulbs with small, cup-shaped flowers in purple, lavender and white on short stems. Excellent for semi-shaded spots and rockery pockets with good drainage. Plants can be left undisturbed for many years. Plant in autumn 5 cm deep and 10 cm apart.

CYCLAMEN

These attractive tuberous plants are usually started from seed sown in summer and early autumn and grown as pot plants. (See Chapter 9.) The tubers will carry over quite well but must be repotted 5 cm deep during late summer or early autumn.

CYRTANTHUS (Ifafa Lily)

A rather uncommon bulb with tubular flowers in cream and salmon on stems 60 cm tall. Plants prefer full sunlight and a friable, well drained soil, either in the open garden or as pot plants. Plant in autumn 7 cm deep and 15 cm apart.

DAFFODIL (Narcissus)

Daffodils are the best known and most adaptable of all spring flowering bulbs. They thrive in all climates except very warm northern districts. They grow well in full sun or semi-shade and prefer a friable, well drained soil to which a mixed fertilizer has been added. The plants respond to liquid feeds when buds appear. Daffodils can be grown in almost every garden situation. On grassy banks they become a permanent feature but are very popular for clump planting to combine with other spring annual flowers such as alyssum, nemesia, pansy, viola and linaria. They are excellent for growing in tubs, pots or troughs outdoors or for bowls indoors.

Daffodils come in a wide range of gold or cream shades and flower forms. New varieties are continually being added to the list. King Alfred is still the most widely grown of the large trumpet types. Fortune, Dr Roseby, Tunis and Russ Holland are also popular. Other varieties readily available include Early Prince, Carbineer, Carlton, Pink Blossom and Hoop Petticoat. Packs of mixed varieties such as Gay Spring are available too. Plant bulbs in autumn (preferably before the middle of May in most districts) 12 cm deep and 10–15 cm apart. Bulbs in pots, troughs and bowls can be planted closer together.

Daffodil

FREESIA

Freesias have been popular spring flowering bulbs for a long time, highly prized for their delightful perfume. The Super Giant strain has larger flowers and a wide colour range in shades of yellow, orange, rosy-red, ruby and blue. The Bergunden strain is very similar but the flowers are more highly perfumed. They are excellent for cut flowers too. Freesias do best in a fairly sunny position but will grow well under shrubs and trees if not too shaded. They need a friable, well drained soil which is not overly rich but with a scattering of mixed fertilizer worked into the soil during preparation. Usually this preplanting fertilizer is all that is necessary. The plants are suitable for massing in beds or on grassy banks but are particularly attractive in clumps or in rockery pockets. Freesias can be left in their permanent position for several years or the corms can be lifted after the plants die for storing and replanting. Plant the corms in autumn at a depth of 7 cm at the same distance apart. Packs of white, yellow and mixed colours are available.

FRITILLARIA

These attractive members of the lily family grow best in cool climates. *Fritillaria imperialis* (Crown Imperial) is the best known. The plants are 60 cm tall with drooping bell-shaped flowers in yellow, orange and bronze. They are best planted in clumps or in rock gardens and should be left undisturbed unless they become overcrowded. Plant in autumn 10 cm deep and 30 cm apart.

GRAPE HYACINTH (Muscari)

Grape hyacinths are one of the most attractive dwarf bulbs for spring, with spikes of bead-like, deep-blue flowers on stems 10–15 cm tall. The plants are very adaptable to soil and will grow in full sun or semi-shade. They are good for naturalizing and ideal for clump planting in the garden or in rock pockets. Plant bulbs in autumn 7 cm deep and 10 cm apart.

HIPPEASTRUM (Fire Lily)

These plants produce magnificent trumpet-like flowers in salmon, red, rose and variegated white and red on strong stems 60 cm tall. The plants prefer friable, well drained soil in a sunny position but will tolerate some shade during the day. They need plenty of water in spring when flower buds appear. Plant the bulbs in winter, keeping the neck of the bulb at the soil surface. Space bulbs 30–40 cm apart.

HYACINTH

New, improved varieties of hyacinth are very adaptable and will do well in all but very warm climates. They produce giant spikes crowded with large flowers in white, yellow, pink, red, light blue or dark blue. The plants prefer a sunny position and fertile, well drained soil for best results. They are excellent for clump planting among spring annuals and are ideal for container growing, both outdoors and indoors. For container growing, select large, plump bulbs of good shape and free of blemishes. Store bulbs in a cool spot in the house and pop them in the refrigerator crisper tray for 3–4 weeks before planting. Plant the bulbs in autumn (April or early May is a good time in most districts) 15 cm deep and 15 cm apart. Container grown bulbs can be planted with the neck at soil level and the spacing reduced to 7 cm. Bulbs are available in separate colours or as assortments.

IRIS, FLAG

The most popular and widely grown iris is the Flag, Bearded or German Iris. These plants grow from rhizomes into large clumps of attractive, grey-green leaves to 60 cm or more tall. The silky-textured, flag-like flowers come in a wide range of colours including yellow, gold, lavender, mauve, purple, shades of brown and white. Many have colour combinations. The flag iris does well in almost any garden situation but prefers a sunny aspect and well drained soil. Add a ration of mixed fertilizer when preparing the soil; an application of lime is recommended on acid soils. Plant in autumn through to winter, keeping the top of the rhizome at soil surface. Space rhizomes about 30 cm apart.

IRIS, DUTCH AND SPANISH

Unlike flag iris, the Dutch and Spanish types grow from bulbs and not from rhizomes. The plants form attractive

Dutch irises make grand cut flowers

clumps and have flowers in a range of colours including blue, bronze, yellow and white on stems about 60 cm tall. The flowers of Dutch iris are larger and rather earlier than the Spanish types. Plant bulbs in autumn 10 cm deep and 15 cm apart.

IXIA (Corn Lily)

These attractive South African bulbs have clusters of bell-shaped flowers in shades of yellow, gold, pink, orange and port wine on stems 60 cm tall. They are excellent for cut flowers. The plants need a sunny position and are best grown in clumps among annuals or between shrubs. Plant bulbs in autumn 7 cm deep and 10 cm apart.

JONQUIL (Bunch-flowered Narcissus)

A close relative of daffodils, jonquils are just as adaptable to climate and soil. They are one of the first bulbs to flower in spring, with small clusters of fragrant, daffodil-like blooms. They are useful for naturalizing on lawns or grassy slopes and are attractive in clumps around shrubs and odd corners of the garden. The plants can remain undisturbed for many years. The most popular variety is Soleil d'Or with orange-red cups surrounded by gold petals. There are many other varieties with cream or white blooms. Plant in autumn 10 cm deep and 10 cm apart.

LACHENALIA (Cape Cowslip)

Attractive dwarf plants from South Africa with 15 cm spikes of tubular, waxy flowers in shades of yellow, orange-red, red or green according to variety. The plants prefer full sunlight and a well drained, friable soil. They are excellent for borders, clump planting, rockery pockets, outdoor pots and troughs and indoor bowls. The variety *tricolor* is yellow, red and green; *quadricolor* is a combination of red, yellow, green and purple; *pendula* has large, red flowers. Other varieties are *aurea* (golden yellow) and *pallida* (lime-green). Plants may remain in the ground for several years or bulbs can be lifted when foliage dies for replanting the following season. Plant in autumn 7 cm deep and 10 cm apart.

LILY-OF-THE-VALLEY (Convallaria)

These dwarf plants grow from rhizomes and have tiny, fragrant, bell-shaped white or cream flowers in late spring. The plants need a shady, moist but well drained position with plenty of organic matter in the soil. Plants can remain undisturbed for several years. Plant rhizomes in winter 3 cm deep and 10 cm apart. Plants can also be grown in pots for indoor decoration, especially in warmer climates where they do poorly in the open garden.

ORNITHOGALUM

There are three widely grown species of this spring-flowering bulb. All three need well drained soil and prefer sunlight for at least half the day. They are best left for several years to form clumps but can be lifted after foliage dies and replanted next season if preferred. The plants do well in warm climates. *Ornithogalum thyrsoides* (Chincherinchee) grows to 30 cm with clusters of papery-white flowers which open from the base upwards. Good for cut flowers which are capable of changing colour when the stems are dipped in dye or coloured ink. *O. arabicum* (Arab's Eye) has papery-white petals surrounding a black centre and the blooms have an aromatic fragrance. *O. umbellatum* (Star of Bethlehem) has waxy, white flowers striped green on the underside of the petals. Plant bulbs in autumn 7 cm deep and 15 cm apart.

RANUNCULUS

Ranunculus, like anemone, is excellent for mass bedding, borders or for planting in clumps or drifts with other spring flowering annuals. The plants grow to 60 cm tall and the semi-double and double flowers come in a remarkable colour range in shades of red, crimson, scarlet, pink, orange, yellow, lemon, cream and white. Bulb packs of the Picasso strain are available as separate colours—red, pink, orange, gold and white—or as mixtures. The plants need full sunlight and a friable, fertile, well drained soil to which organic matter and mixed fertilizer has been added during preparation. They respond to liquid feeds when buds form. You can dig the corms when the foliage dies down but results are usually better from fresh corms each year. Plant corms in autumn 3 cm deep and 15 cm apart. Make sure the corms are planted with the 'claws' downwards. Like anemone, the corms can be started in a shallow seedling tray for transplanting later.

SNOWDROPS (Galanthus); SNOWFLAKES (Leucojum)

Both these dwarf, spring flowering bulbs have dainty, white, bell-shaped flowers with a green spot on the outside of each petal. The plants are excellent for clump planting and need a semi-shaded or shaded aspect for best results. They are excellent for planting under deciduous trees where winter sun can reach them or on grassy banks. Plant bulbs in autumn 7 cm deep and 10 cm apart.

SPARAXIS (Harlequin Flower or Wand Flower)

These bulbs from South Africa need much the same conditions as freesias—a sunny position. They are excellent for naturalising on lawns or grassy slopes. They are ideal for rock gardens and outdoor container growing. The bell-shaped flowers are quite large and come in shades of red, orange or cream with black geometrical markings. Plant in autumn 7 cm deep and 10 cm apart.

TRITELEIA (Star Violets)

Dwarf plants with soft green foliage and star-shaped, lavender-blue flowers on 15 cm stems. The plants become naturalized when grown under trees or shrubs and are attractive in borders, rockery pockets and containers. Plant in autumn 5 cm deep and 7 cm apart.

TRITONIA (Montbretia)

Another bulb from South Africa. The plants are very similar in height and flower form to sparaxis. Flower stems are rather longer and flowers are in shades of orange, pink and red. Tritonias are grown in the same way as sparaxis or freesias. Plant bulbs in autumn 7 cm deep and 10 cm apart.

The delightful Picasso ranunculus is a spring garden favourite

TULIP

In the sixteenth century, single bulbs of the tulip fetched fantastic prices in Holland and other European countries. Tulips have large, beautifully formed, bell-shaped flowers in dazzling shades of cream, yellow, orange, pink, scarlet, red and deep maroon. The plants do best in cool climates but are very adaptable and should be grown more widely in temperate climates. They prefer full sunlight but tolerate semi-shade. Tulips need friable, well drained soil which has been prepared in advance of planting with compost and a mixed fertilizer added. Tulips do not like acid soils so it is best to add a ration of lime in most coastal and highland districts. (See Chapter 4.) When flower buds appear, give liquid feeds of soluble fertilizer to promote long, strong stems and large blooms.

Tulips are best grown in clumps in the garden surrounded by dwarf annuals like alyssum, bellis or violas. They are ideal, of course, for container growing too. As for hyacinths, chill the bulbs in the refrigerator crisper before planting. Plant in autumn (late April or May are best in most districts) 12 cm deep and the same distance apart. After flowers and foliage

Tulips are best grown in clumps

die, lift the bulbs for storing in a cool place and replanting next season. In warmer climates it may be best to buy new bulbs for a good display.

WATSONIA (Bugle Lily)
Watsonias develop into large clumps of strap-like foliage 90–120 cm tall. The flower spikes are even taller with dainty, tubular flowers in pink, salmon, red and white. Flowers are excellent for cutting in late spring when other flowers may be scarce. The plants are useful for background work and can be divided after the foliage dies in late summer or early autumn. Plant in autumn 7 cm deep and 30 cm apart.

SUMMER-FLOWERING BULBS

AGAPANTHUS (African Lily)
These vigorous, bulbous plants are evergreen and grow into large clumps 60–90 cm tall. The large clusters of tubular flowers are blue or white. Agapanthus is a good background plant but will grow almost anywhere—on dry banks in full sun or in shade under trees. Because of their vigorous roots and spreading habit, they crowd out weaker plants and tend to take over the garden. Clumps can of course be thinned and replanted if necessary. Plant from late autumn to early winter. Just cover the fleshy rhizomes and space them 50 cm apart.

AMARYLLIS (Belladonna Lily)
Also known as Naked Ladies because the buds on tall, fleshy, stems appear before the foliage. The large, fragrant, trumpet-shaped flowers are usually pink but also creamy white. The plants grow in full sun or under deciduous trees in semi-shade. They are best left undisturbed for a few years but can be lifted when dormant after flowering. Plant bulbs in late autumn or early winter, just covering them with soil and spacing them 30 cm apart. Brunsvigia is a close relative and is grown in the same way.

BEGONIA (Tuberous)
Tuberous begonias are usually grown in pots in a sheltered area, shadehouse or glasshouse. The magnificent semi-double or double blooms come in shades of red, pink, orange, yellow and white. The plants need a rich, friable, well-drained soil. Tubers can be carried over for repotting the following year. Plant or repot tubers in spring with the top or crown level with the soil surface. Tuberous begonias can also be grown from seed. (See Chapter 9.)

CALLA
This group includes both *Arum* and *Richardia* which are more correctly called *Zantedeschia*. Callas prefer a sunny but damp situation with fairly rich soil. Too much moisture, when the bulbs are dormant in winter, may cause them to rot, so lift them after foliage dies for replanting in late winter. *Z. elliotiana* grows to 60 cm with bright yellow lilies and white-spotted leaves. *Z. rehmannii* is smaller (30 cm) with dainty mauve-pink blooms. *Z. aethiopica* is the Lily of the Nile or Arum Lily. Plant bulbs in late autumn or early winter 10 cm deep and 20 cm apart.

CANNA
A vigorous plant 90–150 cm tall forming dense clumps of brilliant green or bronze leaves from its tuberous roots. The large

flower clusters, with lily-like blooms, come in shades of cream, yellow, orange, pink and red. Some flowers are attractively spotted. Cannas prefer full sun but need damp or even wet conditions. After flowering the stems can be cut down to ground level for regrowth in spring or divided for replanting. Some varieties, however, tend to remain green all year round. Plant tubers in winter or early spring 5 cm deep and 50 cm apart.

CRINUM (Veldt Lily)

The white or pinkish, fragrant flowers of crinum resemble lilies and are borne on stems 60–90 cm tall. Some species are evergreen while others die down after flowering. The plants prefer an open, sunny aspect and do best in warm coastal areas. The unusually large bulbs may be planted with a light soil covering at almost any time of the year, spacing them 30 cm apart.

DICENTRA (Bleeding Heart)

Dicentra is an old-fashioned plant to 25 cm tall with sprays of red, heart-shaped flowers. The plants do best in cool temperate or cold climates and prefer deep, well drained soils, but need plenty of water in spring to mid-summer when flowers appear. The plants are dormant from late summer to spring. Plant the tuberous roots in autumn or early winter 10 cm deep and 60 cm apart.

EUCHARIS LILY

An evergreen, bulbous plant best suited to warm temperate and tropical climates. The plants are suitable for clumps in the garden or in large tubs and need a sunny sheltered position. The white, trumpet-shaped flowers have a sweet fragrance. Plant bulbs in winter to spring, just covering with soil and spacing them 20 cm apart. Flowering is improved if clumps are left undisturbed for a few years.

EUCOMIS (Pineapple Flower)

An unusual bulbous plant growing to 60 cm tall in clumps with large cylindrical flower spikes made up of masses of greenish-white petals tipped with lilac. The flowers are excellent for cutting and last for weeks in water. The plants prefer open sunlight and are best left undisturbed for several years. Give liquid feeds during late spring and summer and water well in hot weather. Plant bulbs in winter or early spring 12 cm deep and 30 cm apart.

GALTONIA (Summer Hyacinth)

A tall (120 cm) bulbous plant from South Africa with clusters of white, drooping, bell-shaped flowers. A good background plant but needs a sunny position. Best in warmer districts as the plants are sensitive to frost. Plant bulbs in spring 10 cm deep and 20 cm apart.

GLADIOLUS (Sword Lily)

Certainly the most spectacular of summer flowering plants, both for garden display and cut flowers. The orchid-like blooms come in a wide range of clear colours and pastel shades. Packs of separate or assorted colours are available from garden stores and nurseries. The plants prefer full sunlight and are best planted in clumps with dwarf summer-flowering annuals. The plants, when well-grown, may reach 150 cm in height so it is best that they are sheltered from strong winds. To prevent wind damage, tie 3 or 4 plants in a clump to a garden stake just before the flower buds appear.

Gladioli are very adaptable to climate and can be grown in hot, mild and cold districts. Flowering takes about 90–100 days from planting and it is best to plant so that flowers appear before or after the extreme heat of summer. They need a well drained, friable soil which should be prepared in advance of planting by adding compost and a ration of mixed fertilizer. In warm districts, planting can be made from May to September so that the plants flower before the extreme heat of summer. In cold districts, it is best to delay planting until August or later. It is good insurance to dust the corms with a fungicide—Zineb, Captan or Dry Bordox—before planting. Plant the corms 10 cm deep, spacing them 20 cm apart.

Thrips are a serious insect pest of gladioli so it is necessary to spray with insecticide every two weeks after the plants reach the four-leaf stage. (See Chapter 17.) In warm weather, always cut flower spikes when the first flowers open. The remaining buds will open indoors if the stems are in water. Corms can be lifted after the leaves

Gladioli are spectacular summer flowers

turn yellow and start to die—generally about 4–6 weeks after flowering. Allow the leaves to dry out, then cut them off close to the corm. Dust the corms with fungicide and store in a dry, cool place until the next planting season.

GLORIOSA (Climbing Lily)

Gloriosa is one of the few climbers of the garden lilies and will grow to 2–3 m tall with orange or red trumpet-shaped flowers in summer. The plants prefer a warm sunny position but tolerate semi-shade. The offsets must be carefully removed from the bulb as the roots are brittle and easily damaged. Plant offsets during winter and early spring 3 cm deep and 30 cm apart.

GLOXINIA

These spectacular tuberous plants are best grown in pots in shadehouses or glasshouses where they are protected from wind and rain. They can also be grown indoors in a well-lit spot, but not in direct sunlight. They need ample moisture and a fairly even temperature (about 20°C). The bell-shaped flowers have a velvety texture and come in many shades of cream, pink, red, blue and purple, many of which are spotted or have contrasting edges and throats. Tubers can be dried off like cyclamen at the end of the flowering season for replanting. (See notes on Cyclamen, Chapter 9.) Plant tubers in winter or spring with the top of the tuber at soil level. Gloxinias can also be raised from seed. (See p. 144.)

HEMEROCALLIS (Day Lily)

These bulbous plants produce clumps of pale green foliage 60–75 cm tall with flowers in cream, orange and red shades. Individual flowers last one day only but new flowers open over a long period. They prefer a sunny position. Plant bulbs in autumn to winter 10 cm deep and 30 cm apart.

KNIPHOFIA

(Torch Lily or Red Hot Poker)

These attractive plants form large dense clumps with long poker-like stems bearing masses of small tubular flowers which open from the base. Colours range from yellow through orange to red. The plants prefer a sunny position and are useful as background plants or for clump planting. They are best left undisturbed for several years. Plant bulbs in autumn or winter 10 cm deep and 60 cm apart.

LILIUM

These are the true lilies. There are many species and varieties and hundred of hybrids with an endless array of flower forms and colours. The most widely grown are the November Lily (*L. longiflorum*) with white trumpet-shaped flowers; Tiger Lilies—*L. tigrinum* (orange), *L. speciosum* (pink) and *L. speciosum album* (white); *L. regale* (large creamy-white, purple-backed flowers) and Golden Rayed Lily of Japan (*L. aurantum*) with large, white flowers spotted purple and striped yellow. Most of the liliums form large clumps which are best left undisturbed. They need half-sun and a friable well drained soil, but respond to generous watering in summer and mulching to keep the soil cool and moist. Plant the scaly bulbs in late autumn to winter 10–20 cm deep (according to bulb size) and 30–40 cm apart. They flower in late spring to late summer according to variety.

LYCORIS
(*See Spider Lily.*)

NERINE
(*See Spider Lily.*)

SOLOMON'S SEAL (Polygonatum)
Solomon's seal is an excellent bulbous plant for cool, shady situations or dappled sunlight. The slender, oval leaves grow in clumps (60 cm tall) from shallow rhizomes and the curving stems bear white, bell-shaped flowers tipped green. Plant rhizomes in late autumn to winter at a depth of 3 cm and 25 cm apart.

SPIDER LILY (Nerine; Lycoris)
These two plants have similar flowers and are usually grown in clumps which are best left undisturbed. They prefer a well drained soil and sunny position. Nerines flower before the foliage appears in the same way as Belladonna lilies. Flowers are pink (*N. bowdenii*) or scarlet (*N. sarniensis*). Lycoris flowers are yellow (*L. aurea*) or red (*L. radiata*). All the spider lilies are excellent for cut flowers. Plant bulbs in winter or early spring, keeping the neck of the bulb above the soil surface. Space the bulbs 15 cm apart.

SPREKELIA (Jacobean Lily)
Sprekelia has large, crimson, orchid-like blooms 10 cm in diameter borne on naked stems 15–30 cm tall. The plants prefer sandy, well drained soil in a situation with half-sun. Plant bulbs in late winter and early spring with necks of bulbs just above surface and space them 12 cm apart.

STERNBERGIA (Autumn Daffodil)
A dwarf, bulbous plant to 20 cm high which needs full sunlight and rather dry conditions. The golden crocus-like flowers appear with the leaves in autumn. Useful little plants for pockets in paving or in clumps on the edge of a bed. Plant in early summer 10 cm deep and 15 cm apart.

TIGRIDIA (Tiger Flower or Jockey's Cap)
These unusual but colourful flowers with broad pink or red petals surrounding

Jillian Wallace, one of the most popular hybrid lilies

Liliums are available in a multitude of forms and colours

smaller, spotted petals are borne on 30–40 cm stems in early summer. Individual flowers last for only one day but there is a succession of blooms. The plants need a well drained, sunny position and are best planted in clumps to remain undisturbed for a few years. Plant late autumn and early winter 7 cm deep and the same distance apart. Plants can also be raised from seed.

TUBEROSE (Polianthes)

The fragrant, white flowers of tuberose are favourites for bridal bouquets. In warm and mild climates the plants will flower at almost any time of the year but summer is the best flowering period. The plants prefer a warm, sheltered position with plenty of moisture in summer. It is best to divide clumps after flowering for replanting. Plant in winter or early spring, just covering with soil and spacing them 20 cm apart.

VALOTTA (Scarborough Lily)

These evergreen bulbous plants grow to 40 cm tall with large orange-scarlet, trumpet-shaped flowers on strong stems. The plants prefer a sunny position but are very adaptable to soil. The clumps are best left undisturbed. Plant bulbs in winter or early spring with the neck of the bulb level with soil surface. Space bulbs 20 cm apart.

ZEPHRANTES

(Autumn Crocus or Storm Lily)

Zephrantes is an excellent bulbous plant to grow for late summer and autumn flowering. The clumps grow to 20–30 cm tall with white (*Z. candida*) or yellow (*Z. citrina*) crocus-like flowers. The plants are evergreen in most climates and are useful for pots, rockeries and border edgings. Plant bulbs in late autumn, winter or early spring 7 cm deep and 30 cm apart. Plants can also be raised from seed.

Flowers of the day lily (Hemerocallis) range from cream to orange and red

PLANTING GUIDE FOR FLOWERING BULBS

Name and Common Name	Planting Season	Planting Depth (cm)	Distance Apart (cm)	Flowering Season
Agapanthus (African Lily)	Late Autumn–Winter	*Note A*	50	Summer
Alstroemeria (Peruvian Lily)	Autumn–Winter	15	30	Spring–Early Summer
Amaryllis (Belladonna Lily)	Late Autumn–Winter	*Note A*	30	Summer
Anemone (Windflower)	Autumn	3	15	Spring
Babiana	Late Summer–Autumn	5	7	Spring
Begonia (tuberous)	Spring	*Note B*	In pots	Summer
Blue Bells (Scilla)	Autumn	7	10	Spring
Brodiaea (Firecrackers)	Autumn	7	15	Late Spring
Brunsvigia	Late Autumn–Winter	*Note A*	30	Summer
Calla	Late Autumn–Winter	10	20	Summer
Canna	Winter–Early Spring	5	50	Summer
Chionodoxa (Glory of the Snow)	Autumn	7	10	Early Spring
Clivea (Kaffir Lily)	Autumn	*Note A*	30	Late Spring
Crinum (Veldt Lily)	Autumn–Spring	*Note A*	30	Summer
Crocus	Autumn	5	10	Late Winter–Early Spring
Cyclamen	Late Summer–Autumn	5	In pots	Late Winter–Early Spring
Cyrtanthus (Ifafa Lily)	Autumn	7	15	Late Spring
Daffodil (Narcissus)	Autumn	12	10–15	Spring

Note A. Just cover with soil.
Note B. Plant with top of bulb at soil surface.

PLANTING GUIDE FOR FLOWERING BULBS

Name and Common Name	Planting Season	Planting Depth (cm)	Distance Apart (cm)	Flowering Season
Dicentra (Bleeding Heart)	Autumn	10	60	Early Summer
Eucharis Lily	Winter–Early Spring	*Note A*	20	Summer
Eucomis (Pineapple Flower)	Winter–Early Spring	12	30	Summer
Freesia	Autumn	7	7	Spring
Fritillaria	Autumn	10	30	Spring
Galtonia (Summer Hyacinth)	Spring	10	20	Late Summer
Gladiolus (Sword Lily)	Late Winter–Spring	10	20	Summer
Gloriosa (Climbing Lily)	Winter–Spring	3	30	Summer
Gloxinia	Winter–Spring	*Note B*	In pots	Summer
Grape Hyacinth (Muscari)	Autumn	7	10	Spring
Hemerocallis (Day Lily)	Autumn–Winter	10	30	Summer
Hippeastrum (Fire Lily)	Winter	*Note B*	30–40	Late Spring
Hyacinth	Autumn	15	15	Spring
Iris, Flag or Bearded	Autumn–Winter	*Note B*	30	Late Spring
Iris, Dutch and Spanish	Autumn	10	15	Spring
Ixia (Corn Lily)	Autumn	7	10	Spring
Jonquil (Bunch–flowered Narcissus)	Autumn	10	10	Later Winter–Early Spring
Kniphofia (Torch Lily)	Autumn–Winter	10	60	Summer
Lachenalia (Cape Cowslip)	Autumn	7	10	Spring

Lilium	Late Autumn–Winter	10–20	30–40	Summer
Lily-of-the-Valley (Convallaria)	Winter	3	10	Late Spring
Lycoris (Spider Lily)	Winter–Early Spring	*Note B*	15	Late Summer
Nerine (Spider Lily)	Winter–Early Spring	*Note B*	15	Late Summer
Ornithogallum	Autumn	7	15	Spring
Ranunculus	Autumn	3	15	Spring
Snowdrop (Galanthus)	Autumn	7	10	Spring
Snowflake (Leucojum)	Autumn	7	10	Spring
Solomon's Seal (Polygonatum)	Late Autumn–Winter	3	25	Summer
Sparaxis (Harlequin Flower)	Autumn	7	10	Spring
Sprekelia (Jacobean Lily)	Late Winter–Early Spring	*Note B*	12	Summer
Sternbergia (Autumn Daffodil)	Early Summer	10	15	Autumn
Tigridia (Tiger Flower)	Autumn–Winter	7	7	Early Summer
Triteleia (Star Violet)	Autumn	5	7	Spring
Tritonia (Montbretia)	Autumn	7	10	Spring
Tuberose (Polyanthes)	Autumn–Winter	*Note A*	20	Summer
Tulip	Autumn	12	12	Spring
Valotta (Scarborough Lily)	Winter–Early Spring	*Note B*	20	Late Summer
Watsonia (Bugle Lily)	Autumn	7	30	Spring
Zephrantes (Autumn Crocus)	Winter–Early Spring	7	30	Autumn

Note A. Just cover with soil.
Note B. Plant with top of bulb at soil surface.

Chapter 11

PERENNIALS, VINES AND CREEPERS

Perennial plants are those that grow for a number of years. This definition is very wide and could cover a vast array of plants, including trees, shrubs, climbers, ferns, bulbs, cacti, succulents and many others—in fact, any plants which are not strictly annuals or biennials. This chapter, however, is confined to flowering perennials and climbing plants for the home garden.

PERENNIALS

Perennials—or more precisely flowering perennials—are defined as plants (mostly with non-woody stems and branches) which die back to the roots or an evergreen crown at the end of the flowering season. They burst into renewed growth the following spring. They repeat this cycle year after year, gradually spreading and increasing in size and number.

The very large group of perennial plants includes hundreds of species and dozens of families so it is difficult to divide them into precise classifications without some overlapping. As we have already seen in Chapter 9, many plants grown from seed as annuals may, under favourable conditions, persist as perennials. Good examples are aquilegia, carnation, chrysanthemum, dahlia, gazania, gerbera, salvia and verbena. Most bulbs and bulbous plants with corms, rhizomes or tubers can also be regarded as perennials. Good examples are agapanthus, calla, canna, iris, kniphofia, lilium, tuberose and watsonia. These and many others have been described in Chapter 10. To avoid repetition, the reader is referred to these two chapters for descriptions.

There are many other perennials which can add interest and colour to the garden. They come in all shapes and sizes, from small prostrate plants suitable for groundcovers or rockery work to large plants several metres tall for backgrounds or planting amongst shrubs. Some grassy perennials such as pampas grass (*Cortaderia selloana*), Nile grass (*Cyperus papyrus*) and New Zealand flax (*Phormium tenax*) have become popular as accent plants around buildings and pools and in pebble gardens.

Herbaceous perennials are those plants which, after flowering in summer, die down in autumn and become dormant in winter. They usually prefer cool conditions and have been used in the traditional herbaceous border or in English cottage gardens. Evergreen perennials are mostly natives of warmer climates and flower in spring. Because they remain green in winter they avoid that completely bare look in the garden. Whether herbaceous or evergreen, perennials can complement shrubs, annual flowers and bulbs. Many provide excellent flowers for indoor decoration, often at a time when annual flowers are scarce. Others have attractive and varied foliage in bronze, yellow-greens, grey and silver shades.

SOIL PREPARATION AND PLANTING

As most perennials will remain undisturbed for several years, it is important to prepare the soil well. Most perennials can be planted in late autumn and winter so it is a good idea to start preparation in early autumn. Dig the bed to spade depth and remove all tree roots and persistent weeds. Add plenty of organic matter in the form of animal manure or compost and a ration of mixed fertilizer such as Gro-Plus Complete at $\frac{1}{3}$ cup per square metre. Slow-acting blood and bone based fertilizers such as Gro-Plus for Camellias and Azaleas or other slow-release fertilizers (Osmocote, Agriform Plant Tablets or Mag-Amp) are also suitable to provide nutrients over a long period. It is a good idea to raise the bed 10–15 cm above the surrounding level too. This will ensure good drainage.

Select plants of the right height and size—dwarf plants in front and taller plants at the back of the bed. Make sure that each plant has sufficient space to develop without crowding or overlapping its neighbours. While a description of some popular perennials is given in this chapter, there are more comprehensive garden catalogues to help you plan your display. While many perennials can be started economically from seeds and cuttings, quicker results will be obtained by using established plants or divisions of good size. These can be purchased from garden stores and nurseries or perhaps you can take advantage of the offers of friends.

Most perennials are planted with the crowns slightly above soil level so that they do not collect water and rot during the dormant season. Plant the crowns so that the roots spread outwards and downwards. After planting, lightly fork the soil between clumps and water well.

MAINTAINING AND DIVIDING PERENNIALS

As the plants grow, cultivate between clumps to keep down weeds or discourage the weeds by mulching with grass clippings, compost, pine bark or leaf litter. Taller plants may need hardwood stakes or circular wire supports (available from garden stores) to protect them against wind. Remove spent flowers regularly during the summer months and cut out spent shoots of herbaceous perennials as they die down. Cut back shrubby, evergreen perennials to a tidy, compact shape after flowering. Removing the spent growth will also make it easier to fork over the surface between the plants. Apply a ration of mixed fertilizer in late winter or early spring each year. Slow-release fertilisers are ideal for this purpose as they can be scattered liberally between and on plants without fear of fertilizer burn. Thinning or dividing perennials is usually necessary when the clumps become overcrowded after four or five years. You will notice that the centre of the clump has become old and straggly so the best parts for replanting are on the outside. It is best to stagger the task of lifting and dividing, and do only a few plants each year. This way, you maintain the overall effect of the bed. After lifting clumps, dig the area well, adding more organic matter and fertilizer for the next few years. You will probably have more plants than you need for replacement but your friends will be grateful for any leftovers.

FAVOURITE PERENNIALS

Abbreviations: H—herbaceous, E—evergreen, FS—full sun, HS—half sun, SS—semi-shade, S—shade.

Height and spread of plant (cm) is shown in brackets, e.g. (20—30).

ACANTHUS (Bear's Breech)

H, FS to S (200—120). Large, glossy-leafed plants with white and purple snapdragon-like flowers in summer.

ACHILLEA (Yarrow or Milfoil)

A. filipendula H, FS (90—30). *A. fomentosa* E, FS (20—30). Both with lacy, finely divided leaves and rounded white, yellow, pink or red flower heads in summer.

ALYSSUM (Basket of Gold)

A. saxatile, E, FS or HS (30—30). Dwarf carpeting plant with soft mounds of foliage covered with brilliant yellow flowers in spring. Good drainage.

ANTHEMIS (Chamomile)

E, FS to SS (40—30). Semi-trailing plant with fern-like foliage and yellow or white daisy-like flowers in spring and summer. Good drainage.

ARENARIA (Sandwort)

E, HS or SS (10—16). Moss-like trailing plant with tiny white flowers in spring.

ARMERIA (Thrift)

E, FS to SS (40—20). Tufts of grassy foliage with globe-shaped heads of white or pink flowers in spring. Good drainage.

ARTEMESIA (Wormwood or Ghost Bush)

E, FS (120—60). Background plant with handsome, divided, greyish white leaves and yellow flowers in summer. Good drainage.

ASTER (Michaelmas Daisy)

E, FS or HS. Tall (120—50); medium (75—50); dwarf (20—25). Daisy-like, white, pink, mauve, lavender or blue flowers in late summer.

ASTILBE (Goat's Beard)

H, HS or SS (60—40). Clumps of attractive foliage with plumes of white, pink and red flowers in late spring. Moist position.

BERGENIA (Saxifraga)

E, SS or S (30—40). Rosettes of large, rounded, glossy leaves with stems of waxy, pink or rose-coloured flowers in winter.

BILLBERGIA (Queen's Tears)

E, SS or S (40—75). Clumps of thin, pointed leaves and green flowers hanging from pink-sheathed stems.

CAMPANULA (Bellflower)

H or E, SS (15—20). Neat dwarf mounds of green covered with blue or white bell-shaped or star-shaped flowers in summer.

CERASTIUM (Snow in Summer)

E, FS (10—50). Dwarf carpeting plant with silvery-grey foliage and masses of white, cup-shaped flowers in late spring. Plants spread rapidly. Good drainage.

Astilbe (Goats beard)

CHRYSANTHEMUM

(*C. maximum*, Shasta Daisy) H, FS (100—75). Large daisy flowers with white petals and golden centres in summer. Protect from strong winds.

CINERARIA

(*C. maritima*, Groundsel or Dusty Miller) E, FS or HS (60—30). Decorative, finely divided, silver foliage with brilliant yellow daisy-like flowers in late summer. Often listed as *Senecio cineraria*.

CONVOLVULUS

(*C. mauritanicus*, Ground Morning Glory) E, FS (30—100). Groundcover and rockery plant with trailing stems and soft, lavender-blue flowers in summer and early autumn. Good drainage.

CORTADERIA

(*C. sellowana*, Pampas Grass) E, FS or HS (150—100). Large tussocky grass plant with silver-white (sometimes buff or pink) plumes on long stems in late summer. Accent plant.

CYPERUS

(*C. papyrus*, Nile Grass) E, FS to SS (250—100). Broad clump of thick stems tipped with green, pendulous, thread-like flower spikes. *C. alternifolius* (Umbrella Plant) is similar but shorter (120 cm) with broader leaves. Moist conditions for both.

DIMORPHOTHECA

(*D. aurantica,* African Daisy) E, FS (60—100). Gay, daisy-like, orange flowers in spring and summer. *D. eklonis* (white flowers with purple centre). *D. barberia* (mauve flowers). Good drainage for all three.

ECHINOPS (Globe Thistle)

H, FS (150—100). Background plant with spiky, grey-green leaves and tall, steel-blue, globe-shaped flowers in summer. Good drainage.

ERIGERON (Fleabane)

E, FS or HS (30—50). Dwarf mat of foliage with small, daisy-like, mauve or lavender flowers in summer and autumn. Good drainage.

ERYNGIUM (Sea Holly)

H, FS or HS (60—40). Thistle-like plant similar to echinops with cone-shaped blue flowers in summer. Good drainage.

EUPHORBIA

(*E. wulfenii*, Yellow Spurge) E, HS (100—100). Compact, rounded clumps of foliage covered with yellow-green or lime-green flower bracts in winter and spring. Good drainage, frost susceptible.

FELICIA

(*F. ammelodes*, Blue Marguerite) E, FS or HS (50—50). Mounds of foliage with dainty, sky-blue, daisy-like flowers in most seasons. *F. angustifolia* (Kingfisher Daisy) is taller (100 cm), with lavender-blue flowers in spring. Good drainage for both.

FESTUCA

(*F. glauca*, Blue Fescue) E, FS or HS (25—25). Small attractive grass clumps with fine blue-grey leaves. Good drainage.

Filipendula

FILIPENDULA

H, HS or SS (100–40). Panicles of pink, mauve or white flowers with attractive, deep cut, fern-like foliage. Prefers moist conditions.

HELIANTHUS (Perennial Sunflower)

H, FS (100—60). Bushy plants with many flowering stems with small sunflowers in autumn. Good drainage.

HELLEBORUS (Winter Rose)

H, HS to SS (50—40). Heart-shaped leaves with lime-green, mauve or purple flowers in winter. Moist conditions.

HEUCHERA (Coral Bells)

E, FS or HS (50—25). Mounds of attractive, scalloped foliage with clusters of small, coral-pink bells in spring and summer. Moist conditions.

HOSTA

(*H. plantaginea*, Plantain Lily) H, FS to SS (25—40). Handsome, variegated foliage with small, bell-shaped, white or lilac flowers in summer. Moist conditions.

HYPERICUM

(*H. reptans*, Gold Flower) E, FS or HS (10—30). Trailing plant with attractive foliage and five-petalled, yellow flowers in late spring. Good drainage.

LAVENDULA

(*L. spica*, Lavender) E, FS or HS (50—30). Attractive clumps of silver-grey, aromatic foliage and lavender-blue flower spikes in early summer. Good drainage.

LIATRIS (Gay Feather)

H, FS or HS (60—30). Clumps of grass-like leaves with tall spikes of rose-purple flowers in summer. Good drainage.

LYCHNIS (Maltese Cross)

H, FS or HS (60—30). Compact clumps (evergreen in mild climates) with clusters of brick-red flowers on wiry stems in late spring.

MECONOPSIS (Blue Tibetan Poppy)

H, FS to SS (100—40). Beautiful large, crepe-petalled poppies in late spring. Good drainage but regular watering.

MESEMBRYANTHEMUM
(Pig Face or Ice Plant)

E, FS (25—75). Drought-resistant carpeting plant with succulent leaves and silky-petalled flowers in late spring. Very wide colour range. Good drainage.

MONARDA (Bergamot or Bee Balm)

H, HS or SS (60—50). Large clumps with many stems bearing pink, mauve or red flowers in summer. Moist conditions.

Liatris

NEPETA (Catmint)

H, FS or HS (25—25). Soft mounds of sage-green foliage (attractive to cats) with dainty sprays of lavender-blue flowers in late spring and summer. Good drainage.

NIEREMBERGIA (Blue Cup Flower)

E, HS or SS (20—25). Dwarf border or rockery plant with fine, lacy foliage and masses of blue, cup-shaped flowers in summer.

OPHIOPOGON (White Mondo Grass)

E, SS or S (25—25). Clumps of green and white, strap-like leaves with white or cream flowers in muscari-like spikes in summer.

PAEONIA (Peony Rose)

H, SS (75—50). Cool climate plant with large, fragrant, ruffled or single flowers in spring. Flower colours in shades of pink, mauve, red and white. Good drainage.

PENSTEMON (Beard Tongue)

H, FS (75—50). Bushy 'gloxinia-flowered' types have spikes of pink, mauve and red flowers in late spring and summer. Good drainage.

PHLOX

(*P. paniculata*, Perennial Phlox) H, FS to SS (60—50). Herbaceous border plant with flowers in every shade of pink, mauve, salmon and white. Flowers late spring to early autumn.

PHORMIUM (New Zealand Flax)

E, FS or HS (150—75). Background or accent plant with green, bronze, red-purple or variegated (green-white) strap-like leaves. Red or yellow flowers on tall spikes in summer. Tolerates tough conditions and salt air.

PLATYCODON (Balloon Flower)

H, HS or SS (60—60). Herbaceous border plant with balloon-like buds which open to star-shaped, violet-blue flowers in summer. Good drainage but regular watering.

STACHYS (Lamb's Ear)

E, FS to SS (20—30). Dwarf clumps of furry, silver-grey leaves for edging or ground cover. Spikes of small purple flowers in summer. Frost susceptible.

STOKESIA (Stokes' Aster)

H or E, FS to SS (60—40). Dense clumps with large lavender-blue flowers (similar to cornflowers) in late spring and summer. Good drainage but adaptable to climate and soil.

THALICTRUM (Lavender Shower)

H, HS or SS (150—50). Mass of fern-like foliage (similar to aquilegia) with tall, shivery stems bearing tiny, lavender flowers on lateral branches in summer.

VERONICA (Speedwell)

H, FS to HS (30—30). Rosettes of glossy, pointed leaves and pink, mauve or blue flower spikes in spring and summer. Good drainage.

VIOLET

(*Viola odorata*) E, HS or SS (20—40). Attractive clumps of rounded leaves with the most fragrant purple (occasionally white) flowers in winter and spring. Good drainage but regular watering. Best replanted every few years.

Note: Refer also to perennial plants described in Chapter 9 and Chapter 10.

VINES AND CREEPERS

'Climbers' is perhaps a better word than 'vines and creepers' to describe this group of plants because most, if not all of them, need support of some kind. Because climbing plants grow vertically, they are good space savers in the garden. They can be used to provide large masses of foliage as a background to shrubs and annuals. Many have attractive and fragrant flowers. Climbers are also useful to cover a bare wall, an ugly fence, an old stump or dead tree. They add more privacy to a corner of the garden. Deciduous vines and creepers growing over a trellis or pergola are ideal for shade in summer but let through winter sun.

TYPES OF CLIMBERS

Vines and creepers cling to their support in different ways. Some, like honeysuckle and wisteria, grow by twining their long stems around the available support—and other plants, if they are growing too close. Others develop thin tendrils or claws which grasp wire or timber to keep the stems

Wisteria is one of the most popular creepers

upright. Good examples are clematis, doxantha (Cat's Claw Bignonia) or vitis (Ornamental Grape). Another group of climbers—ficus, hedera (Ivy) and Virginia creeper—have aerial roots or sucker pads which cling to brick or stone walls, bark of trees and timber. The last group are the woody scramblers like bougainvillea which lean on their supports and bustle their way upwards.

SOIL PREPARATION AND PLANTING

Because most climbers will become a permanent garden feature, it is advisable to prepare the soil well by digging to spade depth and adding organic matter in the form of animal manure or compost. Sites against walls may contain builder's rubble which should be removed and replaced with good soil. Add a ration of mixed fertilizer at $\frac{1}{3}$ cup per square metre. Blood and bone based fertilizers and the slow-release fertilizers are also useful for a pre-planting ration. Climbers, like most plants, need good drainage so make provision for this, especially against brick or stone walls which are likely to trap water. It is a good idea to plant vines or creepers at least 20 cm from a solid wall—30–40 cm is even better if there is room.

Most climbers can be propagated from cuttings, layers or root divisions or you can buy started plants from garden stores or nurseries. Deciduous types are planted in winter but most evergreens are best planted in early autumn or spring. Some quick-growing vines—which are often most useful in a new garden—can be raised from seed. Those which can be raised in this way are antigonon, aristolochia (Dutchman's pipe), cobaea (cup-and-saucer vine), ipomoea (morning glory), mina (*Mina lobata*) and ornamental gourds.

SELECTING VINES AND CREEPERS

It is important to choose suitable plants for your climate and for each situation in the garden. Some creepers spread quickly by self-layering and tend to take over the whole garden if not kept in check. Akebia, doxantha, ficus, hedera, jasmine and wisteria all layer readily. Others, such as bougainvillea, clematis, hardenbergia and solandra, are very vigorous in warm, moist situations and may need cutting back

Dianthus are excellent rockery and edging plants

Calliopsis

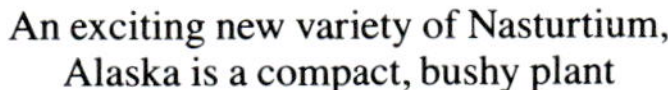

An exciting new variety of Nasturtium, Alaska is a compact, bushy plant

Pulcino is a new variety of Zinnia with a wide range of colours

The popular cyclamen creates a pictur[e] of fragile beauty during autumn an[d] winter

(*Top Left*) Sunflowers are easily grown and give a grand show. This variety is Bronze Hybrid

(*Left*) Sim carnations

(*Bottom left*) Bambini is a new dwarf viola suitable for growing in pots as well as in gardens

Marigold Flamenco is ideal for border[s] and narrow beds

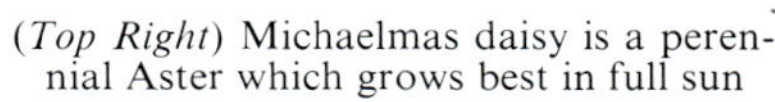

A pleasing blending of colours and forms

(*Top Right*) Michaelmas daisy is a perennial Aster which grows best in full sun

(*Right*) Painted daisy is ideal for special-effect and clump plantings

(*Bottom right*) The dainty Western Australian native *Pimelea sauveolens*

Eucalyptus ficifolia—the spectacular red flowering gum

Boronia serrulata (native rose) is a very fragrant and beautiful plant in bush or garden

(*Top left*) *Leptospermum squarrosum* (Port Jackson Tea Tree) flowers profusely in autumn and winter

(*Left*) *Melaleuca lateritia* is an outstanding native that flowers best in full sun

(*Bottom left*) *Stenocarpus sinuatus* is a spectacular slow-growing tree

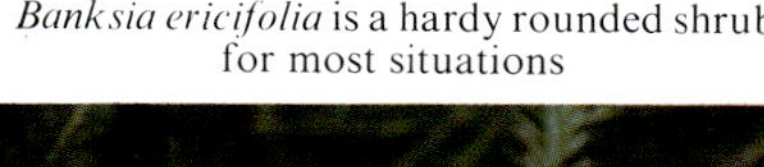

Banksia ericifolia is a hardy rounded shrub for most situations

regularly. A selection of climbing plants is given below with brief notes on the growth characteristics of each.

Abbreviations: D—deciduous, E—evergreen, SD—semi-deciduous, A—annual, FS—full sun, HS—half sun SS—semi-shade, S—shade

AKEBIA

(*A. quinata*) D, FS or HS. Vigorous, twining creeper with divided foliage and fragrant, umbrella-shaped, lime-green to purple flowers in spring. Prune back after flowering and shorten leaders if too vigorous. All climates.

ALLAMANDA

(*A. cathartica*) E, FS. Sprawling climber with large, golden, trumpet-like flowers from spring to autumn. Cut back in winter to keep it within bounds. Warm temperate, tropical.

ANTIGONON

(*A. leptopis*, Coral Creeper) D, FS. Vigorous, twining creeper with clusters of small, bright pink flowers in summer. Plants die down in winter but shoot again in spring. Sow seed in spring. Temperate, tropical.

ARISTOLOCHIA

(*A. elegans*, Dutchman's Pipe) E, HS or SS. Small perennial vine with heart-shaped leaves and quaint purple and white flowers in summer and autumn. Sow seed in spring. All climates but frost susceptible.

ASPARAGUS

(*A. plumosus*, Asparagus Fern) E, FS to SS. Dainty climber with fern-like leaves much in demand for bouquets and flower decoration. Also useful as a container plant with wall support or frame. All climates.

BAUHINIA

(*B. scandens*) E, FS or HS. Small tendril-climber with butterfly-like leaves and shell-pink flowers with red stamens. Temperate, tropical but frost susceptible.

BIGNONIA

Plants listed under the name Bignonia have been reclassified. Please refer to correct name as follows:

B. australis, see *Pandorea pandoreana*
B. cherere, see *Phaedranthus buccinatorius*
B. jasminoides, see *Pandorea jasminoides*
B. linleyi, see *Clytostoma callistegiodes*
B. tweediana, see *Doxantha ungis-cati*
B. venusta, see *Pyrostegia ignea*

BOUGAINVILLEA

E, FS. Large, woody, scrambling creepers with hard, hooked thorns. Plants need full sun and tolerate dry conditions. The showy flower bracts make a dazzling display in spring and summer. Needs strong support and hard pruning if plants are too vigorous. Warm temperate, tropical. The best known species are: *B. magnifica trailii* (bright purple), *B. laterita* (brick-red), *B. Mrs Butt* (port wine), *B. rosea* syn., *B. thomasii* (dusty pink).

CLEMATIS

D, FS or HS. Tendril climbers with showy flowers in white, pink, red or lavender-blue (depending on species and variety). Cool or temperate climates.

Clematis are showy climbers for cool and temperate climates

CLERODENDRON

(*C. splendens*) E, FS or HS. Vigorous creeper with glossy foliage and tightly packed clusters of scarlet flowers in summer. Needs warm, sheltered position. Temperate, tropical.

CLYTOSTOMA

(*C. callistegioides*) E, FS. Previously classified as *Bignonia lindleyi*. Vigorous creeper with masses of trumpet-shaped, lavender-blue flowers in summer. All climates but frost susceptible.

COBAEA

(*C. scandens*, Cup and Saucer Vine or Cathedral Bells) SD, FS or HS. Quick-growing, tendril vine often grown as an annual as plants become ragged in winter. Large bell-shaped flowers—green turning lilac or violet—in summer. Sow seed in early spring. Temperate, tropical.

DOXANTHA

(*D. ungis-cati*, Cat's Claw Creeper) E, FS or HS. Previously known as *Bignonia tweediana*. Vigorous creeper with claw-like tendrils and golden-yellow trumpet flowers in spring. All climates but frost susceptible.

FICUS

(*F. pumila*) E, FS to SS. Aerial root creeper with dense foliage if clipped regularly. Suitable for low walls, fences or natural covering for rocks. Temperate, tropical.

GELSEMIUM

(*G. sempervirens*, Caroline Jasmine) E, FS or HS. Small, attractive twining creeper with pointed leaves and fragrant, bell-shaped, yellow flowers in late winter and spring. All climates.

HARDENBERGIA

(*H. violacea*, Australian Sarsparilla) E, HS to SS. Native, twining creeper with lavender-blue to violet, pea-shaped flowers in spring. All climates except coldest regions.

HEDERA (Ivy)

E, FS to S. Foliage creepers with aerial roots. Very adaptable to climate, soil, sun and shade. Useful for covering fences, walls, banks or tree stumps or as ground

Golden yellow flowers of doxantha crown a fence

cover. Most popular species are *H. canariensis variegata* with large, glossy, green and cream leaves and *H. helix* (English Ivy) with smaller leaves, but there are many varieties with deep greens, silver variegated, yellow variegated and yellow centred leaves. All climates.

HOYA

(*H. carnosa*, Wax Plant) E, HS or SS. Small twining plant useful for trellis or in containers with wall support or frame in a warm, sheltered position. Clusters of pink, star-shaped flowers in summer. New buds arise from old flower spurs. Temperate, tropical.

IPOMOEA (Morning Glory)

E, FS or HS. Vigorous vines for quickly covering walls, fences, banks and batters. Mixed packets of seed are available with a formula mixture of a wide range of bright colours. Early Rose has large rose-pink flowers in summer. All climates except coldest regions.

JASMINIUM (Jasmine)

E, HS or SS. Vigorous, twining or scrambling vine with sweetly-scented flowers in spring or summer (depending on species). Temperate, tropical but frost susceptible. Most widely grown species are: *J. polyanthum* with rosy-pink buds and small starry-white flowers in early spring; *J. grandiflorum* with larger white flowers produced sparsely throughout the year; *J. nitidum* with very large white flowers throughout the year but best suited to tropical climates.

LONICERA (Honeysuckle)

SD or E, FS to SS. Vigorous twining or scrambling creeper which is very adaptable to climate and soil. Fragrant spring flowers are creamy-white, yellow, pink or red depending on species. Plants must be pruned after flowering to keep them within bounds. All climates.

MANDEVILLA

(*M. sauveoleus*, Chilean Jasmine) SD, FS or HS. Handsome vine with heart-shaped leaves and clusters of fragrant, white, trumpet-shaped flowers in summer and autumn. Prune back after flowering to prevent overcrowding. Warm temperate, tropical, but frost susceptible.

MINA

(*M. lobata*) A, FS or HS. Quick-growing vine to 2 m tall, useful for short-term cover of trellis or fence. Vines are covered with sprays of red and yellow flowers in late summer and autumn. Sow seeds in spring. All climates but frost susceptible.

MONSTERA

(*M. deliciosa*, Fruit Salad Plant) E, FS to S. Vigorous climber with aerial roots which cling to masonry, and with large, round but deeply divided leaves. Also useful as indoor plant in large pots or tubs. Arum-type yellow flowers develop into long, cylindrical fruit with a delicious flavour. Warm temperate, tropical climates.

ORNAMENTAL GOURD

A, FS. Quick-growing vines closely related to cucumbers and pumpkins. Ripe fruit, in curious shapes and colours, are dried for indoor decoration or as ornaments. Sow seeds in spring or early summer and grow plants in the same way as trellised cucumbers. All climates but frost susceptible.

Mina lobata

PANDOREA

E, FS or HS. *P. jasminoides* (previously *Bignonia jasminoides*) is an attractive creeper with glossy leaves and white, trumpet-like flowers flushed rosy-purple in spring. *P. pandoreana* (previously *Bignonia australis*) is commonly known as Wonga-Wonga vine. A similar creeper but the white, tubular flowers are in clusters. Warm temperate, tropical climates for both species.

PARTHENOSISSUS

(*P. quinquefolia*, Virginian Creeper or Boston Ivy) D, FS or HS. Previously known as *Ampolepsis quinquefolia*. Vigorous self-clinging creeper, grown for its umbrella-shaped leaves which colour brilliantly in autumn. Ideal for brick walls and stonework. All climates.

PASSIFLORA (Passionfruit)

(*See Chapter 17.*)

PELARGONIUM (Ivy Geranium)

E, FS or HS. Geraniums with long stems for training over walls and fences or for trailing over banks or batters. Also for hanging baskets. Ivy-shaped leaves and white, pink or red flowers in spring and summer. All climates.

PHAEDRANTHUS

(*P. buccinatorius*, Mexican Blood Trumpet) E, FS or HS. Previously known as *Bignonia cherere*. Vigorous creeper with strong stems and clusters of orange-red trumpet flowers in spring and summer. Warm temperate, tropical but frost susceptible.

PYROSTEGIA

(*P. ignea*, Flame Vine) E, FS. Previously known as *Bignonia venusta*. Vigorous, adaptable creeper for covering fences, trellis or outbuildings with brilliant orange, tubular flowers in late winter and spring. Temperate, tropical but tolerates light frosts.

ROSA

(*R. wichuriana*, Rambler Rose) SD, FS. Spectacular spring-flowering roses for trellis or pergola or for weeping standards. All climates. For cultivation notes on roses see Chapter 15.

SOLANDRA
(Cup of Gold or Hawaiian Lily)

E, FS to SS. Sprawling, rampant vine which needs solid support and regular cutting back to keep it under control. Large, creamy-yellow trumpet flowers to 25 cm diameter in spring. Warm temperate, tropical but frost susceptible.

STEPHANOTIS (Madagascar Jasmine)

E, FS or HS. Handsome creeper which needs light support and training. Fragrant, white, trumpet-like flowers in late summer are favourites in bridal bouquets. Warm, sheltered position. Temperate, tropical.

THUNBERGIA

E, FS. Several species are attractive climbers which flower intermittently throughout the year. The best known is *T. alata* (Black-Eyed Susan) with orange petals and black centres. *T. gibsonii* (Golden Glory Vine) has masses of orange flowers. *T. grandiflora* (Sky Flower) has pale-blue flowers. All species prefer a warm, sheltered position. Temperate, tropical but frost susceptible.

TRACHELOSPERMUM (Star Jasmine)

E, FS to SS. Sometimes known as *Rhynchospermum*, this useful creeper has rich, glossy foliage and fragrant, lace-like, white flowers in spring and summer. All climates except coldest regions.

VITIS (Ornamental Grape)

D, FS. Vigorous tendril vine with cool, green summer foliage and brilliant autumn colours. Ideal for training on trellis or pergola, especially to let through winter sun. All climates except hot, tropical regions.

WISTERIA

D, FS to HS. One of the most popular spring-flowering creepers with pendulous clusters of pea-like flowers in white, lavender and lilac (depending on species or variety). Vines may be slow to establish but become very vigorous and long-lived so need strong support. Ideal for covering a pergola for summer shade, winter sun and spring blossom. All climates.

Chapter 12

TREES AND SHRUBS

Trees and shrubs can make an important contribution to the general outline and colour of your garden. They provide an easy-care background for annuals, bulbs and perennials which change with the seasons. They can ensure privacy and protection against winds, screen unsightly views and diminish traffic noise. They offer a leafy canopy on hot summer days and deciduous species let warm winter sun penetrate to indoor and outdoor living areas. From the hundreds of kinds of trees and shrubs available, it is essential to choose those which suit your climate, situation and soil. Plant form, shape and height when mature is very important in deciding their position in the garden, either as single specimens or in groups. Lists of popular trees and shrubs are given in this chapter as a guide but the best plan—especially for new gardeners—is to visit a leading nursery where the plants are displayed. Alongside each kind should be a label which gives the size of the plant when fully grown.

PLANTING

Some plant nurseries still sell open-ground plants, but these are mostly deciduous kinds—roses, flowering fruit trees and others which lose their leaves in winter. Most shrubs and trees are sold in plastic pots or plastic bags. This has made year-round planting possible although the cooler months of autumn or spring are the best. Containers of different sizes also offer the choice of small, semi-advanced, or advanced specimens.

Before planting, check the position to make sure it is suitable for the plant when fully grown. Avoid sites close to sewerage or drainage lines which could be invaded by strong-rooting species and cause future problems—especially with trees like willows, poplars and rubber trees. Also avoid planting trees under overhead wires where they may eventually be dangerous or cause expense and trouble in keeping them lopped. Do any drainage work necessary beforehand as many shrubs and trees resent wet, soggy conditions. (See Chapter 3.) If the soil is a very heavy clay, organic matter in the form of animal manure, leaf mould or compost should be incorporated to improve structure. The addition of coarse sand or gypsum (calcium sulphate) also makes heavy soils more crumbly.

The soil in the container and the soil in the garden should be dark damp (but not wet) at planting time. Dig a wide but shallow hole—about 60 cm wider than the container but not much deeper. Roots will travel faster through soil which has been well broken up. The soil in the container is often a different texture to the garden soil, which may create an 'interface' problem, that is, the two soils fail to merge, causing a root barrier. Add compost or peat moss to the soil removed from the hole and mix it well. On heavy, puggy, clay soils the addition of coarse sand is also recommended.

Now fill the base of the hole with the soil mixture to bring the root-ball of the plant level with the top of the hole—so that the plant is not deeper in the soil than it was in the container. Take the plant from the container and set it at the correct level. If the plant is slightly root-bound (roots tangled round and round), tease the roots away gently, spread them out and cover with soil progressively. Alternatively, hose the root-ball gently to wash away the outside soil so that roots can be separated and covered. Add the rest of the soil gradually, firming it down to exclude air spaces, but avoiding heavy pressure by treading. The leftover soil is used to form a raised ring around the plant. Water well and then spread grass clippings, leaf mould or fibrous compost for a mulch. Keep the mulch away from the stem as it might encourage collar rot or other root rots. The mulch will prevent the surface soil from caking and conserve moisture. Where stakes are considered necessary, these should be driven in outside the root-ball to avoid damage. Trees or shrubs should be tied to the stake with plant ties or lengths of old rag or nylon stocking. Do not use wire which will cut into the stem.

Water thoroughly when needed, making sure that the root-ball is wetted because this may dry out more quickly than the surrounding soil. Avoid using strong powdered or granular fertilizers, either at planting or immediately afterwards. Slow-acting blood and bone based fertilizers such as Gro-Plus for Roses or Gro-Plus for Camellias and Azaleas are safe to scatter on the mulched surface. Alternatively, Agriform Tree Tablets may be placed adjacent to the root-ball prior to filling in around the plant, or slow-release fertilizers like Osmocote or Mag-Amp scratched into the soil surface.

PRUNING

Many trees and shrubs are more or less self-shaping, so pruning is rarely necessary. Others are improved by regular or occasional pruning. Briefly the objects of pruning are:

1. To shape the plant—especially young shrubs and trees—and ensure a balanced framework for the future. For hedges the aim is to promote a dense growth down to ground level.
2. To reduce competition by thinning out crowded growth, reduce competition with plants nearby or prevent obstruction of pathways and light to windows.
3. To stimulate new growth and encourage flowers (and fruits). With flowering shrubs, the removal of spent blooms prevents fruits and seeds from forming and so directs plant energy to further flowering or vegetative growth.
4. To remove diseased or dead wood and cut away unwanted branches such as those which have reverted from a variegated leaf to a plain leaf or from a dwarf form to a tall form.
5. To remove suckers from root stocks on which some shrubs are grafted.

Only sharp tools (secateurs or pruning saws) should be used, because jagged cuts provide a resting place for disease spores and a greater area for the plant to heal or form a callus. (See Chapter 6.) Large cuts should be treated with water-resistant paint or grafting mastic to prevent infection. Always make cuts just above a bud or level with the joint to a larger branch to protect large areas from dieback. The shaping of a tree, shrub or hedge should begin as soon as possible. If early pruning is neglected it is more difficult to achieve a balanced shape later. Shears are suitable for clipping or trimming hedges. While it is not possible to control the exact position of the cut on each stem, most hedge plants develop laterals freely and do not suffer from dieback at the cut ends.

Most slow-growing trees and shrubs need little pruning. Good examples are azalea (*Rhododendron*), camellia, daphne, gardenia and frangipani (*Plumeria*). As a general rule, most flowering shrubs and some flowering trees are pruned after they have flowered, although deciduous types fall into two groups. Those which flower (usually summer or early autumn) on current season's growth (new wood) are pruned in winter. Examples are cassia, hydrangea and lagerstroemia. Those which flower in spring on last season's growth (old wood) are pruned immediately after flowering. Pruning in winter would remove much of the spring blossom. Good examples are flowering peach and flowering almond. Other spring flowering trees and shrubs are best pruned back (if at all) after flowering. Multi-stemmed shrubs like Japanese bamboo (*Nandina*), may (*Spirea*) and barberry (*Berberis*) require little pruning except to remove older canes periodically at ground level to promote new growth.

REPAIRING DAMAGED TREES AND SHRUBS

Very often, a broken branch or stem, if not completely severed, can be repaired by fitting the broken section together using a splint (a piece of dowel stick for small branches or stouter hardwood for larger limbs) and binding the break and splint tightly together with plastic or adhesive tape. Heavy branches may be propped with stakes from below, or better still, secure the branch to the main trunk or stem with rope or garden stakes tied at each point. Pruning the foliage from the damaged branch will reduce its weight and prevent further possible wind damage. Closely-bound branches should mend in three or four months. If a break is beyond repair, saw the branch off cleanly. Make the first cut underneath to about halfway or until the saw binds. Make the second cut about 5 cm in front of the first and take it through until the branch falls. Now make a third cut right through to remove the stub, which is easily handled. In some cases it may be best to remove the branch entirely, cutting close to the trunk. Apply a waterproof paint or grafting mastic to the cut surface to prevent infection and help heal the wound.

Trees which have been blown over can often be salvaged because usually only half the roots are broken and exposed. Trim the broken roots cleanly and paint them with grafting mastic. Lift the tree upright while the soil is still moist, using one or more forked poles for extra leverage. Move them in closer to support the tree as it is raised. A permanent support of a guy wire (fencing wire or clothes-line wire) fixed to one or more stakes on the windward side should be sufficient. Thread the wire through a short piece of old garden hose to prevent damage where it circles the trunk of the tree.

Shrubs and small trees can be moved successfully if handled carefully when dormant. Winter is the safest time for most kinds but warmth-loving shrubs like gardenia and hibiscus are often best moved in very early spring. Dig around the plants so that a wide, rather shallow root-ball is taken. Slide or juggle the plant on to a sack or piece of canvas to carry or drag it to the new site. Take care that the root-ball remains intact.

SELECTED TREES AND SHRUBS

The height to which a tree or shrub grows, whether it is evergreen or deciduous and whether it is grown for foliage, flowers or fruits are important factors in its selection. A summary of the most popular trees and shrubs is given in this chapter. It is by no means complete but may be a useful reference for the new gardener. Further information can be gathered from nursery catalogues or one of the many good reference books which deal specifically with this aspect of gardening. The most practical way of learning about trees and shrubs is to see them actually growing in your own district. Many leading nurseries employ trained horticulturists who can give valuable information to help you make your final selection. In the summary which follows, plants are listed in alphabetical order by their botanical name, followed by their common name.

Abbreviations: E—evergreen, D—deciduous, SD—semi-deciduous, FS—full sun, HS—half sun, SS—semi-shade, S—shade, H—height, W—width. (Measurements shown in metres (m) or centimetres (cm).)

ABELIA (Glossy Abelia)

E, FS or HS, H 2–3 m, W 1–2 m. Arching shrub with small leaves and white to pink, bell-shaped flowers in summer and autumn. All climates except coldest districts. Makes good specimen or hedge plant.

ABUTILON (Japanese Lantern)

E, FS to SS, H 1–2 m, W 1–2 m. Attractive, usually mottled, foliage and pendulous, hibiscus-like flowers in white, yellow, orange or pink (depending on variety) in summer. All climates except coldest districts.

ACACIA

(*See Chapter on Australian Native Plants*)

ACALYPHA (Fiji Fire Plant)

SD, FS or HS, H 2–3 m, W 1–2 m. Attractive, large, multi-coloured foliage plant with insignificant flowers. Another species has green leaves and drooping red flower spikes. Both are frost susceptible and need warm temperate or tropical conditions.

ACER (Maple)

D, FS to SS, H 2–18 m, W 1–15 m. Attractive trees with spectacular autumn foliage for cold and temperate climates. Japanese maples (*A. palmatum*) have many leaf forms and colours (some variegated) and are usually small to medium size trees. Box elder (*A. negundo*) and Norway maple (*A. platanoides*) are larger trees for cold climates.

AGONIS

(*See Chapter on Australian Native Plants*)

The ever-popular azalea

ARBUTUS (Irish Strawberry Tree)

E, FS or HS, H 5–6 m, W 3–4 m. Rounded, densely-foliaged tree with masses of small white blossoms from summer to winter followed by large, rough, multi-coloured (green, yellow, orange) berries. Very adaptable to climate and soil except in tropical areas.

ARDESIA (Coral Berry)

E, HS to S, H 60–90 cm, W 30–60 cm. An attractive small shrub with clusters of white or pink flowers in spring followed by brilliant red berries in autumn and winter. Very adaptable to climate but prefers cool, shaded position.

AUCUBA (Gold Dust Tree)

E, HS to S, H 1–2 m, W 1 m. Large, oval, glossy leaves flecked with gold; may produce red berries when both male and female plants are grown together. Prefers damp, cool, shady position but may scorch in hot, dry climates. Can also be grown as indoor plant.

AZALEA

E or SD, HS to S, H 30 cm–3 m, W 30 cm–2 m. The best known and most useful of spring flowering shrubs. There are several species from dwarf (*A. kurume*) to large *magnifica* or *splendens* types (*A. indica*) with hundreds of hybrids and named varieties in single and double flowers in almost every colour. The evergreen types will grow in almost any climate except the tropics but deciduous kinds prefer cooler conditions. They must have an acid soil and will not grow in alkaline soil. They need good drainage but the shallow roots need to be cool, moist and shaded. Azaleas make excellent plants for pots or tubs.

BACKHOUSIA

(*See Chapter on Australia Native Plants*)

BAECKEA

(*See Chapter on Australian Native Plants*)

BANKSIA

(*See Chapter on Australian Native Plants*)

BAUHINIA (Butterfly Tree)

E or SD, FS or HS, H 5–6 m, W 5–6 m. Spreading small tree or large shrub with twin, butterfly-like leaves and rosy purple flowers in spring. It is deciduous except in warm climates and is best suited to temperate or tropical areas, but will tolerate moderate frosts. There are several other shrub or tree species and a creeper (*B. scandens*).

BELOPERONE

(*See Drejerella.*)

BERBERIS (Barberry)

E or D, FS or HS, H 1–2 m, W 1–2 m. Small, compact, usually spiny, shrubs with yellow flowers in spring and red berries in autumn. Most deciduous species have brilliant autumn foliage. Very adaptable but prefer mild temperate or cold climates. Those with purple-bronze foliage (*B. thunbergii atropurpurea*) make good accent plants. All can be used as hedge plants.

BETULA (Birch)

D, FS or HS, H 6–9 m, W 3–6 m. Slender trees with graceful foliage and pendulous catkins in spring. Brilliant autumn foliage. Silver birch (*B. verrucosa*) with silvery-white bark is the best known. Adapted to cool temperate and cold climates.

BORONIA

(*See Chapter on Australian Native Plants*)

BRACHYCHITON

(*See Chapter on Australian Native Plants*)

BRASSAIA

(*See under Schefflera in Chapter on Australian Native Plants.*)

BROWALLIA

(*See Streptosolen.*)

BRUNSFELSIA (Brazil Rain Tree)

E, HS or SS, H 2–3 m, W 2–3 m. This handsome shrub is also called Yesterday, Today and Tomorrow because the fragrant flowers in spring open deep blue, fade to lavender and then to white on successive days. Adaptable to all climates except coldest districts and prefers well drained soil in a semi-shaded situation.

BUDDLEIA

(Summer Lilac or Butterfly Bush)

SD, FS or HS, H 2–3 m, W 2 m. Fast-growing shrub with long sprays of slightly fragrant, lilac or purple flowers in spring and summer. Adaptable to all climates—evergreen in warm districts and deciduous where colder.

BUXUS (Box)

E, FS or HS, H 2 m, W 1 m. Attractive, tidy, small tree or shrub with small shiny foliage. Often used as a low hedge or as accent plants in formal gardens. Very adaptable to all climates.

CALLISTEMON

(*See Chapter on Australian Native Plants*)

CALODENDRON (Cape Chestnut)

E, FS, H 9–15 m, W 5–6 m. A handsome, well-shaped tree with showy clusters of pink or mauve, orchid-like flowers through summer. All climates except cold districts. Frost susceptible and may need protection when young.

CAMELLIA

E, HS to S, H 3–5 m, W 1–3 m. Camellias compete with azaleas for the most popular shrub or tree. The most widely grown is *C. japonica* which flowers in winter and early spring. There are hundreds of named varieties with single, semi-double or double flowers which range in colour from pure white through pinks and mauves to deep red. Another species is *C. sasanqua* with mostly single flowers in late autumn and winter; it is useful as a background plant or trimmed as a tall hedge. Like azaleas, camellias will not tolerate alkaline soils and need good drainage, plenty of organic matter and a cool root area. Camellias prefer semi-shade or shade as strong sunlight can scorch the blooms. They make excellent tub specimens.

CASSIA

E, FS or HS, H 1–6 m, W 1–3 m. There are several species of these showy shrubs or small trees with yellow pea-like flowers in spring, summer or autumn. Buttercup Tree (*C. corymbosa*) is the most widely grown with masses of yellow blooms in spring. Most species are fast growing and adaptable to all climates except cold districts. A dwarf Australian species, *C. artemisiodes*, is very suitable for hot, dry regions.

CERATOPETALUM

(*See Chapter on Australian Native Plants*)

CERATOSTIGMA

D, FS, H 3 m, W 3 m. Small shrubs with pale green, bronze-tinted leaves and bright blue flowers in summer. Foliage colours in autumn. Cut back hard in winter to promote new growth and flowers. Very adaptable plant to most climates.

CESTRUM

E, FS or HS, H 2–3 m, W 1–2 m. Fast-growing shrubs with showy, tubular flowers in light green, orange or red according to species. The best known is Night Jessamine (*C. nocturnum*) with strongly-perfurmed flowers on warm nights in summer. Temperate and tropical climates.

CHAMAELAUCIUM

(*See Chapter on Australian Native Plants*)

CHORIZEMA

(*See Chapter on Australian Native Plants*)

CINNAMOMON (Camphor Laurel)

E, FS or HS, H 15 m, W 9 m. Vigorous, self-shaping tree with dense foliage with camphor odour and green berries turning black in summer. Takes a lot of space. All climates except coldest districts.

CITHAREXYLUM (Fiddlewood)

E, FS or HS, H 6–9 m, W 5 m. Fast-growing tree with rich green foliage with copper tones in autumn and small sprays of white flowers in summer. Excellent shade tree or as hedge or windbreak in groups. Temperate and subtropical climates.

COLEONEMA (Pink Diosma)

E, FS or. HS, H 1–2 m, W up to 1 m. Compact, delicate foliage with white or pink star-shaped flowers in spring. Good dwarf shrub for accent and rock gardens. Prune back after flowering. All climates.

COPROSMA (Looking-glass Plant)

E, FS to S, H 1–2 m, W 1–2 m. Attractive shrub with glossy green or variegated foliage. Very adaptable and tolerates sea spray. All but coldest climates.

COTONEASTER

E or D, FS to SS, H 1–4 m, W 1–4 m. Many species and varieties from prostrate spill-over shrubs to small trees, all of which have attractive clusters of orange or red berries. Mostly evergreen but some deciduous. Excellent garden shrubs, also useful for hedges and espalier training. Small types for rockeries. All climates.

CRATAEGUS (Hawthorn)

D, FS to SS, H 4–6 m, W 2–3 m. Thorny shrubs or small trees with white or pink rose-like flowers in spring followed by brilliant yellow, orange or red berries according to species. Adapted to all climates except very hot districts.

CUPHEA (Cigar Plant)

E, FS to SS, H 30–60 cm, W 60 cm. Small spreading shrubs with red tubular flowers tipped with ash grey. Excellent for rockeries. Adapted to most climates.

DAPHNE (Sweet Daphne)

E, HS to SS, H 1 m, W 1 m. Charming dwarf shrub with glossy foliage and highly perfumed, waxy, white, pink or red flowers in winter and early spring. Needs good drainage but a mulch to keep roots cool and moist. Prefers morning sun or semi-shade. Temperate or cool climates.

DIOSMA

(*See Coleonema*)

DEUTZIA (Wedding Bells)

D, HS or SS, H 2–3 m, W 1.5 m. Attractive shrubs with long canes and clusters of white or pink flowers in late spring and early summer. Cut back after flowering. All climates except very hot districts.

DREJERELLA

(syn. *Beloperone*, Shrimp Plant)

E, FS or HS, H 1 m, W 1 m. Small soft-wood shrub with yellow or pink, over-lapping, shell-like bracts suggesting a prawn or shrimp. Very adaptable to climate but prefers warm sheltered position.

EPACRIS
(*See Chapter on Australian Native Plants*)

ERICA (Heath)
E, FS, H up to 2 m, W up to 1 m. Heath-like plants in many varieties and forms with needle-like leaves and masses of tubular or bell-shaped flowers in shades of yellow, orange, pink or mauve. They need well-drained, slightly acid soils, as for azaleas, to which they are related. Best in cool temperate or cold climates.

ERIOSTEMON
(*See Chapter on Australian Native Plants*)

ESCALLONIA
E, FS to S, H 2–3 m, W 1–2 m. Glossy foliage shrubs with rose-pink flowers in late spring and summer. Cut back after flowering. All climates except tropical and very cold.

EUCALYPTUS
(*See Chapter on Australian Native Plants*)

EUGENIA
(*See Chapter on Australian Native Plants*)

EUONYMUS
E, FS or HS, H 1–2 m, W up to 1 m. A number of species of shrubs, mostly with attractive variegated foliage, may be clipped to any shape or size or made into a hedge. Also attractive tub specimens. All climates.

EUPHORBIA (Poinsettia)
D, FS or HS, H 3–5 m, W 2–3 m. Poinsettia (*E. pulcherrima*) is the most popular shrub of this genus. The tiny orange flowers are surrounded by showy bracts in deep crimson, pink or yellow. Best suited to subtropical and tropical climates but also in warm, sheltered situation in temperate zones. Frost susceptible.

EXOCHORDA (Pear Bush)
D, FS to SS, H 2–3 m, W 2–3 m. Attractive shrub with pale green foliage and pearly white flowers like apple blossom in spring. Prune lightly after flowering. Temperate and cool climates.

Variegated form of Euonymus

FAGUS (Beech)
D, HS or SS, H 9–18 m, W 9–12 m. Large but slow-growing deciduous tree. For cool temperate and cold climates only. Magnificent foliage trees with many varied forms and colours of leaves. Silver Beech and Copper Beech are the most widely grown.

FICUS (Fig)
E, FS or HS, H 9–15 m, W 6–15 m. Many species of medium-sized to very large trees (Port Jackson Fig and Moreton Bay Fig). Small species are India Rubber Plant (*F. elastica*) and Weeping Fig (*F. hillii*) with dense, pendulous foliage. *F. pumila* is an evergreen creeper (see Chapter 11).

FRANGIPANI
(See *Plumeria*)

FRAXINUS (Ash)

D, FS or HS, H 6–12 m, W 4–6 m. Fast-growing, attractive trees with brilliant autumn foliage for cool temperate and cold climates. Golden Ash, Desert Ash and Claret Ash are the most popular species.

FUCHSIA

E, SS or S, H 1–2 m, W 30–60 cm. Decorative softwood shrubs which flower from late spring to early winter. Flowers are pendulous with a tubular corolla surrounded by sepals in contrasting colours. There are hundreds of named varieties. All prefer semi-shade and a friable, well-structured soil. Most are suitable for pots, tubs and hanging baskets. Most climates except very cold districts.

GARDENIA

E, FS or HS, H 1–2 m, W up to 1 m. Attractive shrubs or small trees with dark, glossy foliage and highly fragrant, waxy, white flowers from late spring through summer. If necessary prune back plants in winter. They need a sheltered, sunny position and do well in all climates except very cold districts.

GARRYA (Silk Tassel or Curtain Bush)

E, SS to S, H 3 m, W 2–3 m. Dense-foliaged shrub with pendulous clusters of greyish-yellow catkins in winter and early spring. Prefers semi-shade or shade and does best in cool temperate and cold climates.

Hibiscus

GINKO (Maiden Hair Tree)

D, FS or HS, H 9–12 m, W 4–6 m. Pale green, two-lobed leaves turning golden yellow in autumn. Cool temperate and cold climates.

GORDONIA

E, FS or HS, H 3–4 m, W 2–3 m. Tall, fast-growing shrub with glossy foliage and large single white flowers with prominent yellow centre from autumn to spring. Flowers fall intact to form a floral carpet beneath. Best in temperate climates but have some cold tolerance.

GREVILLEA

(*See Chapter on Australian Native Plants*)

HAKEA

(*See Chapter on Australian Native Plants*)

HEBE (Veronica)

E, FS to SS, H 1–2 m, W up to 1 m. Small, compact shrubs with dense foliage and showy racemes of white, blue or purple flowers in winter and spring. Plants can be pruned as hedges and are suitable for rockery work. Best suited to temperate and cool climates.

HELIOTROPIUM (Cherry Pie)

E, FS to SS, H 60–90 cm, W 90 cm. Small, fast-growing shrubs with attractive, pale or dark green, wrinkled leaves and fragrant, lavender or purple flower heads in summer. Cut back after flowering. A good plant for a rockery. Adapted to most climates but frost susceptible.

HIBISCUS

E or D, FS or HS, H 2–4 m, W 1–2 m. One of the best summer-flowering shrubs for warm temperate and tropical climates. There are many species (mostly evergreen) and hundreds of named varieties. The magnificent large flowers (some to 20 cm across) are single or double and range in colour from pure white through lemon, yellow, gold, orange, pink, red and maroon. All species are frost susceptible.

In cooler temperate climates select a sunny, sheltered position.

HYDRANGEA

D, SS or S, H 0.5–3 m, W 0.5–2 m. Popular deciduous shrubs for shade and moist situations with large, showy flower heads in summer. Most flowers are in shades of pink or blue but colour can change depending on soil reaction (pH)—alkaline soil for pink, acid soil for blue. In some varieties the flowers are white or greenish and do not change colour with soil. Prune back plants when dormant in winter. All climates.

HYPERICUM

E, FS or HS, H 60–120 cm, W 60–120 cm. Small, twiggy shrubs with deep yellow, buttercup-like flowers in summer. Useful rock garden plants. Adapted to most climates.

ILEX (Holly)

E, FS to SS, H 3–5 m, W 1–2 m. Handsome shrubs or small trees with glossy foliage and white flowers followed by brilliant red berries in winter. Plants may be trimmed to shape and make a dense hedge. Adapted to cool temperate and cold climates only.

JACARANDA

Semi-D, FS or HS, H 9–12 m, W 9–12m. Graceful tree with fine, fern-like leaves which turn golden bronze in winter and with masses of lavender-blue flowers in late spring and early summer. Adapted to all climates except very cold or extreme tropical regions. Young trees may need protection from frost until established.

KOLKWITZIA (Chinese Beauty Bush)

D, HS or SS, H 2 m, W 1.5 m. Attractive shrubs with long arching canes covered with pale pink, trumpet-shaped flowers in spring. Adapted to cool temperate and cold climates.

LAGERSTROEMIA (Crepe Myrtle)

D, FS or HS, H 2–3 m, W 1–2 m. Large shrubs or small trees, often with gold autumn leaves, and showy clusters of pink, mauve or carmine flowers in late summer. Prune hard in winter. All climates except coldest districts.

LEPTOSPERMUM

(*See Chapter on Australian Native Plants*)

LESCHENAULTIA

(*See Chapter on Australian Native Plants*)

LIQUIDAMBER (Sweet Gum)

D, FS or HS, H 6–12 m, W 4–9 m. Tall, conical trees with maple-like leaves in brilliant autumn colours of yellow, orange, red and purple. Adaptable tree but best suited to temperate and cold climates.

LIRIODENDRON (Tulip Tree)

D, FS or HS, H 9–12 m, W 4–6 m. Attractive large tree with fiddle-shaped leaves turning bright yellow in autumn. Flowers are lime-green and tulip shaped. Cool temperate and cold climates only.

LOROPETALUM (Chinese Fringe Flower)

E, HS or SS, H 1–2 m, W 1–2 m. Graceful shrub with rounded leaves and spidery, cream flowers on the outside stems in spring. Temperate and cold climates, but frost susceptible.

LUCULIA

E, HS or SS, H 2–3 m, W 2–3 m. Attractive shrub with rounded heads of fragrant pink flowers in early winter. Cut back after flowering. Temperate climates but frost susceptible.

MAGNOLIA

E or D, FS or HS, H 5–9 m, W 3–6 m. Several species but the deciduous tree (*M. soulangeana*) is the most widely grown. The large, tulip-shaped flowers in white, pink, mauve or purple appear in early spring before the foliage. Suited to temperate or cool climates. The evergreen tree (*M. grandiflora*) has large, fragrant, waxy-white flowers in summer and can be grown in all but the very coldest climates.

MALUS (Crabapple)

D, FS or HS, H 3–5 m, W 3–5 m. One of the most delightful spring flowering trees with blossoms in white, pink or red followed by small, multi-coloured apples used for decoration or apple jam. Prune trees to shape and size. All climates except warm tropical regions.

Magnolia

MELALEUCA
(*See Chapter on Australian Native Plants*)

METROSIDEROS
(Pohutukawa or New Zealand Christmas Tree)
E, FS or HS, H 5–6 m, W 4–5 m. Rounded, self-shaping large shrub or small tree with shiny leaves (grey underneath) and brilliant red, brush-like flowers in summer. Excellent shrubs for exposed seaside conditions and tolerate salt spray, dust and city smog. Temperate and coastal climates only.

MURRAYA
(*See Chapter on Australian Native Plants*)

NANDINA (Japanese Bamboo)
E, FS to SS, H 1–2 m, W 1 m. Multi-stemmed shrub with finely divided foliage, small cream flowers and red berries. Attractive autumn foliage, especially in dwarf variety. (*N. domestica nana*). All climates.

NERIUM (Oleander)
E, FS to SS, H 3–5 m, W 1–2 m. Adaptable shrubs with lance-shaped, rather leathery, leaves and masses of single or double flowers (white, cream, yellow, pink, crimson) from early summer to late autumn. All but cold climates.

NYSSA
D, FS or HS, H 6–9 m, W 4–6 m. Shapely tree with large leaves on horizontal branches. Brilliant autumn colouring. Mild temperate and cold climates.

PARROTIA
D, FS or HS, H 6–9 m, W 4–6 m. Handsome tree with attractive foliage in both spring and autumn. Also reddish-brown flowers in late winter. Mild temperate and cold climates.

PHILADELPHUS (Mock Orange)
D, HS or SS, H 2–3 m, W 2 m. Cany shrubs with sweetly scented white flowers in late spring. Prune back after flowering. All but tropical climates.

PHOTINIA
E, FS to SS, H 2–5 m, W 1–3 m. Foliage plants with brilliant red new growth, often used as hedge plants. All climates except tropics and hot inland areas.

PIMELIA
(*See Chapter on Australian Native Plants*)

PIERIS (Lily-of-the-Valley Shrub)
E, HS to SS, H 1–2 m, W 1–2 m. Small attractive shrubs with sprays of cream flowers in spring. Well drained soil and cool, sheltered position, similar to azaleas. Cool temperate and cold climates.

PITTOSPORUM
E, FS to SS, H 2–6 m, W 1–4 m. Several species of handsome foliage shrubs, some variegated and others with fragrant flowers and attractive berries. Very adaptable to most climates.

PLATANUS (Plane Tree)
D, FS or HS, H 12–15 m, W 6–9 m. Beautiful shade tree with large fan-shaped leaves turning yellow in autumn. Very adaptable to most climates.

PLUMERIA (Frangipani)
D, FS or HS, H 3–4 m, W 2–3 m. Small deciduous tree with fragrant, waxy, white (or pink) flowers in summer. Frost susceptible, warm temperate and tropical climates only.

POINSETTIA
(*See Euphorbia.*)

POLYGALA (Sweet Pea Shrub)
E, FS to SS, H 1–2 m, W 1–2 m. Small, fast-growing shrub with purple, pea-like flowers from winter to early summer. Prune after flowering to keep within bounds. Temperate and tropical climates, but dislikes excessive moisture.

POPULUS (Poplar)
D, FS to HS, H 9–24 m, W 3–12 m. Beautiful deciduous trees with brilliant autumn colours, but rather large and greedy for small gardens. The tall, slender Lombardy Poplar, the Silver Poplar and Cottonwood are the best known. Very adaptable but prefer temperate and cold climates.

PROSTANTHERA
(*See Chapter on Australian Native Plants*)

PRUNUS

D, FS or HS, H 3–6 m, W 1–6 m. This large genus includes not only the flowering plum but also most of the spring flowering blossoms—peaches, almonds and cherries. There are many species and varieties which differ in size, leaf and flower colour, flower form (single or double) and time of flowering. All are pruned after flowering. Generally, this group of beautiful trees prefers full sunlight, good drainage and a cool winter climate. Leading nurseries can advise on kinds and varieties suitable for your district.

PSORALEA (Blue Butterfly Bush)

E, FS or HS, H 2–3 m, W 2–3 m. Fast-growing shrubs with masses of pale blue, pea-like flowers in spring. Prune back by one third after flowering. Frost susceptible. Prefer temperate to warm climates.

PULTENAEA

(*See Chapter on Australian Native Plants*)

PYRACANTHA (Firethorn)

E, FS to SS, H 2–3 m, W 2–3 m. Often confused with hawthorn (which are deciduous), these spiny, evergreen shrubs have clusters of small white flowers followed by yellow, orange or red berries. Good hedge plant. Temperate and cold climates.

QUERCUS (Oak)

D, FS or HS, H 9–18 m, W 5–9 m. Large trees in varying shapes and sizes. English Oak is round-headed and spreading; Pin Oak is conical with red autumn leaves. Temperate and cold climates.

RAPHIOLEPIS

E, FS or HS, H 2 m, W 1–2 m. Compact shrubs with pink flowers in spring followed by black berries. Very adaptable to most climates.

RHODODENDRON

E, SS or S, H 3–6 m, W 2–4 m. Spectacular spring flowering shrubs or trees for mild temperate and cold climates. Need similar conditions (semi-shade, good drainage, acid soil) to azalea, to which they are closely related. Many varieties and flower colours.

RONDELETIA

E, HS or SS, H 2–3 m, W 1–2 m. Attractive

Rhododendrons make a spectacular spring show

Native Hare's Foot Fern *(Davallia)* is ideal for baskets

Hibbertia scandens is a fast-growing, twining plant

The Royal Botanical Gardens, Hobart, is renowned for the cultivation of Tuberous Begonias

Glorious autumn tints dominate this Canberra street

shrub with dark green foliage and rounded masses of pink blossom in late winter and early spring. Frost susceptible. Temperate to warm climates.

ROSMARINUS (Rosemary)

E, FS to SS, H 1–2 m, W 1 m. Attractive small shrub with glossy, aromatic leaves (used as herb in cooking) and pale blue flowers in both early spring and autumn. Good hedge plant. Very adaptable to most climates, except tropics.

RUSSELIA

E, FS or HS, H 1–2 m, W up to 1 m. A stemmy, rush-like plant with scarlet, tubular flowers in summer. Can be tied to a stake or used as spill-over plant for walls and rockeries. Temperate and warm climates.

SALIX (Willow)

D, FS to SS, H 6–9 m, W 6–9 m. Graceful trees suited to wet conditions. Weeping Willow is best know, but may invade drains in small gardens. There are smaller, less vigorous species—Pussy Willow with silvery catkins in spring or Tortured Willow with twisted stems and leaves. All climates.

SCHEFFLERA

(*See Chapter on Australian Native Plants*)

SPIRAEA (May)

D, FS to SS, H 1–3 m, W 1–3 m. Several species of cany shrubs with masses of white single or double flowers in spring. Prune back after flowering to preserve compact shape. Adaptable to all except tropical climates.

Virgilia, a quick-growing small tree

STREPTOSOLEN

(syn. *Browallia jamesonii*) E, FS or HS, H 1–2 m, W 1 m. Cany shrubs with clusters of bright orange flowers in spring and early summer. Needs occasional pruning and thinning. Frost susceptible. Temperate and tropical climates.

SYRINGA (Lilac)

D, FS or HS, H 2–3 m, W 1–2 m. Attractive, suckering shrub with fragrant clumps of flowers in white, pink, red, mauve and purple. Cool temperate and cold climates only.

TAMARIX (Flowering Cypress)

D and E, FS or HS, H 3–9 m, W 2–5 m. Small trees with cypress-like, pendulous foliage and feathery masses of pink flowers in spring and summer. Very adaptable and tolerates dry heat, strong winds and salty soils. All climates.

ULMUS (Elm)

D or SD, FS or HS, H 9–18 m, W 9–18 m. Several species of medium to large trees with attractive foliage in spring and autumn. Golden Elm and Chinese Weeping Elm are widely grown. Cool temperate and cold climates.

VIBURNUM

E or D, FS to SS, H 2.5 m, W 1–3 m. More than 100 species and many more named varieties of these attractive shrubs grown for their fragrant, hydrangea-like flowers and attractive berries. Prune both evergreen and deciduous types after flowering. Most climates except tropics.

VIRGILIA

E, FS to SS, H 4–6 m, W 3–4 m. Fast-growing small trees with fern-like foliage and sprays of mauve flowers in spring. Very adaptable but sometimes short-lived. All climates except cold districts.

WEIGELA

D, HS or SS, H 2–3 m, W 1–2 m. Cany shrubs with white, pink or red trumpet-like flowers in spring. Cut back after flowering. Temperate and cold climates.

WESTRINGIA

(*See Chapter on Australian Native Plants*)

CONIFERS

This is the group of cone-bearing plants. Most are evergreen, grow to a definite (usually symmetrical) shape and almost never need pruning. They vary in size, shape and leaf colour from small prostrate trees or shrubs suitable for tubs or rockery pockets to magnificent specimens 30 metres or more in height.

As a group, conifers are slow growers but respond to attention in watering, mulching and fertilizing. As with other trees and shrubs, slow-acting blood and bone-based fertilizers or slow-release fertilizers are most suitable. Again as a group, conifers are best suited to temperate and cold climates where they are valuable for winter effect. Very few conifers are suitable for tropical climates. There are dozens of species and hundreds of varieties and the reader is advised to consult a nursery catalogue or better still visit a specialist nursery where advanced specimens are displayed. The following is only a brief summary of the most important conifers.

ABIES (Fir Trees)

Tall, pyramid-shaped trees with beautiful foliage and best suited to cool to cold climates with high rainfall.

Silver Fir (*A. alba*) H 15 m. Green/silver foliage.

Colorado White Fir (*A. concolor*) H 15 m. Bluish foliage.

Caucasian Fir (*A. nordmanniana*) H 18 m. Green/silver foliage.

ARAUCARIA

Tall, symmetrical trees but grow rather too large for the average garden.

Monkey Puzzle (*A. araucana*) H 12–15 m. Cool, moist climates.

Bunya Pine (*A. bidwellii*) H 18–30 m. All climates except coldest.

Hoop Pine (*A. cunninghamii*) H 24–45m. Moist coastal climates.

Norfolk Island Pine (*A. heterophylla*) H 18–30 m. Moist coastal climates.

CALLITRIS (Cypress Pine)

Ornamental Australian native trees useful in dry inland climates where many of them grow naturally.

White Cypress Pine (*C. columellaris*)

H 15 m. Dark green foliage.
Port Jackson Pine (*C. rhomboidea*) H 12 m. Olive green foliage.

CEDRUS (Cedar)

Shapely, pyramidal trees with needle-like leaves and upright, barrel-shaped cones. Adaptable trees but best in cool temperate and cold climates. Atlas Cedar (*C. atlantica*) H 15–18 m. Grey-green foliage; var. *glauca* with silver-blue foliage; var. *aurea* with yellow foliage. Deodar (*C. deodara*) H 15–18 m. Green or grey green foliage; var. *aurea* yellowish foliage. Cedar of Lebanon (*C. libani*) H 12–15 m. Similar to *C. atlantica* but a more flattened shape.

CHAMAECYPARIS (syn. *Retinospora*, False Cypress)

Shapely, ornamental, medium-sized or dwarf trees which are useful for garden landscape effects. Need good drainage and do best in cool temperate or cold climates with good rainfall. Lawson Cypress (*C. lawsoniana*) H 12–15 m. Pyramidal growth to ground level with many varieties and foliage colours. Dwarf varieties range from 60 cm to 3 m in height. Hinoki Cypress (*C. obtusa*) H 9–12 m. Flattened, fan-like foliage; var. *crippsii* yellow foliage and smaller tree (6 m). Many dwarf varieties ranging in height from 60 cm to 2 m. Sawara Cypress (*C. pisifera*) H 9–12 m. Similar to *C. lawsoniana* with many leaf forms and colours. Also dwarf varieties 60 cm to 120 cm in height.

CRYPTOMERIA

Japanese Cedar (*C. japonica*) H 12 m. Large stately tree for cool temperate and cold climates; var. *nana* is dwarf variety 1–2 m tall.

Handsome conifers dominate this well planned garden. The blue spruce in the foreground is a particularly fine specimen

CUPRESSUS (Cypress)

Fast-growing, attractive conifers for specimen trees or for screen or windbreak planting. Very adaptable but prefer temperate or cold climates. Funeral Cypress (*C. funebris*) H 12–18 m. Grey-green Weeping foliage. Arizona Cypress (*C. arizonica* syn. *C. glabra*) H 6–12 m. Silver-grey foliage, drought resistant for inland districts. Monterey Cypress (*C. macrocarpa*) H 15–18 m. Fast-growing conifers for hedges and windbreaks. Smaller varieties are *brunniana* (H 12 m, gold foliage); *erecta aurea* (H 8 m, gold foliage); *lambertiana* (H 12 m, green foliage) and *lambertiana aurea* (H 12 m, gold foliage). Italian Cypress (*C. sempervirens*) H 12–18 m, erect pyramidal shape; var. *stricta* is more slender, often called Pencil Pine. Bhutan Cypress (*C. torulosa*) H 12–18 m. Tall, pyramidal shape useful for screens or windbreaks.

JUNIPERUS (syn. *Sabina*, Juniper)

A large group of conifers with interesting shapes and leaf colours, expecially in the dwarf species and varieties. Very adaptable and suited to all climates except tropical areas. Bermuda Cedar (*J. bermudiana*) H 12–15 m. Frost sensitive. Chinese Juniper (*J. chinensis*) H 9–12 m; var. *aurea* with yellow foliage; var. *variegata* with grey and cream foliage. Pencil Cedar (*J. virginiana*) H 12–18 m. Narrow column shape, very adaptable. Alligator Juniper (*J. pachyphlaea*) H 3 m. Erect habit, silver-blue foliage. Creeping Juniper (*J. horizontalis*) H 60–90 cm. Prostrate, creeping habit, blue-green foliage. Good spill-over plant for rockeries and banks. Meyer Juniper (*J. squamata meyeri*) H 2–3 m. Erect habit, blue-grey foliage. Savin Juniper (*J. sabina*) H 1–2 m. Spreading habit, blue-green foliage.

PICEA (Spruce)

Handsome, symmetrical, cone-shaped trees. Best in cool temperate and cold climates. Norway Spruce (*P. abies*) H 9–15 m. Fast-growing, green foliage. White Spruce (*P. glauca*) H 9–15 m. Upturned branches, grey-green foliage; var. *albertina conica* is dwarf (1–2 m). Blue Spruce (*P. pungens*) H 6–9 m. Slow-growing with blue-green foliage; var. *glauca* with blue-grey foliage; var. *kosteriana* with silver-blue foliage.

PINUS (Pines)

Large, fast-growing conifers with needle-like leaves. Pyramidal in shape when young but may lose their symmetry when mature. Very adaptable to climate but generally prefer temperate regions. Canary Island Pine (*P. canariensis*) H 9–15 m. Pendulous, grey-green foliage. All but coldest districts. Cuban Pine (*P. caribaea*) H 18–24 m. Fast-growing pine for coastal and tropical areas with summer rainfall. Aleppo Pine (*P. halepensis*) H 18 m. Drought-resistant, suitable for poor soils in low rainfall regions. Mexican Pine (*P. patula*) H 12–15 m. Fast-grower with pendulous, blue-green foliage. Temperate and warm climates. Monterey Pine (*P. radiata* syn. *P. insignis*) H 18–24 m. Dense, fast-growing pine for shade, shelter or windbreaks. Very adaptable to all but tropical climates.

PODOCARPUS

(*P. elatus*, Illawarra Plum Pine) H 12 m. Attractive, round-headed, native tree with glossy foliage, plum-coloured when young. Adaptable to all but cold and dry inland climates. Yellowwood (*P. falcatus*) is a similar tree from South Africa with blue-green foliage.

RETINOSPORA

(*See Chamaecyparis.*)

SABINA

(*See Juniperus*)

TAXODIUM

(*T. distichum*, Swamp Cypress)

H 15–18 m. Large deciduous tree with green, feathery foliage turning russet-brown in autumn. Useful in wet situations. All climates except very cold and tropical regions.

THUJA

Ornamental conifers with attractive shapes and flattened frond-like foliage of various colours. Best suited to cool temperate and cold climates. American Arbor Vitae (*T. occidentalis*) H 15 m. Fast-growing tree with conical shape. There are many smaller and more colourful varieties including *fastigiata* (H 2–3 m) green foliage; *lutescens* (H 2–3 m) foliage tipped yellow; *hoveyii* (H 1 m) green foliage; *rheingold* (H 1 m) golden foliage; *little gem* (H 60 cm), green foliage. Bookleaf Cypress (*T. orientalis*) H 9–12 m. Fast-growing conifer with dense, compact shape. Smaller coloured-leaf varieties are also available. Western Red Cedar (*T. plicata*) H 12 m. Dark green foliage which droops at the ends; var. *aurea* is smaller with gold-tipped foliage and var. *zebrina* has striped foliage. THUJOPSIS (*T. dolabrata*, False Arbor Vitae) H 9–12 m. Slow-growing conifer with heavy, coarse foliage. A smaller variety (4 m) is *variegata* with cream patches through the foliage.

PALMS

This large family of woody, evergreen plants contains over 1000 species. In their natural habitat, most species are found in tropical and subtropical climates, but can be grown successfully in temperate climates too. They are not really suited to cold climates, unless they are planted in a warm, sheltered position or grown as indoor plants in containers.

Some palms have a single trunk (others have several) with a distinctive crown of leaves or fronds. The fronds may be fan-shaped (palmate) or deeply divided (pinnate). As a group, palms are very ornamental and easy to grow. They add a tropical touch to the patio or garden and are becoming very popular for planting in the surrounds of swimming pools. Most palms can be grown successfully in pots or tubs and many make ideal indoor plants. (See Chapter 14.) Most palms can be grown in containers for five years or more before they become too large. In the summary which follows, the species are listed in alphabetical order by their botanical name, followed by their common name. The height is given at maturity and at five years of age in brackets; for example, H 3 m (1 m).

A SELECTION OF POPULAR PALMS

ARCHONTOPHOENIX

(*A. cunninghamiana*, Bangalow Palm) H 9–12 m (2 m). Graceful, Australian native palm with smooth grey trunk, dense crown of feathery fronds and pendulous bunches of flower spikes and fruit. Good container plant when young.

ARECASTRUM

(*A. romanzoffianum* syn. *Cocos plumosa*, Plume or Queen Palm) H 12 m (2 m). Slender, smooth trunk and a head of arching, blue-green fronds. Prefers full sunlight.

BUTIA

(*B. capitata*, Jelly or Wine Palm) H 6 m (1 m). Short, thick trunk with head of blue-grey fronds and orange-red bunches of fruit. Sunlight or shade. Good container plant.

CARYOTA

(*C. mitis*, Fish Tail Palm) H 6 m (1 m). Multiple trunks with long, yellow-green fronds with toothed, wedge-shaped leaflets.

CHAMAEDOREA

(*C. elegans* syn. *Neanthe elegans*, Parlour Palm) H 2 m (1 m). Delightful pot or tub plant for patio or terrace in semi-shade or indoors. Bright green, papery fronds arranged in a spiral around the thin, dainty stem.

CHAMAEROPS

(*C. humilis*, European Fan Palm) H 5 m (1 m). Clumps of several stems, bearing shiny, deeply-cut, fan-like leaves. Good palm for small gardens and containers, both indoors and outdoors. Tolerates cooler conditions more than most palms and does well in full sun or shade.

COCOS

(*C. nucifera*, Coconut Palm) H 12–15 m (2–3 m). Graceful inclined trunk with rather sparse crown of feathery, yellow-green fronds. Requires tropical, high rainfall conditions.

HOWEA (syn. *Kentia*)

Two attractive species, native to Lord Howe Island: *H. belmoreana* (Sentry or Curly Palm) H 6 m (1 m) has stout ringed trunk with feathery fronds curving inward. Good container plant for indoors or outdoors. Sunlight or semi-shade; *H. forsteriana* (Kentia or Thatch Leaf Palm) H 7–9 m (1 m) is a more slender palm.

Rhapis excelsa

Ideal indoor palm. Mature palms of both species have long spikes of green fruit turning yellow and red as they ripen.

LINOSPADIX

(*L. monostachyus*, Walking-Stick Palm) H 3 m (1 m). Attractive, mid-green fronds are broad and fringed. Blends well with other plants in half sun or semi-shade position. Ideal container plant for shady terrace or indoors.

LIVISTONIA

(*L. australis*, Cabbage Tree Palm) H 6 m (1 m). Australian native palm with rough ringed trunk topped with broad, fan-like fronds with spines along the edges of the base of the stem. Fruit is orange-red turning black. Full sun or semi-shade.

PHOENIX (Date Palms)

Several species including the Date Palm (*P. dactylifera*) which is grown for its fruit. *P. canariensis* (Canary Island Palm) H 6 m (2 m) has arching fronds from ground level for 8–10 years. The fronds then grow from a thick, robust trunk. Heavy bunches of edible, orange berries are produced periodically. Very adaptable but prefers full sun. *P. roebelinii* (Dwarf Date Palm) H 4 m (2 m). Very slow growing but extremely attractive palm with glossy, finely-feathered, arching fronds. A good garden palm, and ideal for container growing (indoors or outdoors). Grows well in sun or shade.

RHAPIS

(*R. excelsa*, Lady Palm or Ground Rattan Cane) H 4 m (1 m). Attractive palm with several slender trunks topped by small fan-shaped fronds. An excellent container plant for indoors or outdoors.

TRACHYCARPUS

(*T. fortunei*, Fan or Chusan Palm) H 9 m (2 m). Slender trunks with persistent fibres and bearing fan-shaped fronds with stout spines at base. Large clusters of fragrant yellow flowers followed by black fruit. Full sun or semi-shade. Like the European fan palm, this species tolerates cooler conditions.

Chapter 13

AUSTRALIAN NATIVE PLANTS

Since the beginning of the 1960s there has been a growing awareness of the value of Australian native plants in gardens. Over the last few years this interest has accelerated into a boom.

This continent has a wealth of native plants bred and conditioned over many years to withstand the vagaries and the harsh realities of our climate.

The present practice of mixing natives and exotics in home gardens is to be commended, as both have much to contribute to garden layouts in form, colour and visual effect.

Many Australian native plants are very adaptable and are easy to grow in gardens, but because Australia is a very large continent with a wide climatic range, some are difficult to grow, except in conditions close to their natural environment. For example, it is almost impossible to grow a plant from the dry, inland regions in a wet, coastal climate and vice versa. So it is important to know the climatic and soil requirements of native plants before attempting to grow them.

Leading nurseries can give you helpful advice on the best species and varieties to grow in your garden. There is also a Society for Growing Australian Plants (SGAP) which will welcome your interest and membership. The Society holds meetings, discussion groups and field days to promote the preservation and cultivation of native flora. It also collects and distributes seeds of some species which may be difficult to obtain through the usual channels.

It is often claimed that Australian plants need little attention because they 'grow wild'. This is not true, and most native plants respond to care in a well tended garden. They are healthier and produce more flowers than their counterparts struggling in their natural surroundings. Admittedly, some Australian plants grow naturally on rather poor soil and resent heavy application of quick acting fertilizers, but most will respond to slow-release fertilizers such as Osmocote, Agriform Plant Tablets, or Magamp. Another factor is root disturbance. Most Australian natives will not tolerate excessive cultivation and possible root disturbance. Many are shallow rooted, so weed around them by hand or use a shallow tool, like a Dutch hoe to chip the surface. Mulching with grass clippings or fibrous compost will deter many weeds. Desiccant weedicides such as Tryquat, Polyquat or Zero are also useful (see Chapter 18). This Chapter does not pretend to be a definitive work on Australian flora. We list the most commonly available trees, shrubs, climbers and ground covers, but there are many others from which to choose.

For general notes on cultivation and care see the chapter on Trees and Shrubs. This also contains the key to abbreviations used here.

SELECTED AUSTRALIAN TREES AND SHRUBS

(Abbreviations same as in previous Chapter)

ACACIA (Wattle)

E, FS or HS, H 1–9 m, W 1–9 m. Australia's national flower. There are over 500 species of this genus ranging from small shrubs to large trees. All are evergreen with yellow flowers in late winter or spring, but the foliage may be feathery, flat or needle-like. All prefer full sun or half sun and are usually fast-growing but short-lived Reliable and adaptable species are:

- Gold Dust Wattle (*A. acinacea*) H 1–2 m, W 1–2 m.
- Cootamundra Wattle (*A. baileyana*) H 3–5 m, W 3–5 m.
- Box-leafed Wattle (*A. buxifolia*) H 2–3 m, W 2–3 m.
- Black Wattle (*A. decurrens*) H 6–9 m, W 3–5 m.
- Sydney Golden Wattle (*A. longifolia*) H 4–5 m, W 2–3 m.
- Queensland Silver Wattle (*A. podalyriifolia*) H 3–4 m, W 3–4 m.
- Golden Wattle (*A. pycnantha*) H 3–5 m, W 2 m.

AGONIS (Willow Myrtle)

E, FS or HS, H 5–6 m, W 3–5 m. Attractive tree with pendulous, willow-like branches and small, white, tea-tree-like flowers in spring. Frost susceptible. Temperate and warm climates.

BACKHOUSIA (Lemon-scented Myrtle)

E, FS or HS, H 2–4 m, W 1–2 m. Attractive shrub with glossy lemon-scented leaves and clusters of small greenish-white flowers in early summer. Frost susceptible. Temperate and warm climates.

BAECKEA

E, HS or SS, H 1–2 m, W up to 1 m. Small attractive shrubs with white flowers, not unlike those of boronia, in spring and summer. Temperate and warm climates.

BANKSIA

E, FS or HS, H 3–6 m, W 2–5 m. Quaint but attractive shrubs or trees with thick, often serrated, leaves and large erect cones of stiff wiry flowers in shades of greenish-white, yellow, orange and red. Prefer well-drained sandy soils in warm coastal climates. Popular species are:

- Scarlet Banksia (*B. coccinea*) H 2–3 m, W 1–2 m.
- Hairpin Honeysuckle (*B. collina*) H 2–3 m, W 1–2 m.
- Heath Banksia (*B. ericifolia*) H 3–4 m, W 2–3 m.
- Coast Banksia (*B. integrifolia*) H 3–5 m, W 2–3 m.

Australia's national flower, the wattle or acacia, is a favourite in many gardens

BORONIA

E, HS or SS, H 60–90 cm, W 30–60 cm. Many species of small shrubs with spicy fragrance and attractive delicate flowers—mostly pink, but one is brown and yellow. All prefer well-drained, slightly acid, sandy soils similar to those of their natural habitat. Mulch with compost or leaf mould to avoid root disturbance and hold moisture. Warm coastal climates are best. The most widely grown species are:

Sydney Boronia (*B. ledifolia*) H 90 cm, W 30 cm.

Brown Boronia (*B. megastigma*) H 90 cm, W 60 cm.

Native Rose (*B. serrulata*) H 75 cm, W 30 cm.

To raise plants from seed, see Chapter 9.

BRACHYCHITON
(Illawarra Flame Tree)

D, FS or HS, H 12–24 m, W 6–9 m. Large, pyramid-shaped tree with glossy, lobed leaves, usually deciduous in spring, and masses of brilliant red flowers in late spring and early summer. Temperate and tropical climates. Other useful trees in this genus are:

Queensland Lacebark (*B. discolor*) H 9–12 m, W 6 m.

Kurrajong (*B. populneus*) E, H 6–9 m, W 4–6 m.

Banksia—a symbol of the Australian bush

CALLISTEMON (Bottlebrush)

E, FS or HS, H 2–6 m, W 2–5 m. Attractive shrubs or small trees with brush-like flowers in shades of cream, yellow, pink and red in spring and early summer. Very adaptable plants to both wet and dry conditions with good tolerance to salty soils. All climates except coldest districts. Many named varieties of the most popular species are available.

Crimson Bottlebrush (*C. citrinus*) H 3 m, W 2 m.

White Bottlebrush or Pink Tips (*C. salignus*) H 5–6 m, W 2 m.

Weeping Bottlebrush (*C. viminalis*) H 6 m, W 5 m.

CERATOPETALUM (Christmas Bush)

E, FS or HS, H 3–6 m, 2–3 m. Small shapely tree with showy pink or red bracts in summer. Prune back after flowering. Prefers well drained soils but water regularly from spring onwards. Temperate to warm coastal climates. To raise plants from seed, see Chapter 9.

CHAMAELAUCIUM (Geraldton Wax)

E, FS or HS, H 2–3 m, W 2–3 m. Attractive West Australian shrub or small tree with needle-like leaves and waxy, pink or red flowers. Prefers well-drained sandy soils. Temperate or warm climate.

CHORIZEMA (Flame Pea)

E, FS to SS, H 1 m, W 1 m. Compact small shrub with heart-shaped leaves and brilliant orange-red pea flowers in winter and spring. Prefers well drained soil but plenty of water in winter. Temperate climate.

EPACRIS (Native Fuchsia)

E, FS to SS, H 1 m, W 30 cm. Heath-like shrubs with sprays of slender, tubular red flowers with white tips. Needs well drained, slightly acid, sandy soil and resents root disturbance. Temperate and cold climates.

ERIOSTEMON (Waxflower)

E, HS or SS, H 120 cm, W 90 cm. Compact shrubs with fragrant foliage and white or pink, star-shaped flowers in winter and early spring. Prefers well drained, slightly acid sandy soils. Temperate and warm climates.

EUCALYPTUS (Gum Tree)

E, FS to SS, H 6–90 m, W 3–12 m. Over 600 species of eucalypt trees are recorded and dominate the Australian landscape. They range in size from small trees to forest giants. Most are fast-growing and very adaptable but it is best to consult your local nurseryman for advice on suitable varieties for your own district. Some have showy flowers while others are suitable for windbreaks, street planting or as specimen trees. Popular species for small gardens are:

Western Australian Flowering Gum (*E. ficifolia*) H 5 m, W 3 m.
Dwarf Sugar Gum (*E. cladocalyx nana*) H 6–9 m, W 5–6 m.
Lemon-scented Gum (*E. citrodora*) H 12–18 m, W 5–6 m.
Scribbly Gum (*E. haemastoma*) H 5–6 m, W 5–6 m.
White Gum (*E. maculosa*) H 6–9 m, W 3–5 m.
Small-leafed Peppermint (*E. nicholli*) H 5–9 m, W 3–5 m.

EUGENIA (Lilly Pilly)

E, FS to SS, H 5–6 m, W 2–3 m. Small, self-shaping tree with glossy foliage and starry white flowers followed by edible berries in shades of pink, red or purple. Frost susceptible. Temperate and warm climates.

GREVILLEA (Spider Flower)

E, FS or HS, H 1–2 m, W 1–2 m. A great range of native shrubs from tropical, temperate and cold climates—there is one to suit every garden. The spider-like flowers in winter or early spring are usually pink or red but there are yellow and orange flowers too. Plants need good drainage, a slightly acid soil and resent root disturbance. Consult a nursery catalogue for the best species and varieties for your district. Silky Oak (*G. robusta*), the largest of the family by far, is a handsome, self-shaping tree which grows to a height of 12–15 m and a spread of 6 m, with fern-like foliage and showy orange flowers in late spring or early summer.

HAKEA

E, FS or HS, H 4–6 m, W 2–3 m. Fast-growing shrubs with rather thick, leathery leaves for screening or windbreaks. Pincushion Hakea (*H. laurina*) has red, globe-shaped flowers in spring. Willow Leaf Hakea (*H. salicifolia*) has bronze-tipped foliage and can be trained as a hedge. All climates except coldest districts.

LEPTOSPERMUM (Tea Tree)

E, FS, H 1–5 m, W 1–5 m. Many species and named varieties of fast-growing shrubs native to Australia or New Zealand. Tiny leaves and white, pink, or red flowers like

Melaleuca leucadendron

peach blossom in spring or summer. Plants prefer well drained soil and resist drought, wind and salt spray. Temperate coastal climates. Some of the most widely cultivated species are:

Coastal Tea Tree (*L. laviegatum*) H 3–4 m, W 3–5 m.
Port Jackson Tea Tree (*L. squarrosum*) H 2–3 m, W 1–2 m.
Lemon-scented Tea Tree (*L. petersonii*) H 4–5 m, W 3–4 m.
Manuka (*L. scoparium* hybrids) H 1.5–2.5 m, W 1.5–2 m.

LESCHENAULTIA (Mirror of Heaven)

E, FS, H 30–90 cm, W 30–60 cm. Delicate small shrub with soft blue flowers. Prefers well drained, sandy or gravelly soil. Temperate climates.

MELALEUCA (Paper Bark)

E, FS or HS, H 2–9 m, W 2–6 m. Shrubs or small trees noted for their showy flowers (similar to those of bottle brush) and decorative, papery bark. Many tolerate swampy conditions, strong winds and salt spray. All but coldest climates. Widely-grown species are:

Bracelet Honey Myrtle (*M. armillaris*) H 5–6 m, W 4–5 m.
Dotted Melaleuca (*M. hypericifolia*) H 2–2.5 m, W 2–2.5 m.
Broadleaf Paper Bark (*M. leucadendron*) H 6–9 m, W 4–5 m.

MURRAYA (Satin Wood)

E, HS or SS, H 2–3 m, W 2–2.5 m. Dense, rounded shrub with glossy foliage and fragrant clusters of white flowers like orange blossom in spring and summer. Frost susceptible. Temperate and tropical climates.

PIMELIA

E, FS, H 1–2 m, W 1 m. Small attractive shrubs with small glossy leaves and dense clusters of cream, yellow or pink flowers in spring to early summer. Temperate climates.

PROSTANTHERA (Mint Bush)

E, FS or HS, H 2–4 m, W 2–3 m. Several species of attractive shrubs with small aromatic leaves and masses of mauve, purple or blue flowers in spring. They prefer good drainage and resent root disturbance. Prune back after flowering to prevent bushes becoming leggy. Cool temperate to subtropical climates.

PULTENAEA (Bush Pea)

E, HS or SS, H 2–3 m, W 1–2 m. Many species of small native shrubs with yellow or orange pea-like flowers in spring. Temperate climates.

SCHEFFLERA (Umbrella Tree)

F, FS to S, H 6–9 m, W 3–6 m. Glossy foliage resembles segments of umbrella. Reddish flowers in spring followed by purple berries in autumn. Used as indoor plant when small. Temperate and tropical climates.

STENOCARPUS (Queensland Fire Wheel)

E, FS, HS, H 9 m, W 3–5 m, Slow-growing but shapely tree with large glossy leaves and unusual orange-red flowers arranged like the spokes of a wheel. Warm temperate and tropical climates.

TELOPEA (Waratah)

E, FS or HS, H 2–3 m, W 2–2.5 m. This attractive and stately shrub has spectacular cones of vivid red flowers (the floral emblem of New South Wales) in late spring. Prefers well drained, slightly acid, sandy soils but with organic matter added. Prune back plants after flowering. Tem-

Silky Oak, (*Grevillea robusta*)

perate climates. To raise plants from seed, see Chapter 9.

THRYPTOMENE (Heath Myrtle)

E, FS, H 1–1.5 m, W 1 m. Several species of attractive, dwarf shrubs with small leaves close to the stems and tea-tree-like flowers (white, pink or lavender) in spring. Prefers well drained, slightly acid, sandy soil. Temperate and cold climates.

WESTRINGIA (Coastal Rosemary)

E, FS or HS, H 1–2 m, W 1.5 m. Attractive, dense shrubs with grey-green foliage and small white or pale lavender flowers. Very adaptable plants which tolerate strong winds, salt spray and dry conditions. Useful as low windbreak or hedge plant. All climates but coldest.

GROUND COVERS

AJUGA (Blue Bugle Flower)

E, HS or SS (20–40 cm). Dwarf carpeting plant with purple-bronze leaves and lilac-blue flowers in spring. An attractive variegated leaf form is also available.

BRACHYSEMA LANCEOLATUM

E, FS (1.5 m–3 m) Spreading, rounded, glaucous foliaged shrub, with red pea flowers. Needs a well-drained position.

BRACHYSEMA LATIFOLUM

E, FS (20 cm 1 m) Attractive prostrate, trailing plant bearing orange-red pea-shaped flowers in spring. Good drainage essential.

CORREA DECUMBENS

E, FS (30 cm–3 m) Spreading shrub carrying erect tubular flowers, red with yellow tips for most of the year. Best flowering in winter.

CORREA DUSKY BELLS

An interesting plant of somewhat variable growth habit. Hardy and frost hardy with pink flowers in winter. It will take heavy clay soils, is a vigorous grower and like all Correas good for attracting birds to the garden.

DAMPIERA DIVERSIFOLIA

E, FS (20 cm–1 m) An excellent ground cover, bearing masses of dark blue flowers over spring and summer.

Telopea or waratah

DAVALLIA PYXIDATA (Hare's foot fern)
A fern with long creeping stems found in the bush on trees or rocks. Excellent for ground covering and hanging baskets.

DICHONDRA REPENS
A prostrate plant producing roots at the nodes similar to grasses. Mainly grown for the dense, kidney-shaped leaves. Needs plenty of sun and ample watering. Insignificant greenish flowers.

GREVILLEA JUNIPERINA
(prostrate form) E, FS (60 cm–2 m) Excellent cover for sloping banks, large expanses without shrubs and for covering rock walks. Yellow spider-like flowers in spring and summer. Dark green, prickly leaves.

GREVILLEA POORINDA (Royal Mantle)
Low growing, very vigorous and hardy carrying red toothbrush-like flowers in winter. Tolerates drought and will accept clay soils.

GREVILLEA TRIDENTIFERA
(*Syn biternata*) E, FS (40 cm–1 m) Bright green, dissected foliage with masses of white perfumed flowers in spring. Suitable for sloping banks and low maintenance areas between shrubs or trees.

HARDENBERGIA VIOLACEA
E, FS or SH (20 cm–1 m) Trailing plant with generally purple flowers. However, there are some white, pink and mauve forms.

KENNEDIA PROSTRATE
E, FS or SS (15 cm–1.5 m) Prostrate and vigorous. Sometimes called the Running Postman. Free flowering with scarlet blooms in spring and early summer. Ideal for banks and ground cover between shrubs.

MYOPORUM PARVIFOLIUM
E, FS or HS (1 m–11 m) A frost hardy plant bearing white star-like flowers in spring and summer. Suitable for sloping banks, cover between trees and shrubs, between paths or along driveways.

SCAEVOLA (Mauve Clusters)
Is a large genus of plants varying from large shrubs to perennials and ground covers. Mauve Clusters is regarded as one of the best ground covers, carrying mauve fan-shaped flowers from late spring to early winter. Will tolerate wet conditions and has attractive, dense, bright green leaves.

VIOLA HEDERACEA
E, FS or SS (10–40 cm) Commonly called the Native Violet, this forms an extensive mat in damp areas and carries typically violet-shaped flowers in white and purple. Suitable as a dense ground cover between shrubs, in rockeries, and will also grow indoors. Useful in baskets.

CLIMBERS

CISSUS ANTARCTICA (Kangarro Vine)
This is the best known Cissus, with medium green toothed ovate leaves to 10 cm long. Bears small flowers and black edible fruits. Suitable for pergola, fence or indoors. Foliage more attractive in shade. Requires ample moisture.

CISSUS HYPOGLAUCA
Dark green leaves with five leaflets emerging from the one point. Bears small flowers and a bluish fruit, a useful foliage climber for shade or indoors. Requires ample moisture.

CLEMATIS ARISTATA
Vigorous climber bearing creamy white flowers 5 cm in diameter in spring. Attaches itself by twisting its petiole around supports. Hardy in well-drained, sunny or semi-shade position. Ideal for fences.

HARDENBERGIA COMPTONIANA
Moderately vigorous, carrying sprays of purplish-blue pea-shaped flowers in spring. Does well in semi-shaded position, excellent for trellises or pergolas. Moderately vigorous with trifoliate leaves.

HARDENBERGIA VIOLACEA
E, FS or SH (20 cm–1 m) Trailing plant with generally purple flowers. However, there are some white, pink and mauve forms. Selected plants may be used as a ground cover.

HIBBERTIA EMPETRIFOLIA
Not really a climber as it needs support,

but will grow behind wire and will produce a pillar of brilliant yellow for much of the year. Flowers about 15 mm in diameter. Can be kept cut and grown as a shrub.

HIBBERTIA SCANDENS

Rapid growing twining plant with large, bright green leaves and yellow flowers 5 cm in diameter. Grows well in most areas except where frosts are heavy.

HOYA AUSTRALIS

This rainforest native has succulent-like leaves and white or pale pink, waxy flowers in globe-shaped clusters of about 8 cm in diameter. Suitable for outdoor in most positions and indoor. Will accept salt spray.

KENNEDIA COCCINEA

Vigorous climber bearing masses of small orange to red, pea-shaped flowers during spring. Requires sun or partial shade, very showy. Can be used as a ground cover in sun.

KENNEDIA MACROPHYLLA

Very vigorous climber with large, light green leaves and red pea-shaped flowers in spring and summer. Full sun to light shade, avoid frosts.

KENNEDIA NIGRICANS

Very vigorous climber with large, dark green leaves, and black and yellow flowers. The most vigorous of the Kennedias, and probably one of the most rampant plants in southern Australia. Prefers full sun to light shade, will not survive heavy frosts. Ideal for a quick cover.

KENNEDIA RETRORSA

Vigorous plant with dark green leaves and purple flowers. It is relatively frost hardy, and will grow in most positions.

KENNEDIA RUBICUNDA

Vigorous grower with dark green leaves and large red flowers. Will accept most situations, but prefers full sun.

MILLETIA MEGASPERMA

The native Wisteria from the rainforests of Queensland and northern NSW—very showy large white to purple pea-shaped flowers in sprays up to 15 cm in diameter. Excellent climber for covering a fence or pergola in milder areas. Will tolerate some frost.

PANDOREA JASMINOIDES

A vigorous climber with large pink, trumpet-shaped flowers with a deep red centre, and dark green, shiny leaves. Plants are hardy and flowers are showy. Grows well on trellis and prefers some shade. There are a number of cultivars available.

PANDOREA PANDORANA

A well-known Wonga vine, a vigorous climbing plant with glossy leaves, tubular flowers ranging in colour from white to brown. There are a number of cultivars available.

PASSIFLORA CINNABARINA

Vigorous creeper with red flowers 8 cm in diameter. Very showy for fence or for pergola and suits most aspects. Will attach to cracks in mortar between bricks. Flowers in spring.

SOLLYA HETEROPHYLLA

Commonly called Blue Bell Creeper. A vigorous Western Australian bushy plant with a twining habit. It produces a number of branches from ground level and carries flowers of blue and occasionally pink in spring and summer, followed by fleshy, cylindrical blue fruits. Very hardy and is suitable for all soils and aspects.

Chapter 14

ROCKERY PLANTS AND GROUNDCOVERS

Hundreds of plants—annuals, biennials, perennials, bulbs, small shrubs and ferns—are suitable for rock gardens. Readers should refer to previous chapters for descriptions of many of them. The success of a rockery depends on careful selection of planting material. If the rock garden is in a sunny position choose plants which thrive in full sun and tolerate heat and fairly dry conditions. For a cooler, damp, southerly aspect you will need shade-loving plants.

A well-planted rockery should not be so crowded with plants that the rocks are completely obscured. A good ratio is to have about two-thirds plants and one-third rocks. Size of plants should also be considered. Large, vigorous specimens soon overgrow smaller plants which quickly give up the unequal struggle. Plants with strong, greedy roots may dislodge rocks or weaken the whole structure. You will achieve a pleasing and balanced effect by selecting prostrate or creeping plants, cushion or mound plants and clump plantings of bulbous plants with strap-like leaves and taller flower spikes.

Rockery plants should be fairly permanent to cut down on maintenance and ensure an attractive display of flowers or foliage for most of the year. Avoid large plantings of annuals. If you rely too much on annuals you will have bare spaces for long periods—but they are useful when tucked into odd pockets for extra seasonal colour.

PERENNIAL PLANTS

When making your selection of permanent plants, make sure you have flowers for each season—many rock gardens are loaded with spring flowers but devoid of colour for the rest of the year. Many flowering perennials make superb rockery subjects—perennial phlox, Stokes aster, shasta or Michaelmas daisies, bedding begonias, delphinium, gazania and verbena are a few suggestions. Thrift (*Armeria*), basket of gold (*Alyssum saxatile*), bellflower (*Campanula*) and blue fescue (*Festuca glauca*) are good mound or cushion plants. Small shrubs such as ceratostigma, pink diosma (*Coleonema*), cherry pie (*Heliotropium*) and veronica (*Hebe*) make good permanent plants too.

FOLIAGE PLANTS

Foliage colour should also be considered. Many perennials have attractive variegated or coloured foliage which will provide a nice contrast in the rock garden even if there is a scarcity of flowers at any particular time. Differences in size, shape and texture of leaves also add interest, so look for plants with contrasting foliage—large or small, plain or ferny, shiny or furry. For silver foliage, try lavender, dusty miller (*Cineraria maritima*), ghost bush (*Artemisia*), dianthus, snow-in-summer (*Cerastium*) and lamb's ear (*Stachys*). Small shrubs with coloured foliage include dwarf *nandina* (red tones in autumn) and *Berberis thunbergii* (purple-bronze). *Hebe armstrongii* has golden, whipcord foliage and resembles a minature conifer. Some of the prostrate junipers are also attractive for a more formal or spill-over effect. New Zealand flax (*Phormium tenax*) is a good foliage plant too, but often grows rather large for a rockery.

BULBOUS PLANTS

Many bulbs and bulbous plants are ideal for rockeries. For small spaces use daffodils, freesias, snowflakes, iris, lily-of-the-valley, dwarf day lilies (*Hemerocallis*) and other small growing bulbs. For larger rock gardens clump plantings of African lily (*Agapanthus*), Kaffir lily (*Clivea*), torch lily (*Kniphofia*) and the iris-like plants—*Dietes* or *Moraea*—are suitable.

HOBBY PLANTS

A rockery can also be useful to display a collection of species or varieties of a particular kind of plant or 'hobby plant'. You could grow a collection of cacti or succulents in a sunny rockery or a collection of ferns in a shady one. Less exotic, but just as interesting, are collections of geraniums, mini-roses, herbs, dwarf bulbs, dwarf conifers or even a collection of ornamental grasses. Many Australian native plants do well in rock gardens and can be used as a small native collection or as part of an Australian native garden.

ANNUALS

Annual plants in rockeries are rarely destined to provide the main display but can be very useful as fill-ins—especially to give a quick cover while the slow-growing permanent plants are becoming established. Dwarf marigolds, ageratum, petunias (especially cascade types), mignonette, Livingstone daisy, pansy, phlox, sweet peas (dwarf Bijou), verbena and viola are all suitable annuals. Always try to have a touch of white in the rock garden to highlight the more colourful plants. Carpet of snow (*Alyssum*) is a good standby and has a profusion of white flowers during most of the year. There is also a Rockery Mixture available in seed packets. It contains about twelve kinds of seeds of plants suitable for vacant spots in the rock garden.

ROCKERY MAINTENANCE

Your rockery must be well drained. See Chapter 3 for methods to ensure good

Rocks and plantings skilfully arranged

Alaska nasturtium makes an excellent groundcover.

Marigold Rockery Mixed, a dwarf blend useful for groundcover

drainage. Some plants growing in rockeries cannot tolerate damp conditions and will brown off at ground level during wet, humid periods. To help prevent this from happening, spread a layer of pebbles or pine bark to keep the plants from direct contact with the soil. This layer of mulch will help to suppress weeds as well. On the other hand, rockery pockets (like pots, tubs and other containers) tend to dry out faster than beds in the open garden. Always water rockeries thoroughly—especially in hot, dry weather—to make sure that moisture penetrates the soil evenly and deeply. Slow-release fertilizers such as Osmocote or Mag-Amp are the most suitable for use in rock gardens. They can be scattered on plants or the surface soil and washed in when the garden is watered. Because these fertilizers are released slowly, there is very little chance of fertilizer burn to plants. For small rockeries with easy access, the water soluble fertilizers such as Thrive, Aquasol and Zest are useful too. A rockery may become overcrowded. If and when this happens, prune back the plants so the rocks are not completely smothered. Plants which have grown too large for their situation should be removed and replaced with smaller ones. Once established, rock gardens are usually trouble-free and easy to maintain but need a general clean up periodically to keep plants in check and get rid of weeds.

GROUNDCOVERS

A grass lawn is the most popular groundcover but there are some situations where grass just will not grow or is difficult to maintain. Such situations include shady areas under trees, steep banks or batters which are difficult to mow, rocky outcrops, and damp, soggy spots with poor drainage. The answer to these problem areas is groundcover. Gravel, pebbles, quartz chips and pine bark can be considered as groundcovers. The materials can be used alone or can be combined with plants. Very often the best solution to a damp, shaded area is a mulch of one or other of these materials softened by shade-loving plants such as ribbon grass (*Ophiopogon*), plantain lily (*Hosta*), saxifraga (*Bergenia*), bugleflower (*Ajuga*) or native violet (*Viola hederacea*).

Most groundcover plants are prostrate perennials which spread rapidly by above ground stolons or underground rhizomes. When established, groundcovers need little attention and maintenance. An annual dressing of fertilizer—slow-release fertilizers are the most suitable—and occasional pruning to keep the plants from getting out of hand is about all that is necessary.

It does pay dividends to prepare the site thoroughly before you establish your groundcover. If possible, start preparation in winter and keep the soil cultivated through the spring months to destroy weeds—especially perennial weeds with persistant underground parts. The desiccant weedicides Tryquat, Polyquat and Weedex, and the new weedicide known as Zero (Glyphosate), are very useful here because none of these materials has a residual effect on the soil. Early summer is a good time for planting. Close planting will give a quick cover (assisted by scattering a mixed fertilizer, such as Gro-Plus Complete, before planting). You may need to continue with some weeding until the groundcover is well established and forms a dense mat.

On a bank or sloping site, prostrate shrubs can be used to bind the soil and form a weed-resistant cover. One of the best native shrubs is *Grevillea biternata* with shiny, green foliage and clusters of creamy yellow, lightly-scented flowers. Another attractive, shrubby groundcover is shore juniper (*Juniperus conferta*) which has tiny, grey-green leaves and grows to 60 cm in height. Creeping juniper (*J. horizontalis*) is not quite so tall (30 cm) but spreads over the ground covering 2–3 metres. Bulbous plants like agapanthus, clivea and hemerocallis are good, low-maintenance soil binders for steep banks too. Climbing plants are very suitable for groundcovers on sloping or flat sites. Ivy is one of the most popular. There are many varieties of this adaptable plant. Creeping boobialla (*Myoporum parviflorum*) is an attractive groundcover plant which forms a dense green mat with white, tubular flowers followed by purple berries. Other climbing plants which are useful for soil binding or covering rock faces are jasmine (*J. polyanthum*), star jasmine (*Trachelospermum*), Australian sarsparilla (*Harden-*

bergia), creeping fig (*Ficus pumila*), Virginian creeper (*Parthenocissus*) and ivy geranium (*Pelargonium*). Many perennials make suitable groundcovers, among them bugleflower (*Ajuga*), chamomile (*Anthemis*), ground morning glory (*Convolulus*), pig face (*Mesembryanthemum*), snow-in-summer (*Cerastium*), lamb's ear (*Stachys*) and violets (*Viola odorata*). Many of these have been described in Chapter 11.

Other widely-grown groundcovers are Japanese spurge (*Pachysandra*), with glossy foliage and greenish white flowers; London pride (*Crassula*), a shrubby succulent with oval leaves and small, pink flowers; Spanish shawl (*Schizocentron*), a dense mat of short foliage, small carmine flowers; periwinkle (*Vinca*), glossy green or variegated leaves, blue or mauve flowers, and thyme (*Thymus*) of which there are many forms, all with aromatic leaves. Lemon-scented thyme (*T. citrodorm*), and wild thyme (*T. serpyllum*), are both prostrate, creeping species. Creeping mint (*Mentha requienii*) is also a good carpeting plant with aromatic leaves. Lippia (*L. nodiflora*), which forms a loose mat of small leaves with clusters of tiny white flowers, is a very popular groundcover in South Australia. Most of these groundcovers will tolerate shading but will not stand heavy traffic. You will need to lay stepping stones or discs of sawn hardwood in areas where people are likely to walk.

Davallia (rabbit's foot fern) used as a groundcover

Ivies make attractive groundcovers all the year round

Dichondra is an excellent, dense groundcover, easily grown from seed

Chapter 15

GARDENING IN CONTAINERS

Growing plants in containers is an interesting hobby and a great boon to people who live in flats or home units and do their gardening on a windowsill or balcony. Potted plants are very decorative, especially if they are grown in attractive containers. They can be used to soften and beautify large paved areas like patios and courtyards or can create a focal point in the garden. Perhaps the best thing about container-grown plants is they can be moved about from one place to another, providing they are not too heavy. This way you can give plants a suitable microclimate or show them off to their best advantage. The upsurge in interest in container gardening is due to the very wide range of tubs, pots, troughs and hanging baskets now available, better potting mixtures and a wide choice of plants and varieties to grow in them. Almost any plant can be grown in containers—flowering annuals, bulbs, ferns, creepers, shrubs and trees, as well as more practical plants like vegetables and herbs.

SITUATION AND SOIL

To be a successful container gardener you must choose the right plant for your situation. Home unit balconies are often windy so anything you plant should be able to stand up to the breezes. The amount of sunlight is very important and will also influence your choice of plants. Sun-loving plants—which includes vegetables—need at least 4 to 5 hours of sunlight each day to grow successfully, so check the amount of sunshine before spending money on plants that may not be suitable.

Containers must have free drainage, otherwise your plants will drown. Most pots and tubs have one or several drainholes 1–2 cm in diameter. Cover these with a piece of broken tile or flat stone to stop soil running out. There is no need for a whole layer of crocks, coke pieces or gravel at the bottom of the container providing the soil mix drains freely. Proprietary potting mixtures are available from garden stores and nurseries. These are open, porous mixes which are very satisfactory and have the added advantage of being free from weed seeds, soil pests and diseases. Special potting mixes like Orchid Compost, African Violet Mix and Bulb Fibre (for growing bulbs indoors) are available too.

Ordinary garden soil is often unsuitable for pot culture because it does not drain well and tends to set hard. But you can make your own potting mixture by using one-third average garden soil, one-third compost, peat moss or vermiculite and one-third coarse river sand. The sand must have large particles—almost fine gravel—to keep the mix open and free-draining. If your garden soil is very sandy you should increase the quantity of moisture-holding material and reduce the quantity of coarse sand. For most home-made mixes you should add 30 g of mixed fertilizer such as Gro-Plus Complete to each bucket of mixture. You can also feed potted plants as they grow. When potting up most plants, don't be tempted to put a small plant into a large pot with the idea of saving yourself some work. Plants do not thrive in over-large containers—some prefer to be crowded. It is best to move a plant into a slightly larger pot when the previous one fills with roots.

WATERING AND FEEDING

Whatever you decide to grow in your pots, remember that container-grown plants have a restricted root system and cannot forage for moisture as they would do in the open garden. On hot summer days, daily watering may be needed—perhaps twice a day if the plants are in full sunlight. Always water thoroughly—not just a sprinkle. Leave a margin between soil level and the rim of the container. Fill this space slowly with water until

Pot-grown carrots thrive in a sunny spot

it weeps out of the drainage holes. This way, you restore the soil to field capacity (as much water as the soil can hold). A mulch of grass clippings, compost, coarse gravel, pebbles or pine bark helps to reduce evaporation and cools the surface soil.

Good drainage and frequent watering also means loss of plant nutrients. Regular, small amounts of fertilizer are needed to keep plants growing strongly. The water-soluble fertilizers such as Thrive, Aquasol or Zest are suitable for regular liquid feeds. Use at half-strength for tender plants. Slow-acting blood and bone-based fertilizers or slow-release fertilizers—Osmocote, Mag-Amp or the numerous brands of plant tablets—are also suitable to provide nutrients over a long period. Whatever fertilizer you choose, always use it according to the manufacturer's directions. Too much fertilizer for potted plants can be disastrous, especially if the soil becomes too dry.

PESTS AND DISEASES

Container-grown plants are not immune to attack by pests and diseases. Grubs, bugs, blights and mildews must always be guarded against. A few plants on a balcony can often be kept clear of caterpillars, snails and other leaf-eating pests by picking them off by hand but safe insect sprays are available which are ideal for controlling most pests of pot plants and indoor plants. A few pellets of snail bait at the base of potted plants will control snails and slugs.

FAVOURITE SHRUBS FOR TUBS

Camellias and azaleas with all-year-round foliage and exquisite flowers make excellent tub specimens. They prefer partial shade or filtered sunlight. They need acid soil, so most proprietary potting mixes, which contain a lot of peat moss, are excellent to grow these plants successfully. Growing these acid-loving plants in tubs offers a good solution if your garden soil is alkaline, which it may well be in drier inland districts.

Hydrangeas make magnificent tub plants for shady situations, although they do look rather bare in winter. They give a tremendous flower display in summer and can be brought indoors when flowering. Flower colour depends on whether the soil is acid or alkaline—blue in acid soil, pink in alkaline. If you want your hydrangeas to be a particular colour, grow them in a tub and treat the soil accordingly. (See Chapter 4.)

Fuchsias—and there are dozens of varieties—are dependable flowering shrubs for tubs, pots or hanging baskets. Flowers in white, pink, red and purple are produced over a long period. Gardenia, with handsome glossy leaves and waxy-white fragrant flowers, makes a good tub specimen. Plants prefer full sun or half sun in a warm, sheltered spot. In the right position gardenias will flower from spring to autumn. Daphne is another neat, evergreen shrub with exquisitely perfumed pink, red or mauve flowers in late winter and spring. Plants need good drainage and are often more reliable in a large pot or tub than in the open garden. They prefer morning sun but shade for the rest of the day.

Geraniums will provide the brightest colour on a sunny terrace or patio. Grow them in tubs, large pots or window boxes; with plenty of sun they will flower from early spring to late autumn. The ivy-leaf trailing varieties are ideal for hanging baskets. Bougainvillea is another sun-lover, especially for warm northern climates. Although a creeper, it can be pruned to shape in a tub. It gives a magnificent display of flower bracts in summer. Citrus trees are both ornamental

Rhododendron is a favourite for tub culture

and useful tub plants for outdoor living areas. The smaller citrus trees like kumquats and limes are easy to grow in tubs but lemons, oranges and mandarins need pruning to keep them within bounds. (See Chapter 16.) Japanese bamboo (*Nandina*), with its lacy foliage, makes a good container plant and can be kept in bounds by pruning. True bamboo, which belongs to the grass family, gives an oriental effect for tub culture but only dwarf varieties up to 2–3 m tall should be chosen.

FLOWERING BULBS FOR POTS

Bulbs are the easiest plants to grow in pots. Favourite spring flowering bulbs for this treatment are daffodils, jonquils, hyacinths, bluebells, lachenalias, freesias, tritelias and tulips. Daffodils, hyacinths and tulips are ideal for growing in bulb fibre indoors. In cool districts, lily-of-the-valley can be planted several to the pot and will be a great source of enjoyment either indoors or outdoors. Permanent plantings of large evergreen bulbous plants like agapanthus are suitable for large tubs around swimming pools, barbecue areas or sunny patios. The strap-like leaves are attractive all year round with large clusters of blue or white flowers in summer when outdoor living areas are most used. Clivea, with orange-red blooms in winter followed by attractive red berries, is a good substitute for shady areas. Hippeastrums are popular container-grown bulbs in Europe and favourites for their large red or red-and-white striped, lily-like flowers in spring. They are ideal for pots on balconies, patios and courtyards and can be brought inside as temporary house plants when they flower. Many other bulbs and bulbous plants can be grown this way. Hymenocallis (one of the spider lilies), eucharis, eucomis (Pineapple Flower) and liliums are good examples. Gloriosa (Climbing Lily) can be grown in a hanging basket. For further information on flowering bulbs see Chapter 10.

COLOURFUL ANNUALS

Annual flowers make a colourful display in pots or hanging baskets and are invaluable for brightening a balcony, terrace or patio. Potted annuals are also useful for special occasions like parties or weddings which are planned well ahead. In cold areas, a small glasshouse or glass frame is helpful for raising early annuals which can be moved out to vantage points in the garden or brought indoors when they begin to flower.

For spring flowers, plant cineraria, lobelia, nemesia, pansy, polyanthus, primrose, primula, schizanthus, sweet pea (dwarf Bijou) and violas. These can all be raised from seed or seedlings planted in late summer to autumn. Use proprietary potting mixes or make up your own mixture as suggested earlier. Annuals for summer pots include petunia (especially Cascade for hanging baskets), phlox, nasturtium (Gold Jewel and Red Jewel are good, trailing basket plants) and verbena. All need a sunny position but sheltered from wind. Marigolds flower over a long period in containers—use dwarf varieties for small pots and troughs and larger ones for tubs where they will make a dazzling display. Celosia makes a striking pot plant for hot summer weather. Dwarf varieties in scarlet or gold can be combined or planted separately. Bedding begonias are delightful in quite small pots and make a good show planted in a strawberry pot. Impatiens (Busy Lizzie) is an excellent pot plant for shady areas and flowers almost all year round in colours of pink, salmon and mauve. It is easily raised from seed and grows quickly. The more glamorous hybrids with double, rose-like flowers or handsome variegated foliage must be raised from cuttings which strike readily.

As a change from annual flowers in pots, ornamental chilli has interesting, multi-coloured fruits and kochia (Mock Cypress) makes a nice, symmetrical little plant with soft green foliage. Ornamental basil, with purple-bronze, aromatic leaves is another attractive foliage plant which is ideal for container growing. Most herbs, except the very tall ones, can be grown successfully in pots. Many vegetables, especially dwarf varieties, are ideal for pot culture too. Refer to Chapter 9 for detailed information on both vegetables and herbs.

The aromatic herb oregano grows well in a garden pot

INDOOR PLANTS

Indoor plants have become very popular, especially with people living in home units or flats who do not have a garden. There is really no such thing as an 'indoor plant', although many can be adapted to living indoors. All plants have evolved to grow outside so it is not surprising that the most successful indoor plants are those from tropical rainforests where they grow in the shade of large trees and rambling vines in a warm, humid atmosphere. When you grow these plants in your home you should try to give them similar conditions. Your indoor plants will need good light but not direct sunlight. If there is enough light to cast a shadow (test this by holding your hand up against a piece of white paper) there should be enough for most indoor plants. Temperature should be fairly even (about 20°C) without extremes, if possible. High humidity is also preferred, so in very warm dry weather stand pots in trays of water to increase the amount of water vapour around the plants.

The easiest plants to grow indoors are those with glossy leaves—philodendron, rubber plant (*Ficus elastica*), umbrella tree (*Schefflera actinophylla*), cast iron plant (*Aspidistra*), dragon plant (*Dracaena*), prayer plant (*Maranta*), mother-in-law's tongue (*Sanseveria*) and *Diffenbachia*. Two others with attractive flowers as well are flamingo flower (*Anthurium*) and *Spathiphyllum*. These are the toughest and most reliable indoor plants, so start with these before graduating to the more exotic kinds which are available from specialist nurseries.

There are many flowering indoor plants too—African violet (*Saintpaulia*), tuberous begonia, calceolaria, cyclamen, gloxinia, *Primula obconica* and polyanthus. These require more light and care than the hardier foliage plants. Most of them are described in Chapter 9. Flowering chrysanthemums and dwarf poinsettias can also be purchased for long-lasting flowers indoors.

Most indoor plants will grow better if they are rested outside in a shady, sheltered spot periodically (never in full sun after being indoors). If you have a balcony they can be rested there—three weeks indoors and three weeks outdoors is a good system. If you have a few extra pot plants, some can be resting outside all the time. If there is no sheltered place outside, rotate your plants in the best growing spot beside a well-lit window. Plants dislike being left in dark closed rooms. They need fresh air and light to grow. If you are at work all day and your home is closed, try to arrange some ventilation for your plants. Don't pull down the blinds or draw the curtains, but let the light in. Indoor plants need regular watering. This may mean daily watering in hot, dry weather but perhaps only once a week when it is cooler. Water only when the soil feels dry to touch. Don't water plants if the soil feels damp. In cold weather, it is best to use lukewarm water as cold water from the tap will chill the soil and the roots. Most indoor plants react to stress by drooping their leaves. This can mean they are too dry, too wet, too hot, too cold or too dark. You will have to assess the situation to decide which is the real problem.

Fertilize indoor plants to keep them growing steadily. Small doses of liquid feeds (Thrive, Indoor Aquasol or Zest) at half strength every 3–4 weeks should be sufficient. Otherwise use slow-release plant tablets or Osmocote according to directions. Dust plants regularly and gently. Use a clean cloth or duster for glossy-leafed plants but a soft paint brush or old shaving brush for those with furry leaves. It is a good idea to put your plants outside when there is a gentle rain falling. This will wash them free of dust and leach the soil of any build-up of salts from hard water. Plants which have grown too large for their containers can often be rejuvenated by cutting off the tops to let them shoot again from the base. Rubber plant and dieffenbachia can be given this treatment.

ORCHIDS

Orchids make lovely container plants. In mild temperate climates like Sydney, the pots or tubs can be placed under trees in filtered light. Orchids thrive in warm subtropical and tropical areas but need glasshouse conditions in cold districts. When growing orchids in containers it is best to use the special mixtures of orchid compost. These proprietary mixes do not contain any soil and the roots of the plants can move through them freely. Orchids are epiphytes (a plant attached to another plant but not a parasite) and under natural conditions they grow in the debris of bark and dead leaves of trees. It is a good idea to place a layer of crocks, coke pieces or pine bark at the bottom of the container before adding the compost.

Some species of orchid are very easy to grow. The crucifix orchid (*Epidendrum*) can be grown in the open garden in mild climates but also makes an attractive pot plant for a sunny position. There are hundreds of species of tree orchids (*Dendrobium*), many of them Australian natives. The Sydney rock lily (*D. speciosum*) and tongue orchid (*D. linguiforme*) are the best known and very adaptable to

Cymbidium orchids make excellent container plants

climates which are frost-free. Other tree orchids do best in warm tropical areas but need glasshouse treatment where the climate is cooler. Cymbidiums are the most popular orchids for mild climates where the temperature does not drop below 5°C. They make ideal plants for large pots and tubs which can be moved to a favoured position for flowering. They need light shade in summer but full sun in winter when the flower spikes are forming. Flowers last for many weeks. There are hundreds of varieties in shades of white, yellow, pink, red, brown and green.

Slipper orchid (*Paphiopedilum* syn. *Cypripedium*) is a reliable orchid for a small pot (they like to be crowded) and blooms in winter or early spring. Dancing ladies (*Oncidium*) are also delightful orchids for pots or hanging baskets. Natives of Brazil, they need a warm, subtropical climate but can be grown in a shadehouse or glasshouse in cool areas. Angel orchid (*Coelogyne*) has pure white flowers with golden throats in autumn and winter. They are very adaptable to climate, providing there are no frosts. They like to be crowded in the pots and prefer rather dry conditions when flowering. Chinese ground orchid (*Bletilla*) is another adaptable orchid for pot culture. They are dormant in winter and flower in spring. *Vanda* and *Cattleya* are probably the showiest of all orchids but need a warm, humid, tropical climate or glasshouse conditions in cooler areas.

Most orchids respond to supplementary

Dendrobium, a favourite Australian orchid

feeding in addition to the compost in which they are growing. The soluble orchid foods or liquid feeds of Thrive, Aquasol or Zest can be used every 2–3 weeks during the growing period. Do not fertilize orchids when they are dormant.

CACTUS AND OTHER SUCCULENT PLANTS

Cactus plants and other succulents make ideal pot plants for a sunny windowsill, balcony or patio. They are easy to grow and require very little attention. Too much sun and heat will scorch leaves and care must be taken to avoid this. They prefer a dry atmosphere so they need plenty of fresh air in humid climates. For pot culture, a suitable mixture consists of coarse sand or fine gravel with compost, leaf mould or peat moss added. This mixture holds water but is free-draining. Although these fleshy plants can go for long periods without water in their natural habitat, the pots should be watered when the mixture is dry to the touch. Avoid heavy feeding because this promotes rapid, soft growth and plants may rot. Slow-release plant tablets or a scattering of Osmocote (used according to directions) are possibly the most suitable fertilizers to use.

The spines of cactus plants are very sharp and may be difficult to remove from the skin so care must be taken in handling the plants. People with sensitive skins are advised to consider those succulents without spines such as *Kalanchoe*, hen and chickens (*Echeveria*), stonecrop (*Sedum*), jade plant (*Crassula*) and *Sempervivum*. Most of these quaint, fleshy plants have attractive flowers. Crab cactus (*Zygocactus*) and orchid cactus (*Epiphyllum*) are also spineless with jointed, flattened stems and showy flowers in winter and spring. Both are ideal for pots or hanging baskets.

FERNS FOR POTS AND BASKETS

Some of the smaller ferns of the dozens of different species available can be grown in containers indoors, or more often in a sheltered patio or courtyard. Most are grown in pots or hanging baskets but some, like staghorns and elkhorns, can be wired to wooden uprights or boards.

Ferns as a group need cool, moist conditions. When indoors, they are best in full light from a window facing south or where summer sunlight can be excluded by curtains. Free air movement is essential as ferns resent still, dry air and are prone to attack by insects (aphids, mealy bug and scale) under poorly ventilated conditions. (See Chapter 17.) It is best to bring potted ferns indoors for decoration for short periods (1–2 weeks) only and then return them to a shady, sheltered spot for the same length of time.

A suitable soil mixture for pots and baskets of ferns consists of equal parts of garden loam, sand and peat moss or leaf mould. Potting up is done in late winter or early spring when new fronds appear. Pots or baskets should not be too large as the plants prefer to be crowded. Wire baskets are usually lined with moisture-retaining fern bark or sphagnum moss. Ferns must be kept damp at all times although they require less water in winter when growth is slower. Ferns need little feeding and dislike concentrated fertilizers. The safest method is to use water-soluble fertilizers. Apply them at half-strength every 3–4 weeks during warm weather.

Maidenhair fern (*Adiantum*) is probably the most popular and widely-grown fern for pots and baskets. The fronds are finely cut on long wiry stems—a favourite for flower arrangements and bouquets. There are over 200 species of maidenhair fern which differ in leaf shape and size. Rabbit's-foot fern (*Davallia*) has lacy fronds curving downwards and furry, creeping rhizomes. It is an ideal plant for baskets and tolerates warm conditions providing it is kept moist.

Fish bone fern or sword fern (*Nephrolepis*) with bright green upright fronds is perhaps the easiest of all to grow. It tolerates tough conditions—even full sunlight—and spreads rapidly. Brake fern (*Pteris*) is easily propagated from new crowns which develop from the rhizomes. A handsome fern with curled or crimped, many-lobed fronds. Bird's nest fern (*Asplenium*) is a clump of shiny, undivided fronds which may grow to one metre long. It is an epiphyte and may be grown on pieces of old tree or timber but is equally at home in a pot filled with sand

Ferns provide handsome decoration indoors or outside

and leaf mould or peat moss. Propagate by quartering the plant and repotting each section, which will regain the full rosette shape.

Staghorn or elk's horn fern (*Platycerium*) are epiphytes which can be attached to a number of surfaces, including stone or brickwork, cork or timber. Popular in courtyards and ferneries where there is semi-shade. Tree ferns (*Cyathea*) are the giants of the fern family and are best grown in a shady position out of doors. They can be grown in large tubs or in a corner bed of a courtyard. Several species are popular, including black tree fern (*C. medullaris*), New Zealand or silver tree fern (*C. dealbata*) and Tasmanian tree fern (*Dicksonia australis*). All species need semi-shade and plenty of moisture.

POTTED PALMS

Most palms can be grown in pots or tubs for several years before they become too large. Like many other plants grown in containers, palms prefer their roots to be crowded, so it is best to pot them on to a larger container only when the previous one is filled with roots. Good drainage and an open, porous soil mixture are essential. Water regularly and feed with water soluble fertilizers, slow-release plant tablets or Osmocote. For descriptions of the most popular palms, see Chapter 12.

Chapter 16

ROSES

Roses have always been a favourite flower in the garden. They were cultivated by the ancient Babylonians, Greeks and Romans, and no other flower has received more attention from gardeners through the ages. Modern varieties, with their superb colour range, flower form and fragrance make them irresistible for garden display or for cut flowers. Roses are extremely adaptable to both climate and soil. In cool, temperate and cold climates, roses have successive flushes of bloom during warm weather, but in warm subtropical or tropical climates they flower all year round. Many of the roses sold by florists during winter are grown under glasshouse conditions.

There are so many different types of rose that there is one to suit any situation in the garden, always providing there is sufficient sunlight and good drainage. Hybrid tea or bush roses can be used in garden beds, generally with low-growing winter or spring flowering annuals, perennials or bulbs to give colour to the garden during their dormant period. Floribunda roses make good borders or low hedges while climbing roses can be trained on walls, fences and pergolas. Standard roses or weeping standards, which are grafted on to tall root stocks, are excellent for borders or accent plants. They provide lots of cut flowers and allow other annual flowers (and even vegetables) to grow beneath them. Standard roses should be supported with a sturdy stake as the standard may be rather brittle and subject to buffeting by winds. The hardwood stakes should be driven into the soil before the rose bush is planted to avoid root damage.

SITUATION AND SOIL

Roses need a sunny position to grow and flower well. They should not be grown too close to other shrubs or trees which will compete with them for light, moisture and nutrients. Good drainage is necessary too. Raising the bed 15–20 cm above the surrounding level will usually provide sufficient drainage on heavy soils. Roses are very adaptable plants and can be grown on both sandy and clay soil. While they tolerate clay soil better than most plants, they do not prefer or need a clay soil to grow well. The ideal soil is a loamy topsoil with good structure and a clay subsoil which will give an even supply of moisture, providing the clay is well drained and allows excess water to move away from the root zone. Sandy soils, which hold moisture badly, should be improved by adding plenty of organic matter as described in Chapter 3.

Because roses are long-lived plants it is well worth while spending some time and energy in preparing the soil well before planting. It is best to add the organic matter (animal manure, compost or spent mushroom compost) and dig it into the topsoil about 4–6 weeks before planting. If the soil is naturally acid, lime can be added at the same time. Use about $\frac{2}{3}$ of a cup per square metre on sandy soils and double this quantity on heavy soils. As roses cannot make use of fertilizers until growth commences there is no need to apply them during the soil preparation stage. Have the soil in a good crumbly condition for planting by digging it over again about two weeks before planting to break up any clods and mix the organic matter through the soil more evenly. Always cultivate heavy soils in the dark damp condition so that they crumble easily.

PLANTING

Roses are planted during their dormant season. Rose plants are sold bare-rooted at garden stores and nurseries from late May to August. In mild climates, June is a good month for planting because the plants are then completely dormant. In colder areas—Victoria, Tasmania and the eastern highlands—planting can be delayed until July or even August. It is also possible to buy container-grown roses for year-round planting but the vast majority of roses and a much wider selection of varieties are available as bare-rooted plants.

Bare-rooted roses dry out quickly and when the roots have dried the plants will very often die. When buying plants check that the stems, which may be bright green or reddish brown in colour, are smooth and free of any wrinkling, especially at the top of the stem. Such wrinkling indicates that the plants have been allowed to dry out at some stage. Don't buy rose plants until you are ready for planting. The bed should be finally prepared and well settled so you can start planting when you bring the plants home. If you are unable to plant that day, unwrap the plants and heel them into a shady spot in the garden. Cover the roots well with soil and water thoroughly. This way you can 'hold' plants until they can be put into their permanent positions. When you have a number of roses to plant, cover them with a wet bag or stand them in a bucket of water.

The planting hole should be large enough to allow the roots to spread out naturally without bending them. Holes should be not less than 30 cm in diameter and about 20 cm deep. Make a mound of crumbly soil on the bottom of the hole on which to rest the plant and spread the roots. The height of the mound is adjusted so that the bud union (where the scion is

Weeping standard rose

budded onto the root stock) is slightly above or at soil level. Now, cover the roots with 7–10 cm of damp crumbly soil and firm it down well to prevent pockets of air remaining around the roots. Then add half a bucket of water gently, so that soil is not washed away from the roots. Let the water drain away and then fill in the remainder of the soil without firming. The leftover soil is used to form a raised ring around the plants to help direct later watering to the plant roots. Keep the surface soil free and moist by scattering compost, leaf mould or grass clippings around the plant. Later watering will depend on the weather. Check newly planted roses once a week and if soil is dry give about half a bucket of water. After planting use only a slow-release type of fertilizer such as Yates Rose Food, Agriform plant tablets or Osmocote as the plants are dormant and will not utilize it until growth commences in early spring.

GENERAL MAINTENANCE

WATERING, MULCHING, WEEDING

When rose plants are established, heavy watering encourages the roots to go deeper so that plants can tolerate longer periods of dryness. This approach is preferable to frequent light sprinklings which promote shallow roots close to the surface. During the main growing period from spring to autumn a mulch of compost, leaf mould or grass clippings around the base of the plants will help to conserve moisture and control weeds. When rose bushes lose their leaves in autumn, little or no watering is required except in warm, tropical climates where roses behave more or less as evergreen plants. Contact, desiccant weedicides like Tryquat, Polyquat or Weedex are also useful in controlling weeds among rose bushes if they are grown alone without annual plants. They can be sprayed right up to the stems of the roses providing they do not contact leaves or green, sappy stems. These weedicides have no residual action in the soil and break down quickly into harmless compounds.

FERTILIZING

Fertilizers are necessary for vigorous growth and good sized, quality flowers. After planting time use only slow release

American Pillar rose

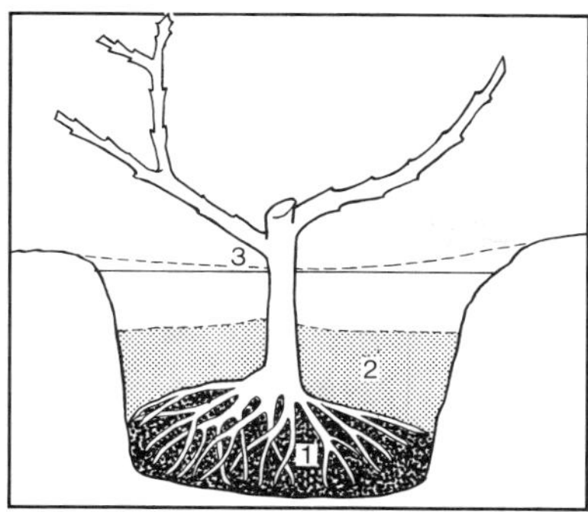

1. Mound of loose soil to support roots.
2. 7-10cm crumbly soil well pressed down.
3. Bud union level with, or slightly below ground level.

Correct method for planting a rose bush

fertilizers as recommended above. An application of mixed fertilizer can be given in late December or January to encourage an autumn flush of blooms. Fertilizers containing blood and bone, such as Gro-Plus for Roses, are ideal.

With established plants, apply fertilizer in late winter or early spring and again in late summer. This can be scattered around the plants (but not too close to the main stem) and lightly raked into the soil. If the soil is mulched, the fertilizer can remain on the surface as the nutrients will soon wash down through the mulch to the soil below. Always apply fertilizers when the soil is evenly moist and water well afterwards to disperse the nutrients safely to the root zone. Alternatively, slow-release fertilizers such as Osmocote, Agriform Plant Tablets or Mag-Amp are suitable.

CUTTING BLOOMS

With young rose bushes, do not cut flowers with long stems as the plants need as many leaves as possible to develop into vigorous bushes. Cut blooms with short stems only. With older bushes, cutting flowers with short stems leads to tall, leggy growth, so make the cut more towards the base of the stem to encourage new growth to come from eyes or buds where the stem is thicker and sturdier. Always make cuts about 6 mm above an eye and slanting back slightly behind the bud. Roses keep best if cut late in the afternoon and placed in a bucket of water overnight for arranging in the morning.

PRUNING

Summer pruning is often called 'summer trimming', and is usually done in late January or early February. It consists of a tidy-up, removing any dead branches or those showing dieback by cutting back to a bud about 5 cm below the dead section. Any unproductive stems which have

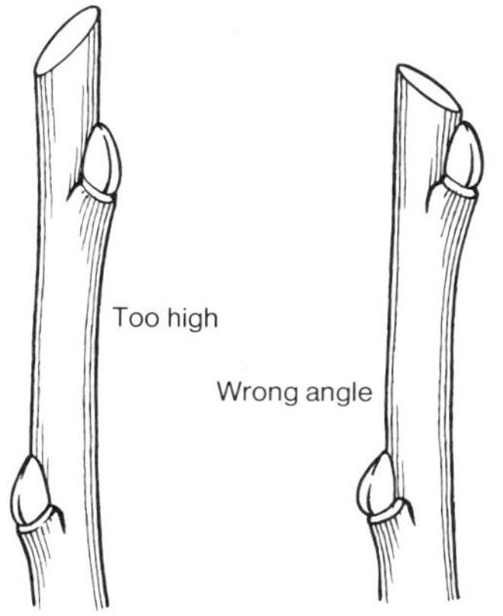

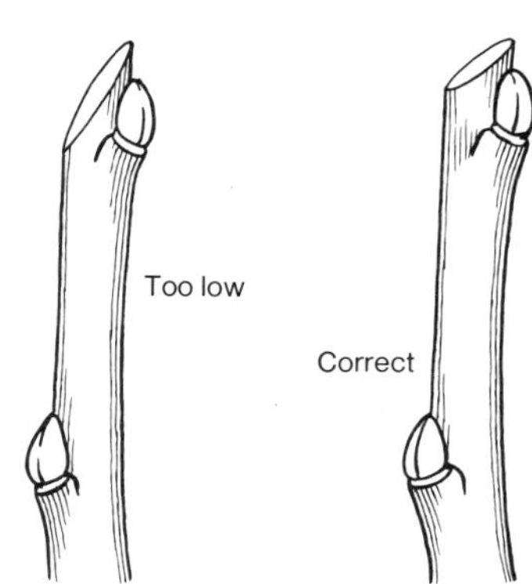

How to prune roses. Make sure the secateurs are clean and sharp to avoid bruising. The cut should be made just above a bud pointing in the direction you wish the new shoot to grow

not produced good flowers or new shoots can also be removed. 'Water shoots' are tall, vigorous, sappy shoots which suddenly develop from the crown of the plant (above the bud union) or from an old cane—they should not be mistaken for 'briar shoots' (which have very small leaflets and arise from below the bud union). Always leave the water shoots because they are important in forming the future framework of the bush. They can be cut back when the wood is fully matured in winter.

Winter pruning is done in late July or early August in most districts but may be delayed until late August or early September in cold climates. The objectives of pruning are to remove dead, old or diseased wood, shorten back healthy branches to promote new growth (flowers are borne on new wood) and to keep the bushes a suitable size and shape. Make sure that the secateurs are clean and sharp and use a pruning saw for cutting thick, woody stems.

For dwarf or bush roses, cut out all dead, yellowing and diseased wood. Also discard thin or weak stems, those which rub against each other or are too crowded. Leave the strongest stems and shorten these back by one-third of their length, cutting to a bud pointing in the direction in which you wish the stem to grow. Floribunda or polyantha roses are usually cut back rather harder than bush roses, shortening the stems by one-half. Standard roses are pruned in the same way as bush roses but cut each stem to an outward pointing bud to avoid too much centre growth and retain a neat shape. Climbing roses require slightly different treatment. Some gardeners give them a light pruning in winter and their main pruning after flowering in spring, removing old canes and dead wood and shortening the remaining canes. Rambling or *wichuriana* roses, most weeping standard roses and Dorothy Perkins types flower in spring on wood grown the previous year. Canes are cut out after flowering, but if this is too drastic they can be cut back half-way to stimulate new growth.

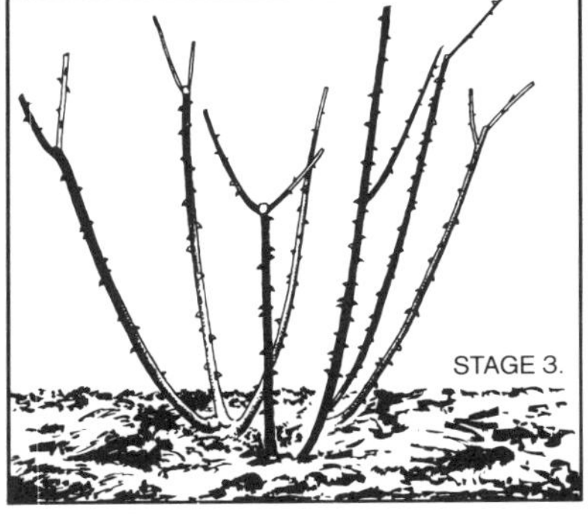

Correct cuts for rose pruning

PESTS AND DISEASES

Rose bushes can be attacked by a number of pests and diseases. Aphids (there are

Efficient use of space enhances this Gold Coast garden

Rhododendrons flank the path above these well-proportioned steps, whilst Lobelia adds a touch of contrast

Kennedia retrorsa is a climber or vigorous ground cover

Grevillea juniperina needs a sunny, well-drained position

Blue-bell creeper *(Sollya heterophylla)* flowers in spring and summer

Pandorea jasminoides is ideal to cover trellis or pergola

Hybrid tea roses make a delightful decoration

several kinds which attack roses) are the worst pests. Colonies of green, brown or pink insects cluster together on young, sappy shoots and flower buds. Thrips, also sap-sucking insects, are just visible to the naked eye and damage both buds and flowers. Red spider, a small mite, infests the lower surface of leaves, causing yellowing or browning and premature ageing. Some caterpillars chew holes in leaves and others roll leaves together. There are also several scale insects which attack roses.

The most serious diseases of rose plants are black spot and powdery mildew. Black spot causes small blackened areas, yellowing leaves and premature leaf fall. Humid weather with heavy night dews favours its spread. Powdery mildew produces a white powdery growth on leaves, stems and buds. It is most active in warm humid weather.

Pests and diseases can be effectively controlled by dusts or sprays of insecticides and fungicides. The recommended chemicals for control are given in Chapter 17. Many gardeners will prefer to use an all-purpose rose dust or rose spray to control both pests and diseases rather than use individual chemicals. There are a number of such all-purpose sprays and dusts available.

Chapter 17

FRUIT TREES AND FRUIT PLANTS

Wherever you live in Australia—sunny Queensland, the Apple Isle or any place in between—you can grow garden-fresh fruit. Fruit trees and fruit plants are not difficult to grow. Very often, a suitable microclimate exists or can be created in your garden. This means that the most unlikely fruits can be grown outside their natural climate—a passionfruit or grape vine in the cool highlands, a pineapple on a sunny patio in Sydney, or an avocado in a sheltered site in Melbourne.

Once upon a time, fruit trees and fruit plants grew in nearly every Australian garden. Nowadays gardens are smaller so it is difficult to find space for this very rewarding activity. But most home owners can grow two or three trees, especially citrus, which take up little room. Many fruit plants—passionfruit, grapes and trailing berries—are great space savers because you can train them on a fence or trellis. The aim of this chapter is to give background information on kinds and varieties of fruit together with recommendations for planting and management. Pest and diseases of fruit trees and fruit plants are dealt with in Chapter 17.

CITRUS TREES

Citrus (lemons, oranges, grapefruit, mandarins, limes and kumquats) have never lost their popularity. The vitamin-rich fruit can be picked progressively over a long time. This makes them ideal for home growing. But they are ornamental as well as useful and make attractive trees with dark green, glossy foliage and fragrant blossom in spring.

Climate, Aspect, Soil Citrus trees do well in all warm and mild climate zones. Providing frosts are not severe, trees will tolerate cool conditions. They also thrive in hot, dry, inland districts with irrigation. Citrus require a sunny position, preferably facing north and protected from strong winds. They are most successful on sandy or loam soils and dislike clay soils or those with a heavy subsoil. Heavy soils become over-wet and drain poorly, leading to root rot problems. If drainage is poor, build the bed up 25–30 cm above the surrounding soil. Improve the texture of clay soils by adding generous amounts of sand and organic matter. (See Chapter 3.) Most varieties of citrus trees are now budded on *trifoliata* root stock which is more tolerant to wet feet than common lemon stock.

Planting Advanced citrus plants are available from nurseries and garden stores. They usually come in large plastic pots or flexible plastic bags, so planting time is less critical than in the past when 'bare root' or open-ground plants were sold. Make sure the trees have not been in the container too long, as they may have become root-bound and root-bound trees often fail to make satisfactory growth. The ideal tree is one year old from budding. It is best to plant trees in early autumn or early spring. This way, young trees avoid the effects of both winter cold and summer heat.

Make the planting hole rather shallow, but 30–50 cm wider than the container. Tip the tree from the pot or cut away the plastic bag. Gently tease out any roots which are slightly pot-bound. For planting depth, keep the bud union (the knee-like joint where the tree has been grafted) at the same level or about 2.5 cm deeper than soil level in the container. For trees balled in hessian, keep the root-ball intact and cut the ties after the tree is in position. Then fold back and trim off the hessian around the bottom. Pack damp, crumbly soil, to which some sand and compost has been added, around the tree and water well. Mulch the soil with dry grass clippings, compost or leaf mould, but keep it 5–7 cm away from the stem.

Grapefruit

An acceptable and simple method of applying fertilizer at planting is to use Agriform Tree Tablets, a tightly compressed block of fertilizer which releases nutrients slowly and will continue to feed the tree for up to eighteen months. Do not use powdered fertilizers at planting as damage may occur. If using powdered fertilizers, wait for 4–5 weeks when blood and bone-based fertilizers like Gro-Plus for Camellias and Azaleas or slow-release fertilizers can be scattered around the tree in moderation. Slow-release fertilisers like Osmocote or Mag-Amp are also suitable. Do not over-water newly planted trees—a good drink every week is sufficient in dry weather.

Pruning and Management Citrus trees tend to be self-shaping and so need little pruning. If growth is overcrowded, thin out the stems after fruiting because flowers and fruit are carried at the ends of branches. Don't thin oranges and grapefruit severely, but mandarins can be shortened back to the second or third shoot down the branch. Lemon trees are taller and less compact so prune them back to keep them at a manageable height. Sappy water shoots of lemon and grapefruit should be cut away unless they can improve the tree shape. Any shoots below the bud union must be removed too. Old citrus trees can be 'skeletonised' in early spring by cutting back to the main branches. They will take a year or two to recover and bear again.

Feeding Feeding roots of citrus are located at the 'drip-line' underneath the outer foliage, so do not cultivate deeply in this area. When applying fertilizer, scatter it around the 'drip-line' and not close to the trunk. Use a complete fertilizer containing about 10 per cent nitrogen. There are several brands of citrus fertilizer including Gro-Plus for Citrus. A well-grown, mature citrus tree should be given 450–500 g of fertilizer each year. Apply two-thirds of this amount in late winter or early spring (July-August) and the balance in late summer (January-February). Water trees well both before and after fertilizing. Fruit drop is a common citrus problem, associated with irregular or uneven watering, especially when young fruits are forming. Lack of fertilizer (or too much of it) can aggravate this condition. Feed trees as suggested above and water them regularly through spring and summer.

LEMON

Lisbon and Eureka (also known as Sweet Rind) are the most popular varieties. Meyer is rather sweeter in flavour and the trees are more tolerant of cool conditions. Lemons bear overlapping crops for picking throughout most of the year.

ORANGE

The Valencia orange is the most widely grown. It is a reliable cropper with fruit for picking from spring right through to the following autumn. The Washington Navel is an excellent quality orange and usually seedless. It bears from late autumn to spring. It is not as consistent a bearer as Valencia and is more liable to fruit drop.

GRAPEFRUIT

Marsh Seedless is the most popular variety; a good cropper which ripens from early winter for picking over some months. The Wheeny variety also bears well but fruits have an acid, lemon-like flavour

An orange tree is ornamental as well as useful

and many seeds. Thompson is a good variety for dry inland districts. Unfortunately, grapefruit are attractive to the fruit fly. This pest can be a problem, especially in coastal areas of Queensland and New South Wales. (See Chapter 17 for methods of control.)

MANDARIN (TANGERINE)

Best early varieties are Imperial and Unshiu. Both are ready to pick in late autumn. A good late variety is Emperor. Ellendale, another late variety sold as a mandarin, is more correctly a 'tangor'—a mandarin-orange hybrid. The tangelo—a mandarin-grapefruit hybrid—has large, juicy, mandarin-like fruits, the best variety of which is Seminole.

KUMQUATS

These small attractive trees are suitable for the open garden or for tub specimens. The small, bitter fruit makes excellent jams, jellies and liqueurs. Marumi (round fruit) and Nagami (oval fruit) are the best varieties.

LIMES

The lime or sweet lime is a small, many-branched tree with green, thin-skinned fruit about 5 cm in diameter. Tahitian is the most popular variety available from leading nurseries.

DECIDUOUS FRUIT TREES

These summer fruiting trees lose their leaves in winter, so can be used where

Eureka lemon

Meyer lemon

summer shade or winter sun is needed in any part of the garden. Most trees have a magnificent display of spring blossom and some, especially pear trees and persimmons, have attractive autumn foliage. Because they are winter dormant, deciduous fruit trees can be grown in districts with cold winters and severe frosts, although late frosts in spring can damage flower and leaf buds. Another problem is fruit fly in warm coastal areas. Fruit ripens in summer when these pests are most active. Apples and pears are also attacked by codling moth which is, perhaps, harder to control than fruit fly. (See Chapter 17.)

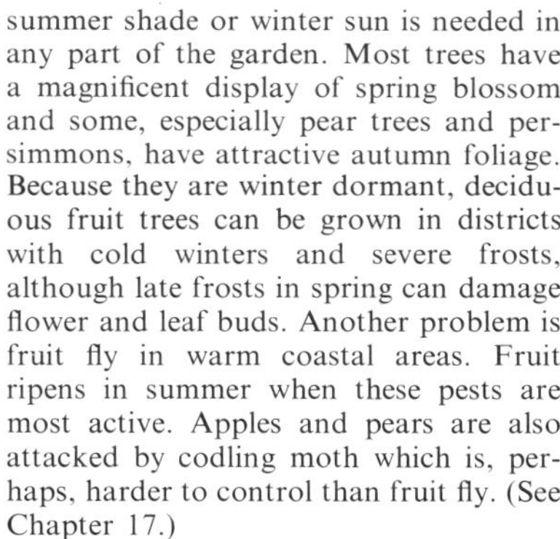

Aspect and Soil Deciduous fruit trees prefer an open, sunny position but will grow on a variety of soils. Like citrus trees, they need good drainage and will not tolerate heavy or water-logged soil.

Planting Plant all kinds and varieties of deciduous fruit trees in winter—June or July would be the best months in most districts. Unlike citrus, they are sold as 'bare root' trees, rather like rose bushes. Although the trees are dormant, roots must not dry out.

Dig shallow holes for planting but break up further soil at the bottom. Very wide holes are not necessary—for most trees a diameter of 60 cm is ample. Plant trees rather deeper than they were in the nursery but keep the bud union well above soil level. Spread roots carefully on a mound of soil, placing the longest and strongest in the direction from which prevailing winds will blow. Fill the hole with crumbly topsoil and tread it firmly. Now add a bucket of water to settle soil around the roots and fill the hole with dry soil to ground level. No fertilizer is needed now (trees cannot use it when they are dormant) but when growth starts in spring, scatter a mixed fertilizer such as Gro-Plus Complete or Gro-Plus for Roses around the tree at the rate of $\frac{1}{3}$ cup per square metre. Chip or rake fertilizer into the topsoil and water to field capacity.

Pruning Pruning aims to regulate growth of branches, allow light to enter the framework and encourage flowers and fruit. Most deciduous fruit trees are pruned to a 'vase' shape but many can be trained as an espalier, which requires a frame support. Generally, trees are pruned in winter, and usually hard for the first few years, to make a sturdy framework and to shape the tree. Later, the main concern of pruning is to remove dead and diseased wood, ingrowing branches, thin overcrowded growth and to cut back leaders or main limbs.

Each kind of fruit tree has its own habit of growth. Fruit may set on current spring growth, on wood formed the previous year or on fruiting spurs which may bear for several years, so it is important to know the bearing wood for each kind before pruning. Brief notes are given in this chapter but detailed information on

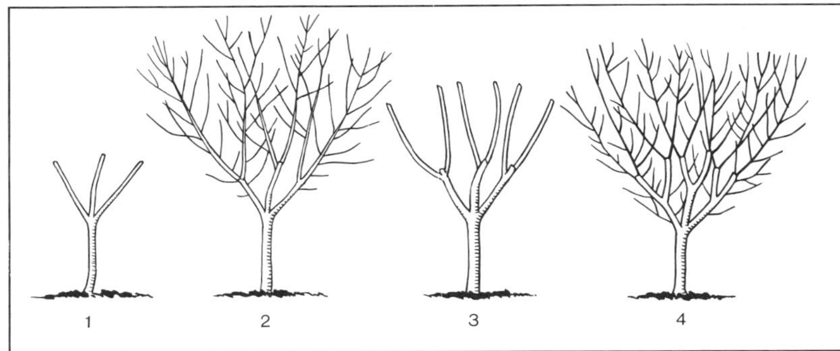

Deciduous fruit trees are pruned in winter to an open-centred vase shape. This shape forms a sturdy framework. Diagrams show: (1) tree pruned after first-year growth; (2) second-year growth; (3) tree pruned after second-year growth; (4) third-year growth

pruning each kind of fruit tree is available from the Department of Agriculture in your own state.

Feeding Fruit trees, like other plants, require feeding. On deep fertile soils it may be some years before fertilizers are needed, but in most gardens, yearly applications of fertilizer will improve growth and fruit yield. A mixed fertilizer which contains about 10 per cent nitrogen is recommended. For young trees, apply fertilizer at the rate of 0.5 kg per tree for each year of age up to bearing. Scatter the fertilizer in late winter or early spring when growth commences. On very poor soils, supplement this by a further application of 250 g per tree in mid-summer (December–January). Slow-release fertilizers used in late winter are also suitable. Animal manures, if available, are a useful way of supplying plant nutrients.

APPLES

Apples grow best in cool to cold climates with a mild summer and cold winter. Most varieties are self-sterile so two varieties are needed for pollination. Some nurseries can supply two or three varieties budded on to one root stock to solve this problem. Apple trees bear fruit on spurs and two-year old or older laterals. Fruiting wood is encouraged by allowing laterals to develop on terminal branches to remain uncut until buds form. Then shorten them if necessary.

Three popular early varieties are Willie Sharp, Early McIntosh and Red Gravenstein, any two of which will cross-pollinate. Late maturing varieties, however, have better quality fruit. The best of the reds is Delicious, an excellent eating apple with a characteristic flavour, and Jonathan, good for eating and cooking. Granny Smith is the most popular green apple, excellent for cooking and pleasant eating when fully ripe. Any two late varieties will cross-pollinate. The development of dwarf apple trees (and pears) for commercial orchards now seems possible. These dwarf, early-maturing trees should have a place in the small home garden too.

APRICOTS

Apricots are very adaptable. They grow well on the coast, in the cool highlands and in hot inland areas. All varieties are self-fertile. Apricots bear fruit on laterals produced the previous year and on spurs which often bear for several years. Strong water shoots develop on some varieties. Cut these back in mid-summer to produce fruiting laterals and spurs.

Glengarry is the most popular variety in coastal districts. It is a good cropper, maturing fruit in mid to late spring. Trevatt, the leading variety for canning, is recommended for cool highlands and inland districts. Morocco is a later variety but keeps rather better than Trevatt.

CHERRIES

A highland climate with mild summers and cold winters is best for cherries. Soils

Apples thrive in cool districts

must be deep and well drained. Trees grow to a large size, so are not well suited to small gardens. All varieties are self-sterile so two varieties are needed for pollination. Cherries bear fruit on spurs growing on two-year old or older wood. After initial tree shaping, little pruning is needed. Pruning, when necessary, should be done in autumn as trees are subject to 'gumming'. Cuts heal quicker then than when trees are fully dormant.

Early varieties are Burgesdorf and Early Lyons. Both have dark red fruit and will cross-pollinate. Ron's Seedling is a dark red, mid-season variety pollinated by Early Lyons. Napoleon and Florence are mid-season, white flesh varieties which will cross-pollinate. St Margaret is the leading late variety with red flesh. Use Florence or Black Boy, another very dark-skinned late variety, as pollinators.

FIGS

Figs, like apricots, are very adaptable to a wide climate range—cool, warm or hot. Trees flourish on the more humid coast but fruit may ferment on the tree in wet seasons. Figs are self-fertile and fruit ripens over a long period. Birds can be a problem—nylon netting over the tree is probably the best answer. Figs bear fruit on current season's growth. Trees usually form a well balanced framework so little pruning is needed.

Purple (Black) Genoa has dark skin and red flesh. White Genoa, pale green skin and creamy flesh, and Brown Turkey, a prolific cropper with dark-skinned fruit, are popular dessert varieties. White Adriatic, green-brown skin and yellow flesh, is a smaller fig but excellent for jam-making.

MULBERRY

Mulberry trees grow to about 6 m with large leaves and small fruit—about blackberry-size but rather longer. They grow well in subtropical, temperate and cool climates and are often grown as ornamentals for summer shade. Mulberries grow easily from cuttings and usually bear fruit in the second year. Apart from initial shaping, little pruning is needed. Top growth and long leaders on older trees can be cut back for easier picking.

Black English is the best variety for cool districts. Hicks, a variety of white or Chinese mulberry, is more suitable for warm climates. The fruits of both varieties make a pleasant, fresh dessert and are excellent for jam or jelly. Leaves of Chinese mulberry are a favourite diet for silkworms if the younger members of your family keep them.

PEACHES AND NECTARINES

Some varieties of peaches and nectarines (which are really a smooth-skinned peach) grow best in subtropical and temperate climates with a mild winter, while others have a high 'chilling requirement' and need a cold winter. Selecting varieties suited to your area is important. All peaches and nectarines are self-fertile, with the exception of J.H. Hale which requires a pollinator flowering at the same time. There are many peach varieties—white or yellow flesh, both clingstone and freestone with a range of inbetweens. Peaches and nectarines bear fruit on laterals produced the previous summer. Laterals fruit for one season only so there must be plenty of new ones coming on each year for continuity of cropping. Prune back trees each winter to encourage new growth but don't remove too much wood carrying buds for next spring.

In warm coastal climates, early varieties should mature before Christmas to escape damage by fruit fly. Watts Early (white freestone), Cardinal, Hiland and Maygold (all yellow flesh, semi-freestone) are recommended. For cool, highland and cold inland areas, varieties must flower late to avoid frost damage. The best varieties (in order of maturity) are Early Becky (white freestone), Starking Delicious (yellow freestone), Redhaven (yellow semi-freestone), Halehaven, J.H. Hale and Blackburn (all yellow freestone). J.H. Hale needs Halehaven or Blackburn as a pollinator.

The best nectarine is Goldmine (white freestone). It is mid-season in maturity and does well in all districts.

PEARS

Like apples, good quality pears require a mild summer and cool to cold winter. China pears do well in warm coastal districts and are used mainly for cooking. The trees are attractive with spring blossom and coloured autumn leaves. Pruning pear trees is the same as for apple trees. Trees take five years or more to bear but they are very long-lived.

Williams (Bartlett) is the most popular variety for eating and cooking. Pick fruit when firm and ripen indoors. Williams is self-fertile but usually crops better with another variety as a pollinator. Beure Bose and Packham's Triumph bear high quality fruit which keep well. Fruit is left to ripen on the tree. Williams is the best pollinator for both varieties.

PERSIMMON

Persimmons, like apricots and figs, are adaptable to a wide climate range. Unlike most deciduous fruit trees, they tolerate moist conditions and heavy clay soils. These attractive, spreading trees grow to about 5 m in height. They are ornamental, too, with brilliant autumn foliage. The salmon-pink fruit is picked when coloured but still firm. Ripen fruit indoors until quite soft before eating. Persimmons have a unique texture and flavour. After the tree has been shaped, little pruning is needed. Fruit is produced on the current season's wood. Fruit on the topmost branches can be picked with a long-handled rake. As most varieties are harvested in autumn or early winter, they escape fruit fly attack. The variety Yemon (Nightingale) is a compact tree suitable for small gardens. Larger varieties are Dai Dai Maru, Hachiya and Tanenashi. Persimmons sucker freely and you can establish new trees by digging these and replanting them when dormant in winter.

PLUMS

There are two types of plums. The European plum has a high chilling requirement suited to cool climates: Japanese plums have a low chilling requirement and grow best in warmer areas. The cherry plum is very similar to the Japanese type. While good drainage is recommended, plum trees usually tolerate heavy soil and moist conditions better than other stone fruits. Neither European nor Japanese plums (with the exception of Santa Rosa) are self-fertile and require a pollinator belonging to the same group. Cherry plums are usually self-pollinating. After the plum tree is shaped it needs little pruning. Fruit is carried on two-year old laterals (European type) or one-year old laterals (Japanese and cherry type) and on spurs which crop for a few years. Angelina, President and Grand Duke are the best European varieties. Any two will cross-pollinate. Early varieties of Japanese plums are Wilson (yellow flesh) which is pollinated by Santa Rosa (red flesh). Santa Rosa in self-fertile—if you only want one plum tree, this is it. Mid-season Japanese varieties are Narrabeen (yellow flesh) and Mariposa (dark red, blood plum). These two will cross-pollinate but Santa Rosa can be used as pollinator for both.

VINE FRUITS

GRAPES

Grapes are grown traditionally on a two-wire trellis but they can be trained over pergolas, arbours and screens. The vines are best suited to inland climates with a dry summer and cool to cold winter, and to the winter-wet, summer-dry, coastal districts of Victoria, South Australia and Western Australia. Fruit splitting and disease are often problems in coastal districts of New South Wales and southern Queensland.

Grape vines are planted in winter about

Wattle brings golden charm to garden or bushland

Well-placed flowering pot plants soften the hard lines of this brick wall

Basket lobelia is a recent introduction. Masses of flowers give a pleasing effect on a sunny verandah

Bottlebrush (*Callistemon*) combines beautifully with geraniums and lobelia to make an attractive colour feature

Daffodils are adaptable and grow well in the garden or in containers

3 m apart. The basic principle of pruning is that fruit buds are borne on one-year old wood which arises from two-year old wood. Water shoots grow from wood older than two years and are not fruit-bearing. Cut the strongest cane on the plant to two buds. Train growth from these in both directions along the bottom wire of the trellis or lateral support. Next winter prune these main arms back to sturdy wood leaving three buds on each. The following winter, prune back these six secondary arms leaving two buds at the base. Continue this process each year.

Golden Muscat is the most suitable variety for humid, coastal districts. European types are best for inland regions and southern states where summer is dry. The best mid-season varieties are Black Muscat and Gordo Blanco. Late varieties are Waltham Cross and Purple Cornichon (Black Lady's Finger). All varieties are self-fertile.

PASSIONFRUIT

This evergreen, perennial vine is very prolific. The dark green leaves and white and purple flowers are also attractive. Passionfruit do well in subtropical and temperate climates which are free of frost. They need a sunny aspect and prefer light soils with good drainage. Passionfruit are self-fertile and you can start by sowing seed from a good cropping vine. Started seedlings and grafted plants are also available from nurseries and garden stores. For trellis growing, wires at 1.2 m and 2 m should run north-south for maximum sun. Vines can be trained on fences, the railings of balconies or terraces and on pergolas. Provide wires or wire netting for the tendrils to cling to. One well-grown vine will give sufficient fruit for the average family but if more than one is required plant vines 2.5 m apart.

Passionfruit respond to generous feeding. Apply a complete fertilizer with at least 10 per cent nitrogen when growth starts in spring. Use 0.5 kg per vine and follow up with light side dressings of a nitrogen fertilizer (Sulphate of Ammonia or Nitram) every 3–4 weeks through summer. Regular watering is needed because roots are quite shallow. Mulching is useful too, but keep mulch away from stem as this may favour collar rot.

Vines planted in spring often give a light crop in autumn but will bear well the following summer. After fruiting, cut the vines back if they are too dense. This allows better air movement and encourages new laterals on which fruit is formed. Vines may become weak and spindly after four or five years so it pays to have new ones coming on as replacements. Fruit colour changes from green to purple when ripe and fruit usually falls. Gather fallen fruit every day in summer. The banana passionfruit is a close relative grown in the same way. It has pink flowers and banana-like fruit with a soft yellow skin.

CHINESE GOOSEBERRY

Chinese gooseberries or Kiwi fruit, as they are often known, are becoming more popular with home gardeners each year. The fruit is 5–7 cm long and covered with short, brown bristles. Flesh, which is light green with dark, soft seeds, can be used fresh or frozen, or in jam, pickles or chutney.

The Chinese gooseberry is a vigorous, deciduous vine suited to mild or temperate climates with warm summer months and freedom from frost. The fibrous roots are shallow so regular watering is needed from spring to early winter when fruit ripens (usually May or June). Chinese gooseberries have male and female flowers on separate vines so you must buy one of each. Nurseries sell them in pairs although one male plant will provide enough pollen for several females. Female vines take four or five years to bear and keep on bearing for at least twenty years.

Vines, which are planted in winter, can be trained on a 2 or 3-wire trellis 2 m high. Space vines 3 m apart. Most home gardeners train the vines over a strong pergola about 2.5 m high and 3 m square. Wires spaced 60–75 cm apart on the sides and top provide support for the twining laterals. Fruit is formed on the first three to five buds of current season's growth. Each winter prune laterals back to two or three buds beyond previous season's crop. If overcrowded, cut some laterals out completely. In summer, shorten back growth if too vigorous. Many varieties, which vary in shape and size, have been selected. Hayward, with large fruit of good keeping quality, is the most popular.

TROPICAL FRUITS

AVOCADO

This handsome, evergreen tree grows to a height of 9 m. The pear-shaped fruit—green or green turning black—contains cream-coloured, butter-textured flesh surrounding a large, oval seed. Although of tropical origin, avocado trees can be grown in sheltered positions in temperate climates. The trees prefer deep, well drained soil and need regular watering in summer. Seedling trees are unreliable and may never bear fruit, so buy grafted plants from nurseries. The trees are self-shaping so no pruning is necessary. Scatter a complete fertilizer beneath the drip line of the tree when flowering commences in spring. Repeat the treatment two or three times during summer and early autumn.

Avocados have a peculiar sex life. Flowers function as female for a few hours, close and then re-open as male flowers the following day, or vice versa. Fortunately, pollination of the variety Fuerte is not so critical because these sex changes overlap in different flowers on the one tree. Sharwill is mid-season and flowers later, but can be pollinated by Fuerte or Haas, a late variety. Haas, in turn, is pollinated by Sharwill but not by Fuerte which has finished flowering. In the Sydney area, Fuerte ripens fruit from April onwards for several months. Fruit does not ripen well on the tree, so pick when the fruit is fully formed and the gloss on the skin fades, then ripen indoors.

BANANAS

Bananas are ornamental as well as useful. They will grow well outside their tropical environment providing they are planted in a sunny spot in a sheltered garden. Older plants will even tolerate light frosts. Bananas are shallow-rooted and need fertile, well drained soil with regular watering in summer.

Start your banana tree from a vigorous sucker with a large, round base. Trim off roots (these will not grow again) and reduce the 'top' by one quarter. Plant the sucker in spring with its base 20–25 cm deep. It will produce fruit in fifteen to eighteen months but may take longer under cooler conditions. Many new suckers will develop, so gouge these out but keep the strongest to replace the parent plant. Strip off any dead or wind-shredded leaves. A banana bunch is right for cutting when fruits have lost their angular shape. Still green, you can ripen them by hanging the bunch upside down in a warm place indoors. Protect bunches which form in cool weather with a blue plastic 'bunch cover'.

Most varieties grow to 6 m tall but Cavendish is shorter at 3 m. Popular varieties are Cavendish, Williams (more tolerant to cool weather), Lady Finger, Sugar and Grand Michel. Friends and neighbours are a good source of planting material. In districts where bananas are grown commercially, home gardeners must apply to the Banana Industry Protection Board for a permit to plant. Only two varieties—Lady Finger and Sugar (both resistant to banana aphids)—may be planted.

MANGO

The mango is an attractive evergreen tree with slender, dark green, leathery leaves. Trees may grow to 9 m under favourable conditions. Mangoes grow best in tropical and subtropical regions but can be successful in coastal districts of New South Wales and warm inland districts such as the Murray River Valley. A sunny, sheltered, frost-free aspect is best. They grow on a wide range of soils but need good drainage in the topsoil. The trees flower in spring and the large oval fruit are ready for picking in late summer and autumn.

You can start your mango tree by germinating a fresh seed from a fully ripe fruit. Place it in a large pot filled with sand and peat moss. One seed may produce several shoots. Retain one shoot only and transplant the seedling when 20–30 cm tall. Started plants are also available from nurseries. Apart from training the tree to four or five main limbs, pruning is rarely needed. Thin inside branches if too crowded. Mangoes respond to fertilizer in small doses. Use 0.5 kg of a complete fertilizer, such as Gro-Plus for Citrus, for each year of growth in four applications—October, December, February and April. Trees start bearing in three or four years. Fruit is ready to pick when the skin turns from green to yellow or orange. The Common variety has long yellow-green fruit with fibrous, sweet flesh. Kensington has a bright yellow skin with a pinkish blush.

PAPAW

Papaw or Papaya, like other tropical fruit, does best in warm, humid climates with good summer rainfall, but if you choose a sunny, sheltered site, this attractive tree will succeed in frost-free temperate climates. A well drained loam or light soil is preferred.

Papaws have male and female flowers on separate trees but there are some bisexual trees with both. Fruit from bisexual trees is long rather than oval and the trees are not as tolerant of cool conditions. Hybrid bisexual trees are available from nurseries. Seeds from fruit of female trees germinate easily in punnets or pots. Transplant seedlings when 20 cm tall. You cannot tell whether seedlings are male or female, so it is wise to set four or five plants in a group spaced about 1.5 m apart. Odds are you will have at least one male tree for pollination. Unwanted male trees are removed, unless you want to keep them as ornamentals.

In favourable conditions, female or bisexual trees will bear within fifteen months but take longer where it is cooler. Trees bear well for about five years so it is best to have younger ones coming on as replacements. Harvesting fruit from older trees is difficult too—even with an extension ladder. Trees which develop three or four lateral stems can be kept at a lower height by cutting out the main stem. Pick fruit when fully coloured, but if weather is cool, pick when fruit is showing a tinge of yellow and ripen indoors. Flowering takes place over several weeks so fruit at different stages of development will be on the tree at the one time. Most papaws have bright yellow or orange flesh but there are some with red flesh.

PINEAPPLE

Pineapples belong to the Bromeliad family. They are one of the few fruit plants which grow in a small space. Pineapples are easy to grow in a warm tropical climate. They are successful in a large pot or tub on a sunny terrace or patio too. Soil must be well drained with plenty of organic matter. Acid soils with a pH 5.0–6.0 are best and you may need to add sulphur to achieve this level. (See Ch. 4.)

Starting a plant from a 'top' (the leafy part of a fruit) is an easy way. Allow the top to dry for a day or two and plant about 5 cm deep in spring. It may take two years to produce the first fruit. After fruiting, prune back to the strongest sucker for the next crop. One plant will bear for several years. Surplus suckers can be used for new plants. Feed plants with a mixed fertilizer in spring each year. Scatter 90 g of fertilizer around each plant. Slow-release fertilizers are also suitable for pot culture. In cool areas, encourage fruiting by using calcium carbide. Dissolve 55 g of material in 5 litres of water and pour 50 ml into the centre of each plant.

LITCHI

Litchi, also known as lychee nut or Chinese nut, is a compact, evergreen tree growing to 12 m. Trees are slow-growing but they are very long-lived. They start bearing at four to five years old. The fruit is round or oval with a red, leathery and rather knobby skin. The white, juicy flesh is very sweet and surrounds a single dark-brown seed. Like other subtropical trees, the litchi needs warm, moist conditions, especially at flowering and for several weeks afterwards, to ensure fruit setting.

Trees from seed are variable so it is best to buy grafted plants. Spring or late summer planting is best. Prune young trees to a strong framework and later remove branches to shape the tree. Fertilize trees in the same way as described for mangoes. Fruit ripens over a four to five week period in December or January. Cut off the small branches with fruit clusters attached every few days. This ensures that new growth for the next crop is on the outside of the tree. Brewster is the best known variety. Others are Kwai Mi, Haak Ip (Blackleaf) and Mauritius.

CUSTARD APPLE

Custard Apple or Sugar Apple is a small, semi-deciduous tree growing to 6 m. It is suited to warm, humid districts in coastal Queensland and northern New South Wales. The fruit is heart-shaped, very knobby with a diameter of 5–8 cm. Flesh is custard-like with many seeds.

It is best to buy grafted plants of the

Cherimoya or Peruvian custard apple because plants from seeds are unreliable. Set plants in early spring or summer. Prune young trees to a vase shape over the first three to four years. Apply fertilizer each year in small doses as for mangoes. Flowering starts in October or November and may continue until January or February, so fruit ripens over a long period. Pick fruit when skin turns a greenish-cream colour. They may take a few days to soften up for eating. African Pride is the most widely grown variety. It is a compact tree, bearing fruit (March–June harvest) after three to four years. Pink's Mammoth is a larger tree with better quality fruit (April–July harvest) but takes six to seven years to bear.

GUAVA

Guavas are not really tropical fruits and can be grown in warm temperate, frost-free climates. Yellow guava, an evergreen tree, is rather hardier than citrus and grows to 5 m tall. The yellow, oval or pear-shaped fruits are about 5 cm in diameter. Strawberry guava is a taller tree but fruits are smaller, purplish-red in colour with a strawberry-like flavour. Pineapple guava or Feijoa is rather more adaptable to cool conditions and will tolerate light frosts. Fruit is oval, about 8 cm long with a greenish-yellow waxy skin and pineapple-like aroma and flavour. Guavas can be eaten fresh but are more popular as jam or jelly. Fruits of all guava trees ripen in late summer and are susceptible to fruit fly attack.

TREE TOMATO

This is a small umbrella-shaped tree growing to 3 m tall. The stem is quite brittle so it is best to support it with a stout stake or post. Tree tomato is not strictly a tropical fruit and grows well in temperate, frost-free districts. It prefers a sunny, sheltered aspect. You can raise tree tomatoes from seed but cuttings are easy too. Shorten plants at a height of 1 m to encourage three or four branches to form the frame. The egg-shaped, purplish-red fruits are 5 cm long with many seeds. The fruits resemble tomatoes in appearance and flavour. They are used fresh or cooked.

BERRY FRUITS

STRAWBERRIES

Strawberries are very adaptable to climate and soil, so it is not surprising they are grown from tropical Queensland to Victoria and Tasmania. The plants need a sunny position, a well drained soil with good structure, regular feeding and watering.

Most home gardeners grow strawberries in a section of raised bed in the vegetable garden with plants spaced at 30 cm each way. Surface mulching between plants will prevent weeds, maintain an even soil temperature and prevent moisture loss in summer. A mulch keeps fruit clean, too. Grass clippings, compost, leaf mould, straw, sawdust, wood shavings or pine bark are all suitable for mulching. Black polythene sheeting makes an excellent mulch for raised beds, and fruiting is earlier because of the warmer soil. Spread the polythene and cut a small slit for each plant, making a depression in the soil below to direct water to roots. Strawberries do well in large pots, tubs or barrels with holes cut in the sides. There are a few non-running varieties suited to this kind of growing.

Prepare soil with animal manure, if available, or compost plus a ration of mixed fertilizer, such as Gro-Plus for Citrus, at $\frac{1}{3}$ cup per square metre.

You can start plants any time between April and August, but early plantings will give fruit in October or November. Remove old dead leaves and trim straggly roots before planting. After flowering commences, give plants liquid feeds of Thrive, Aquasol or Zest every few weeks. Many runners will develop, so pick these off progressively. Plants will bear well for about three seasons. Start a new bed in autumn of the third year. The Department of Agriculture in your own state will recommend suitable varieties for your district. Always ask for plants grown under a Certification Scheme. These are certified free of virus diseases.

Seed of an interesting European wild strawberry is now available. The plants produce fruit in the first season after sowing of seed and heavy crops of medium-sized, flavourful fruit are borne on neat, bushy plants. Ideal for rockeries and pots on balconies. Sow autumn and spring.

RASPBERRIES

Raspberries grow best in cool temperate or cold climates. Districts where apples or cherries grow are ideal. The bushes need deep, well drained soil with lots of organic matter. Prepare and fertilize the soil as for strawberries and set out the dormant canes 60 cm apart in rows 2 m apart.

Raspberry bushes bear fruit on one-year old wood, which is cut back to ground level each winter. Some 'ever-bearing' varieties have an autumn crop on the current season's wood. Thin the strongest of the new canes to 15 cm apart to prevent overcrowding and top them slightly for a manageable height. The canes can be tied together loosely with twine. Raspberries continue to bear for many years. Pick fruit when ripe and well-coloured. Fruit is delicious as a fresh dessert. Raspberries freeze well, too, and make excellent jam or jelly. Widely grown varieties are Lloyd George, Exton Late, Red Antwerp, Williamette, Malling Promise, Malling Jewel and Everbearer. Check with your Department of Agriculture for varieties suited to your district.

TRAILING BERRIES OR BRAMBLES

Loganberry, Boysenberry and Youngberry are the main trailing berries. They are hybrids derived from the Dewberry or Trailing Blackberry. Loganberries prefer a cool to cold climate similar to that for raspberries, but boysenberries and youngberries are more adaptable to warm climates, providing winter months are cool. They grow on a variety of soils, but drainage must be good.

Propagate these berries by cuttings or rooted tip-layers. Plant these 2 m apart underneath a two-wire trellis. Like raspberries, they bear fruit on one-year old wood. Canes produced the previous summer are tied to the top wire and then cut back to ground level after fruiting. New canes are tied to the bottom wire as they grow to keep them tidy and transferred to the top wire when the old canes are cut away.

Prepare the ground well by adding animal manure or compost plus a mixed fertilizer, as for strawberries. Each year, apply fertilizer such as Gro-Plus for Citrus when growth starts in spring. Slow-release fertilizers are also suitable. Pick fruit when well-coloured (red for loganberries, purplish-black for boysenberries and youngberries). Well-grown trailing berries will bear for up to fifteen years. A single vine may yield 5 kg of fruit each year.

GOOSEBERRIES AND CURRANTS

Both fruits grow on small bushes which are suited to cool to cold climates. The English or European gooseberry is started from cuttings in winter and planted 1.5 m apart each way. Gooseberries bear fruit on one-year old wood which is cut out after fruiting. The bush is trained as a vase-shaped, small tree so some pruning is needed. Prepare soil as for raspberries, adding manure or compost plus a mixed fertilizer. Roaring Lion is the most widely grown variety.

Black Currants and Red Currants need a similar climate and soil. Propagate from cuttings and plant at 1.5 m apart. Generally, currants are grown as a many-stemmed bush, which may need thinning when too crowded. They bear fruit on one-year old wood and red currants have fruiting spurs as well. Widely grown varieties of black currant are Carter's Black Champion, Dunnet's and Black Naples. Main red varieties are Fay's Prolific and La Versailles.

NUT TREES

Nut trees are not widely grown in Australian gardens but some are quite ornamental as well as useful. A short summary is given as a guide to how and where the most popular kinds grow.

ALMOND

A deciduous tree, to 6 m tall, suited to winter-wet, summer-dry climates. Dry inland districts with cool winters such as the irrigation areas of southern New South Wales, Victoria and South Australia are ideal. Almonds have similar soil and management requirements to peaches which are close relatives. Nuts are enclosed in a fleshy husk. This dries at maturity and splits open or is easily separated. Many varieties have been selected but two are needed for cross-pollination.

HAZELNUT

Hazelnuts or Filberts grow on small, much-branched deciduous trees to 5 m tall.

Fruit

What to look for	Nutrition and kilojoules	Storage	Preparation	Method of cooking
APPLES Fruit should be the true variety colour, with skin free of bruises. Large apples do not keep as well as smaller fruit.	Fair source of vitamin C and dietary fibre. 222 kilojoules per 100 grams.	Keep in vented plastic bag in refrigerator.	Tart and sharp tasting apples are best for cooking. Remove stalk, wash and dry well. Peel only if necessary.	As fresh juice or cider. Superb raw, pureed, in a tart or strudel, preserved as apply jelly. Use for fritters or bake.
APRICOTS Firm, plump, fully developed fruit with a bright apricot colour. Avoid soft or shrivelled fruit.	Fair source of vitamin C, vitamin A and dietary fibre. Some iron. 188 kilojoules per 100 grams.	Keep in unsealed plastic bag in refrigerator for 2–3 days. Will deteriorate quickly at room temperature.	Wipe over, cut and remove stone.	As a snack. Use in fruit salad and jam. Cooking draws out the flavour. Serve with ham, lamb and duck.
AVOCADOS Generally glossy and hard when unripe. When ripe skin colour is dull, and a toothpick easily pierces flesh at stem. Hass variety has rough dark skin.	Fair source of vitamin C, riboflavin and dietary fibre. Some iron, thiamin and niacin. Fair source of polyunsaturated fat. 674 kilojoules per 100 grams.	Ripen at room temperature, then store in refrigerator.	Remove stone, discard skin. Slice flesh as required. Lemon juice will stop discolouration.	Use mashed on bread and sprinkle with lemon juice. Fill with seafood and dressing. Ideal accompaniment to smoked fish and as a soup. A great ice cream.
BANANAS Best eating quality will be bright, medium sized fruit, yellow to gold in colour, well rounded and free of bruises.	Fair source of vitamin A, vitamin C and dietary fibre. Some iron and thiamin. 364 kilojoules per 100 grams.	Store at room temperature to continue ripening process. Skin will blacken if refrigerated.	Simply peel, or if baking on the barbecue, slightly slit the skin. Lemon juice will prevent discolouration.	Sliced with cinnamon and cream. Ingredient in cakes, biscuits, desserts. Blend with milk for a nourishing drink. Bake on barbecue. Use for fritters.
CHERRIES Firm, fresh, bright uniform coloured fruit, with green stems. Use the taste test.	Fair source of vitamin C and dietary fibre. Some vitamin A. 265 kilojoules per 100 grams.	Keep in unsealed plastic bag in refrigerator to stop from drying out. Highly perishable. Eat soon after harvest.	Wash and remove stem. May be stoned before serving.	Use fresh, or as a tart filling. Blend stoned cherries for fruit sauce. Combine with walnuts and chicken in salad. As a soup.
GRAPEFRUIT Firm and heavy fruit. Skin should be smooth and bright yellow in colour.	Excellent source of vitamin C. 155 kilojoules per 100 grams.	Can be kept outside refrigerator in cool place. Keep in refrigerator crisper for longer storage.	Wipe over and peel. Use grapefruit knife to segment.	Popular as juice or served in halves for breakfast. Serve spiced and grilled as entree. In salads, mix with prawns and mayonnaise.
GRAPES Select bunches of uniform shaped berries, smooth and plump with natural bloom not rubbed off. Stems should be green with fruit firmly attached.	Some vitamin C, iron and thiamin. 276 kilojoules per 100 grams.	In vented plastic bag in refrigerator. Use as quickly as possible.	Wash, dry and remove stems. Pips may also be removed.	Great as a snack. Serve with cheese or pate, and in fruit salad. Combines well with duck, quail and sole.
LEMONS Firm and heavy fruit. Skin should be clean with fine texture. Choose lemons tinged with green for jam making.	Good source of vitamin C. Some calcium and iron. 134 kilojoules per 100 grams.	Keep in cool place, can be home cured for longer storage. Juice can be frozen for use at later date.	Wipe over. Juice, slice for decoration. Cut into wedges and dip in parsley to serve with fish.	Use as a meat tenderiser (mix with mustard to coat meat before baking) and in sorbets. Helps stop apples and bananas from discolouring.
MANDARINS Firm and heavy fruit. Skin should be glossy with a strong orange colour. Heavy fruit gives high juice yield.	Excellent source of vitamin C. Some vitamin A, thiamin and calcium. 193 kilojoules per 100 grams.	Can be kept outside refrigerator in cool place for short time. Keep in crisper for longer storage.	Wipe over and peel. Best eaten raw.	Ideal for lunch box. Use in the same way as an orange, as a crystallised fruit, or in sorbets.

NECTARINES Smooth, plump and highly coloured fruit with no skin blemish. Avoid hard, dull and immature fruit.	Good source of vitamin C. Fair source of vitamin A and dietary fibre. Some iron and thiamin. 260 kilojoules per 100 grams.	Bruise easily. Handle with care. Refrigerate fruit that is riper. Use as quickly as possible.	Wash, cut and discard stone.	Delightful as a snack. Combine with roast beef, cheese and wholemeal bread as a sandwich. Enjoy with cereal or ice cream.
ORANGES Firm and heavy fruit. Skin should be glossy with a fine texture. Colour does not indicate maturity.	Excellent source vitamin C. Fair source dietary fibre. Some vitamin A, thiamin and calcium. 188 kilojoules per 100 grams.	Can be kept outside refrigerator in cool place for short time. Keep in crisper for longer storage.	Peel before eating, slice whole for salads, halve for juicing, quarter and freeze as a snack.	Serve with meat, rice. Use to flavour puddings, breads, biscuits, desserts and in fruit salad, marmalade. Ideal as juice.
PAPAWS Select well coloured fruit. Skin should not be shrivelled or dull, and have no ripe rots or bruising. Aroma is good indicator of ripeness.	Excellent source of dietary fibre. Good source of vitamin C. Fair source of iron. Some riboflavin and niacin. 381 kilojoules per 100 grams.	Ripen at room temperature. Keep ripe fruit in refrigerator. Use as soon as possible.	Wipe over. Slice as required and remove seeds.	Use as meat tenderiser. Serve as accompaniment to smoked beef. Use in fruit salad, or with yoghurt and honey as dessert. Lovely water ice.
PASSIONFRUIT Select full heavy fruit with smooth dark purple skin. Avoid withered fruit.	Excellent source of vitamin C. Fair source of vitamin A. 172 kilojoules per 100 grams.	Keep in plastic bag in crisper of refrigerator. Pulp may be frozen for later use.	Wipe over. Cut in half. Remove pulp and use as required. Discard skin.	Use in fruit salad and fruit punch. Serve as topping over ice cream, pavlovas and flummery or as a fruit sauce. Include in icings.
PEACHES Firm fruit which is just beginning to soften, with a peachy smell. Avoid bruised or under developed fruit.	Fair source of vitamin C and dietary fibre. Some vitamin A, iron, niacin. 172 kilojoules per 100 grams.	Keep in unsealed plastic bag in refrigerator. Will deteriorate quickly at room temperature.	Wash and discard stone. If peeled, use lemon juice to prevent discolouration.	Pies. Top with cinnamon and butter and lightly grill. Eat fresh with cereal, ice cream, cream or yoghurt. Use in compotes and mousses.
PEARS Pears ripen from the inside out after harvesting. Test for ripeness by applying gentle pressure at the stem area. Avoid immature fruit.	Fair source of vitamin C and dietary fibre. 234 kilojoules per 100 grams.	Store firm pears in vented plastic bag in refrigerator. Ripen at room temperature.	Wash and dry. Remove stalk for cooking. Peel only if recipe calls for it.	Eat raw with cheese and walnuts. Preserve. Serve with roast lamb, smoked fish or ham. Poach in vanilla syrup and coat with chocolate. Bake in wine.
PINEAPPLES Skin colour not a reliable guide, but in winter select fruit with quarter yellow colour and no soft spots. Look for fresh deep green leaves and pleasant aroma.	Good source of vitamin C. Fair source of dietary fibre. Some vitamin A, thiamin. 218 kilojoules per 100 grams.	Keep in cool place or in refrigerator. Refrigerate before serving, if desired.	Remove leaves. Make slanting cuts downwards between eyes, and slice as required, using a stainless steel knife.	Fruit salads, upside-down cakes, with ham in salads or sandwiches. Serve with bacon as hors d'oeuvres, or with cheese or sausages. Use in Chinese cookery.
PLUMS Firm, bright and fully developed fruit, with no sign of wrinkling.	Fair source of dietary fibre. Some vitamin A, vitamin C and thiamin. 247 kilojoules per 100 grams.	Ripen at room temperature. Then refrigerate and use as soon as possible.	Wash, cut and discard stone.	Jam, compotes. Serve chilled with camembert or blue cheese. With ice cream. Bake with roast lamb for added flavour. Use as a snack.
STRAWBERRIES Fruit should be clean and brightly coloured with no sign of soft spots or mould. Look for green stem cap and avoid fruit with white or green areas.	Excellent source vitamin C. Fair source dietary fibre. Some iron. 155 kilojoules per 100 grams.	Keep in refrigerator. Very perishable. Use as soon as possible.	Hull and wipe over.	Preserves, jams, tarts. Puree for fruit sauce. Combine with pineapple. Add to fruit salad. With cream or yoghurt. In fruit punch.

Table reproduced by courtesy of the NSW Department of Agriculture

A walnut tree is a long-term investment.

They are suited to cool climates. The shell is hard and woody but the smooth, brown kernel separates easily when cracked. Trees may take several years to bear.

MACADAMIA

This handsome native tree, also known as Queensland Nut, grows to 9 m tall. It is suited to warm, humid climate zones but grows quite well in temperate coastal districts such as the Sydney area. The round white kernels are enclosed in a hard, woody shell. Nuts mature during late autumn and winter and fall when ripe. Seedling trees are extremely variable so it is best to buy grafted trees from nurseries.

OLIVE

The olive is included in this section because the fruit is useful for pickling and the trees are ornamental as well. Trees are evergreen and grow quickly to about 6 m tall. They are suited to temperate climates and withstand hot, dry summers and cool winters with some frosts. The leaves are dark grey (silver underneath) and the white flowers in spring are followed by green or black fruit in summer. Best varieties are Sevillano, Manzanillo, Mission and Verdale. Grafted plants from nurseries should be planted in well drained, sunny situations in autumn or spring. Pruning is rarely needed unless the trees become too large.

PISTACHIO

The pistachio nut is a small deciduous tree to 6 m tall with compound leaves composed of 3–5 leaflets. It is suited to temperate climates with dry summers and cool to cold winters. The fruit is oval-shaped, about 2 cm long, containing the nut with a thin, woody shell which splits open when ripe. The kernel is smooth, light green and rich flavoured.

WALNUT

The common walnut is a shapely, deciduous tree growing to over 15 m unless restricted by pruning. It is suited to cool to cold climates but is susceptible to late spring frosts. Nuts split open or are separated easily when mature. Many varieties are available. Two varieties are desirable for effective cross-pollination.

Chapter 18

PESTS, DISEASES AND WEEDS

You need not be an expert on pests, diseases and weeds to have a clean, healthy garden which will grow top quality flowers, fruit and vegetables. Effective and relatively non-toxic sprays and dusts are available to control most pests and diseases and to eradicate most weeds. The quick reference chart at the end of the chapter tells you what pests and diseases attack your plants, how to identify them (or their symptoms) and what chemicals to use.

It is important to know the normal, healthy appearance of plants so that problems caused by pests or diseases can be detected quickly. Early recognition of a problem gives you a good start in overcoming it. But it is not always a pest or disease which makes a plant unhealthy.

Inadequate watering may be the cause. If you allow soil to dry out to a depth of 5–10 cm below the surface, subsequent light watering, however frequent, will not let the water soak in to reach plant roots. The best way to overcome this is to water thoroughly for several hours until the soil is evenly moist throughout its depth. In other words, the soil should reach field capacity. (See Chapter 3.) Excessive watering and bad drainage can also be the cause of unhealthy plants, especially in the case of trees and shrubs. Overly wet conditions can impair root efficiency by preventing air (for respiration) from reaching them.

Low levels of nutrients or lack of fertilizer, and in some cases a deficiency of trace elements, may be the cause of poor growth. But soil analysis is complicated and does not always give a cut-and-dried answer. The simplest method is to use a mixed fertilizer which contains the major nutrients (nitrogen, phosphorus and potassium) plus animal manure, if available, or compost. (See Chapter 4.) However, never use fertilizers in excess of the quantities recommended.

HOME GARDEN CHEMICALS

There is a wide range of chemicals available for ridding the garden of pests and diseases. Some of these are specific in their action, while others are broad spectrum chemicals and will control more than one pest or disease. Multi-purpose sprays are also available. These contain one or more insecticides to control pests combined with one or more fungicides to control diseases. These mixtures, for example Capthion or Lane's Spraytox, are recommended, especially when there is some doubt as to the cause of the problem.

Whatever garden chemicals are used, it is important to stress that they should be used with care. Fortunately, chemical manfacturers have discarded many of the more toxic insecticides for garden use. Most garden chemicals can be used with perfect safety if directions are followed and a few simple precautions are taken:

1. Read directions carefully before using and use only for the purposes stated on the label.
2. Avoid contact of spray (especially in concentrated form) or dust with skin.
3. Avoid breathing fumes from sprays or dusts.
4. Avoid spraying or dusting on windy days—a calm day is best.
5. Avoid eating or smoking when spraying or dusting.
6. Rinse spray equipment thoroughly after use and wash face and hands with soap and water.
7. Store sprays and dusts out of reach of children or in a locked cupboard.
8. Do not harvest vegetable and fruit crops earlier than the witholding period (see tables for insecticides and fungicides).

Green vegetable bug

DUSTS OR SPRAYS?

Whether you use dusts or sprays, it is important to apply them so that all parts of the plant are covered. Dusts do not need mixing and are simple to apply. There are several small dustguns available at prices around the $15 mark (1983) which can direct the dust from above, below or beside the plant. But dust is difficult to apply to large plants and is easily washed off by rain or overhead watering.

Generally, spraying gives a better and more even coverage of the plant surfaces and therefore more effective control. A small knapsack sprayer which can be carried over the shoulder is probably the best for the home garden. The Rega Junior Pak (4-litre) is ideal and will last for many years. The Rega Uni Spray is a less expensive bucket-type sprayer with a long spray rod with adjustable nozzle and about 3 m of plastic tubing. For small balcony gardens, pot plants and glass or shadehouses a Rega Continuous Atomiser will prove adequate. For correct strength of sprays, always follow the manufacturer's directions exactly. Don't be tempted to put in a little extra for good measure. Most garden chemicals can be used in the same container or spray equipment providing the container, hose and nozzle are washed out thoroughly *immediately* after use. Extra care must be taken to wash hormone weedicides from containers or spray equipment.

PESTS

Garden pests are generally members of the insect world—beetles, bugs, caterpillars, aphids and thrips—but also include a tiny red spider or mite. Snails and slugs are not insects either but probably do as much damage as other pests. Pests are controlled by insecticides, miticides and snail baits. Before we describe them in detail we will divide them into two broad groups according to how they feed.

1. *Chewing insects* actually eat plant tissues—leaves, stems, buds, flowers or fruits. Beetles, caterpillars, codling moth, cutworms, grasshoppers, fruit fly (grubs or larvae) are included in this group. They are controlled by contact insecticides or stomach poisons. Snails and slugs are usually included in this group too because they are controlled by stomach poison baits in the same way as chewing insects.

2. *Sap-sucking insects* are those which suck sap from young shoots, flower buds, leaves and stems. They do not actually eat plant tissue so a contact or systemic insecticide (absorbed by plants into the sap stream) is needed to destroy them. The sap-suckers include aphids, thrips, jassids, mites and several kinds of bugs and scale insects. Most sap-suckers feed by inserting their sharp mouthparts (or beaks) into plant tissue and extracting the sap. This kind of feeding causes collapse of plant cells, destruction of tissues and wilting. Many virus diseases of plants are transmitted from one plant to another by sap-sucking insects.

CHEWING INSECTS

Caterpillars There are many types of caterpillar which are usually the larval (caterpillar or grub) stage of moths such as cabbage moth, potato moth, cabbage white butterfly or tomato moth. The moths lay their eggs on the underside of leaves. The larvae or caterpillars hatch from the eggs and then feed on the leaves or fruit. Caterpillars can be controlled by Carbaryl, Malathion, Thiodan or Derris. Dipel HG is a bio-insecticide which controls leaf-eating caterpillars on cabbages and other crucifers.

Armyworms and Cutworms Another kind of caterpillar is the armyworm or cutworm, which is brown or green (sometimes striped) and mostly feeds at night. These caterpillars cut through the stems of seedlings or transplants. Drench around plants with spray-strength solutions of Carbaryl, Thiodan or Lebaycid.

Beetles, Weevils, Grasshoppers, Crickets Beetles and other insects which chew leaves and stems can usually be controlled by sprays of Carbaryl, Malathion or Thiodan, or by dusting with Derris. Black beetles are a serious pest in lawns but also attack other plants. The adult beetles do some damage in spring but most damage is caused in mid-summer to early autumn. Carbaryl or Malathion will give some control but one of the most effective insecticides is Chlordane. One application in October will control black beetle, lawn grubs, ants and mole crickets. Chlordane is more toxic and has a greater residual effect than most garden chemicals, so it must be used with caution. It is best to prevent children and pets playing on lawns sprayed with Chlordane until the chemical has been well watered into the soil.

Codling Moth Codling moth may cause serious damage in apples, pears and quinces. The moths lay their eggs on leaves and the developing fruit as flowering finishes. When the eggs hatch the caterpillars burrow into the fruit. Apply insecticide sprays (Carbaryl or Lebaycid) at petal fall and then at 14-day intervals until 4–6 sprays have been applied. All infested fruit should be removed and destroyed.

Fruit Fly Fruit fly attacks most kinds of summer fruit, including peaches, plums, nectarines, apples, pears and citrus. It is also a serious pest of tomato, capsicum and eggplant, especially in coastal and some inland areas. The wasp-like fly lays eggs in the developing fruit which are most susceptible to attack a few weeks before harvesting. The eggs hatch quickly and the larvae (maggots) burrow through the fruit and make it inedible. Sprays of Rogor or Lebaycid, both systemic insecticides, prevent the eggs from hatching or kill the young larvae as they emerge. Spray must be applied 3–4 weeks before harvest and then at 1–2 weekly intervals. With tomatoes, use Rogor at weekly intervals (Rogor has a 7-day withholding period). Pick all ripe fruit and fruit showing colour, then spray and harvest again in seven days. Regular spraying against fruit fly has largely superseded the poison baits and lures such as Dak-pots, which were previously recommended. Garden hygiene is also important in controlling fruit fly. All infected fruit should be gathered and destroyed by burning or boiling. These measures are compulsory in most States of Australia. In South Australia, householders suspecting the presence of fruit fly should advise the Department of Agriculture and Fisheries which will take the necessary steps to identify the pest and eradicate it.

Bean Fly Bean fly may be a serious pest of French beans (both dwarf and climbing) in warm, subtropical and tropical areas. The small, adult fly lays eggs on the leaves and the larvae or maggots tunnel into the young stems which swell and break. Spray plants with Rogor three days after emergence and again four days later. Continue sprays at weekly intervals until blossoming.

Leafminers Leafminers are the larvae of a small fly which tunnel through the leaves leaving scribble-like, white markings, especially in cinerarias, nasturtiums, marguerites, spinach and silver beet. As the grubs are difficult to reach with contact sprays, systemic insecticides—Metasystox, Lebaycid or Rogor—give best control.

Borers Borers attack fruit trees and a number of ornamental trees and shrubs. The grubs tunnel into the trunk or branches, leaving a mass of sawdust or gum oozing from the hole. Probe the hole with a piece of wire and inject a strong solution of a contact insecticide—Carbaryl, Malathion, Thiodan or Chlordane. Spraying the trunk and branches in November and again in December will help prevent re-infestation.

Snails and Slugs Proprietary baits such as Baysol or Defender are very effective against snails and slugs. Scatter the baits (as directed on the package) where seeds have been sown or seedlings transplanted. These pests are more active in cool, wet weather, so it is wise to spread baits generously in these conditions, especially under shrubs and hedges where they shelter and breed. Some plants, especially begonias, nasturtiums and lilies, are favourite breeding spots.

SAP-SUCKING INSECTS

Aphids Aphids are small, soft-bodied insects which usually cluster on young shoots and flower buds or underneath leaves. There are many different species which vary in colour—yellow, bronze, green, brown, pink, grey and black. Aphids attack fruit trees (including citrus), roses, camellias, chrysanthemums and other ornamentals, a wide range of vegetables, flowering annuals, bulbs and even weeds. Woolly aphids are another type which give a white fleecy appearance on the branches of apples, pears, hydrangeas and other shrubs. Aphids also transmit virus diseases such as broad bean wilt (on broad beans and sweet peas), potato leaf roll and mosaic virus of stocks. Some weeds are alternative hosts for virus disease. Because aphids are small and often collect on the underside of foliage, they are frequently undetected. Small colonies multiply rapidly and can develop into a heavy infestation in a matter of days. Systemic sprays are most effective in controlling aphids. Metasystox or Lebaycid are excellent for shrubs, ornamentals and flowers because they persist in the sap for several weeks. For vegetables, Rogor (also systemic) and Malathion are recommended because of their 7-day withholding period. Pyrethrum spray or Derris dust also give some control but have little residual effect and plants can be re-infested in a few days.

Jassids Jassids are tiny, quick-flying insects often called leaf hoppers. They are usually green or brown and their feeding causes mottling of the leaves. They are very common on marigolds, and can also transmit virus diseases. Control as for aphids.

White Fly A small white-winged, sap-sucking fly which has become very prevalent in recent years. They usually attack annuals and vegetables, especially tomato, bean and vine crops. Control as for aphids.

Thrips Thrips are small insects about 1 mm long and just visible to the naked eye. They vary in colour from white through yellow and brown to black. Thrips attack the flowers, fruit and foliage of vegetable crops and ornamental plants. Roses, fruit trees, azaleas, gladioli, tomatoes, onions and beans are regular victims of thrip invasion. They also feed on a wide range of weeds. During hot weather, weeds dry up and the insects migrate to more attractive plants. Certain kinds of thrips transmit spotted wilt virus which may seriously affect tomato, lettuce and dahlias in summer. Thrips are often more difficult to control because the eggs are laid inside plant tissue and the pupae and adults often feed on unopened flower buds. This prevents sprays and dusts from reaching the insects. Regular spraying with Rogor, Malathion or Thiodan every week or two will control them. Pyrethrum spray is recommended for small pot plants and indoor plants.

Bugs These large insects, like the green vegetable bug and the bronze orange bug, are often called shield bugs because of their shape and tough exterior. They can be serious pests in summer. The former is bright green in colour and about 1 cm long in adulthood, but more rounded and black and white or black and red in younger stages. It attacks beans, tomatoes, potatoes, sweet corn, vine crops, grapes, sunflowers and other ornamentals. The other, which is orange-brown or black when mature and twice as large, attacks citrus trees causing wilting of the young shoots and flower stalks.

The Rutherglen bug is a smaller grey-brown insect which damages a wide range of plants and may reach plague proportions in hot, dry summers. Lace bugs are small, soft-bodied insects with rather large lacy wings. They cling to the underside of leaves and their feeding causes mottling and bronzing. They are serious pests of azaleas.

Bugs can be controlled by many insecticides. Carbaryl, Thiodan and Malathion are good contact sprays. Rogor, Lebaycid and Metasystox are effective systemic sprays.

Mealy Bugs Mealy bugs are small insects covered with a white mealy coating; some have white hairs attached to their bodies. Heavy infestations can occur on citrus, daphne and other ornamental plants. Orchids and ferns, especially in shadehouses, can become infested too. They may also attack bulbs in storage and the roots of some plants such as polyanthus, liliums and callas. Malathion sprays will control mealy bugs, but use at half strength on soft plants in shadehouses. This also applies to indoor pot plants. For root-infesting mealy bugs drench soil with Malathion or Metasystox at spray strength.

Red Spider, Mites Red spider attacks a wide range of fruit trees (apples, pears, peaches), vegetables (tomato, beans, vine crops) and ornamentals (azalea, roses, marigolds). Symptoms of red spider are bronzing or dull grey mottling of the leaves. In heavy infestations leaves may drop. The tiny pinkish-red mites cluster on the underside of the leaves, often producing a mass of fine webbing. Red spider has become an important pest because the use of insecticides has decreased the number of natural insect enemies, such as ladybirds. Tomato mite and red-legged earth mite (a winter pest of peas and other vegetables) are other pests in this group. They can be controlled by Kelthane, a specific miticide, or by the systemic insecticides Metasystox, Lebaycid and Rogor. Sulphur or lime sulphur, both of which are often used to control fungus diseases, will also give some protection against mites.

Scale Insects The most common scale insect is white wax scale. This insect is easily recognised by the presence of large patches of waxy material along the stems and shoots of citrus trees, gardenias, pittosporum and other ornamental shrubs. The wax covers the insects which are feeding on the sap. Spray trees with white oil (Albarol) in December and again in January before the young insects have secreted their waxy covering. Malathion or Rogor added to the white oil will increase its effectiveness. Red scale, brown scale and white louse scale are related insects which are not such serious pests. They can be controlled in the same way.

DISEASES

Plant diseases are caused by parasitic organisms called 'plant pathogens'. The pathogen may be a fungus, a bacterium, a virus or a nematode.

FUNGUS DISEASES

Most plant diseases are caused by fungi which may be carried in the seed, in the soil or on other plants or weeds. They spread quickly, especially in warm, humid weather, by means of microscopic spores which are distributed by wind or water. Rain or water-splash favours their spread. Fungus diseases fall into four main groups: mildews, rusts, leaf spots, stem and root rots.

Mildew Powdery mildew is a fungus which spreads a white or ash-grey film over the upper and lower surfaces of the leaves of plants—usually the older leaves. Powdery mildew is a common disease of roses, crepe myrtles, dahlias, zinnias, calendulas, sweet peas and vine crops. The systemic fungicide Benlate will control this disease. Karathane is another specific fungicide for powdery mildew. Dusting or spraying with sulphur is also effective.

Downy mildew is often more widespread in younger plants and is recognised by downy, whitish tufts of spores and mycelia on the underside of the leaves.

Downy mildew is common on grapes, vine crops, cabbage and other crucifers, onions, lettuce and stocks. Spray with Zineb, Captan or Mancozeb. Regular spraying is necessary during rainy weather.

Rust Rust fungus is easily identified by the many orange or red pustules on leaves or stems, which break open and release masses of spores. Rust is a common disease of calendulas, snapdragons, geraniums, gerberas and beans. In recent years it has become a serious disease of poplar trees. The most effective sprays to control rusts are Zineb and Mancozeb. Good control is also obtained with sulphur, lime sulphur (Harola) and copper oxychloride (Dry Bordox).

Leaf Spots Leaf spots are easily seen and may be serious on roses (black spot and Anthracnose). Leaf spots and blights are also common on tomato, potato, capsicum, carrot, parsnip, beet and silver beet, polyanthus, iris and many shrubs. Leaf spots are usually more serious in wet conditions. Sprays of Zineb, Captan, Mancozeb, lime sulphur or Dry Bordox will control most leaf spots. The systemic fungicide Benlate also gives good control.

Stem Rots and Root Rots The causes of stem rots, root rots and collar rots are not easily determined but most are due to fungus pathogens which attack the conducting tissues of the plant, causing it to wilt and finally collapse. Sclerotinia is a widespread fungus which attacks many soft-stemmed plants, beans, lettuce, nemesias, linarias and other annuals. The stems rot and the fungus then forms small, hard fruiting bodies called sclerotia which fall to the ground. These develop a mycelium and later small toadstool-like structures which shed spores to restart the cycle. Petal blight of azaleas has a similar life cycle. Benlate is a suitable spray to control this kind of fungus.

Some fungus pathogens attack the stems or crowns of plants at ground level. These fungi are more active in damp conditions and poorly drained soils. Root rot or collar rot is common in delphiniums, carnations, gerberas, strawberries, cabbage and other crucifers. Fruit trees and shrubs may also be attacked. While dampness and poor drainage encourage these root and crown rots, some control can be achieved by drenching the soil around the plants with Zineb, Mancozeb, Benlate or Dry Bordox.

'Damping off' disease of seeds and seedlings has already been mentioned in Chapter 5. The fungus causing this disease is soil-borne and most active under damp cold conditions, so it is wise to dust seeds with a fungicide—Zineb, Captan or Dry Bordox—before sowing. This protects the seeds during germination. Damping off can also occur after seedlings have germinated. If seedlings fall over at soil level, drench the area with Zineb, Captan or Dry Bordox at spray strength.

BACTERIAL DISEASES

Bacterial diseases are not common in the home garden, which is fortunate because there are virtually no chemicals available to control them efficiently. Black rot of cabbage and other crucifers is seen occasionally and there are also a few bacterial leaf spots of tomato and zinnia and a leaf and pod spot of beans (halo blight). None of these diseases is usually serious. Many bacterial diseases are seed-borne but careful attention to hygiene in seed production and treatment by seed companies makes this source of infection unlikely.

VIRUS DISEASES

Virus diseases are often found in the home garden. Spotted wilt of tomatoes, necrotic yellows of lettuce and broad bean wilt of broad beans and sweet peas can be quite devastating. Other virus diseases, like those causing striping in tulips and stocks and the greening of aster flowers, do not usually affect the vigour of the plants. Some other virus diseases gradually reduce the vigour and productiveness of the plants they attack, especially perennials. Good examples are passionfruit virus, crinkle leaf of strawberries, leaf-roll virus of potatoes, spotted wilt of dahlias and mosaic virus of orchids. No chemicals will control the virus diseases of plants, so garden hygiene is most important in controlling infection. All plants suspected of virus infection should be removed and destroyed by burning. It is worthwhile to remove all weeds from the garden and surrounds because many weeds are alternative hosts for virus diseases. Lastly, virus diseases are transmitted from one plant to another by sap-sucking insects —aphids, jassids and thrips. If you control these pests, the battle against virus disease is almost won.

NEMATODES

Nematodes or eelworms are minute soil-inhabiting worms, some of which are useful in decomposing organic matter while others are parasites attacking the roots of plants and causing large swellings or galls. They are more prevalent on sandy soils than heavy soils. Tomatoes, beetroot, carrots, lettuce, cabbage (and other crucifers), carnations and gardenias are susceptible to attack. Control by using Nemacur granules, incorporating them into the top layer of soil and then watering. Do not grow susceptible plants in that part of the garden for one or two seasons. Continuous cultivation during summer is also a useful method of reducing nematode populations. This means leaving the soil completely free of plants (including weeds) from spring to autumn. Another important nematode is the leaf nematode of chrysanthemums. Leaves show large, triangular dead patches and die off from the base upwards. The leaf nematode is often prevalent from late summer to early autumn when accompanied by extended rain periods. Control by spraying with Metasystox or Lebaycid when the lower leaves show damage.

NON-PARASITIC DISEASES AND THEIR SYMPTOMS

Poor growth of plants is not always caused by parasitic organisms. Environmental factors either in the atmosphere or in the soil may be the reason for unhealthy plants. These so-called diseases would more correctly be described as physiological disorders.

The influence of the atmosphere in which plants grow is most important. Excessive cold often causes purple or red pigments to develop in the leaves of roses and many plants. Chlorosis (yellowing) of leaves occurs in others. In other cases, flower buds do not form properly or flowers are not pollinated. Low temperatures are the main cause of poor pollination in tomatoes and capsicum, especially in spring and early summer. This often leads to misshapen fruit called 'catface'. Vine crops often fail to set fruit at low temperatures or in cloudy weather due to the absence of bees.

On the other hand, excessive heat, which is often combined with a dry atmosphere, causes wilting, scorching of leaves (tip burn) and other tender tissues (sunscald of tomato and capsicum), blossom drop in tomatoes and capsicums and faulty pollination in beans and sweetcorn (pollen blast).

Wind is another atmospheric factor. Cold winds slow down growth and hot winds increase moisture loss from plants and soil. Strong winds can cause direct mechanical injury to leaves, stems, flowers and fruits. Insufficient light results in soft, spindly growth especially in seedlings and may reduce flowering and fruiting in older plants.

In industrial areas of cities and towns, chemical pollutants such as sulphur dioxide and the exhaust fumes of motor vehicles produce hydrocarbons and oxides of nitrogen which can have toxic effects on plants. Fluorine, ozone and ethylene are other gases which may cause yellowing of leaves or interfere with flowering and fruit formation. Dust and smoke from factories can cause damage by blocking the stomates (breathing pores) of leaves and forming a film which prevents light from reaching the leaf surface. Some plants are more resistant to gases and dusts in the atmosphere, so gardeners who are unfortunate to live near these sources of pollution should restrict their plants to those which can best resist it. Your local nurseryman will be able to assist you with your selection.

The influence of soil factors on the growth of plants has been mentioned in the opening paragraphs of this chapter and also in chapters 3 and 4. The importance of adequate watering cannot be stressed too strongly. Insufficient water not only causes wilting but a disturbance in the even supply of moisture can be responsible for blossom fall in sweet peas and tomatoes and early fruit drop in citrus trees. Lack of water can also lead to the accumulation of salt (mostly as chlorides) in the soil, especially in gardens in inland irrigation districts or those exposed to salt spray near the coast. Excess chlorides in the soil cause severe leaf scorch, especially on the leaf margins, and may eventually result in death of the plant. Susceptible plants are beans, cabbage, cauliflower, lettuce, onion, tomato, citrus trees, peaches, apricots, bananas and grape vines.

On very acid soils, all the important plant nutrients—nitrogen, phosphorus, potassium, calcium, magnesium and sulphur—are usually in short supply. (See Chapter 4.) A common disorder in tomatoes is blossom-end rot in which the fruit becomes sunken and blackened. This condition is caused by lack of calcium in the developing fruit. Blossom-end rot is aggravated by moisture stress in very hot weather so regular watering and mulching the surface will help control the problem—but over-watering should be avoided because root absorption may be less efficient. An application of lime or gypsum (calcium sulphate) to the bed before planting will lessen the incidence of this disorder. Tip-burn of cabbage and blackheart of celery are similar kinds of calcium deficiency.

A deficiency of magnesium shows symptoms of yellowing between the leaf veins—especially on the older leaves. It is not uncommon on acid soils. Susceptible plants are citrus trees, apples, bananas, grape vines, beetroot, tomato, and members of the cabbage family.

The trace element, molybdenum, is also deficient on acid soil and causes a disorder called whiptail in broccoli, Brussels sprouts, cabbage and cauliflower. The leaf blades are narrow and the margins of the leaves thickened and distorted. The application of water soluble fertilizers—all of which contain molybdenum—will usually overcome the problem. Molybdenum deficiencies have also been recorded in lettuce, tomato and vine crops when grown on acid soil.

On alkaline soils, other elements are likely to be unavailable. This shortage applies particularly to iron and manganese, lack of which causes chlorosis in young leaves of azaleas, rhododendrons, camellias and other acid-loving plants. Deficiencies of manganese can also occur in citrus trees, beetroot and tomatoes, especially in inland areas or on heavily limed soil. On the other hand, excess manganese (manganese toxicity) has been found to cause yellowing in crops of beans and pineapples on acid soils on the coast of New South Wales and Queensland.

Boron is another trace element which may be deficient on alkaline soil. Symptoms are usually associated with growing tissue such as shoots, buds, fruits and storage roots. Susceptible plants are apple (internal browning), broccoli and cauliflower (hollow stems), beetroot (heart rot), silver beet and celery (stem cracking), swedes and turnips (brown heart).

Deficiencies of copper and zinc are not common but have been recorded on very acid, sandy, coastal soils and also on fertile, inland 'black earth' soils. Symptoms occur in the younger leaves which are chlorotic and stunted. The disorder has been called little leaf or rosette in apples, peaches, apricots and citrus trees. Bananas and maize (which includes sweetcorn) are also susceptible to zinc deficiency.

While trace elements are an interesting facet of plant nutrition, deficiencies of these are not common in a home garden situation where organic matter and mixed fertilizers are used. In addition all the water soluble fertilizers—Thrive, Aquasol and Zest—contain trace elements in small but balanced quantities. Because trace elements are required in such small amounts by plants, their correct application is essential. The use of specially formulated fertilizers is much safer and easier than applying trace elements separately or making up your own mixtures.

PESTS AND DISEASES OF FRUIT TREES

Apples, pears and quinces are attacked by codling moth and fruit fly. Aphids can also do damage and woolly aphid can be a serious pest of apples. Stone fruits (peaches, nectarines, plums and apricots) are attacked by fruit fly but not by codling moth. Green and black peach aphids are often a problem, especially in inland districts. (See previous sections under relevant headings for control of these pests.)

Black spot (apple scab) and powdery mildew of apples are controlled by sprays of Benlate. Apply at pink bud stage (early October) and then every 2–3 weeks until after full bloom. Bitter rot of apples can

Roses make excellent formal and informal plantings

ROSE BLACK SPOT

PHOTINIA POWDERY MILDEW

PEACH FRECKLE

CODLING MOTH

ROCKMELON DOWNY MILDEW

WHITE FLY

APHIDS

COTTON CUSHION SCALE

COLEUS MEALY BUG

AZALEA PETAL BLIGHT

HARLEQUIN BUG

BEAN RUST

(See pp. 273-83 for information about pests and diseases)

be checked by sprays of Zineb or Captan. Apply every two weeks from early November onwards.

With stone fruits, a clean-up spray is advisable while the trees are still dormant in June or early July. Spray with Dry Bordox and white oil or lime sulphur before bud burst. Any dried-up fruit (mummies) should be collected from around the base of the tree and burned. Spray stone fruit with Zineb to control freckle and rust in September and repeat at monthly intervals. Both peach leaf-curl and shot hole of stone fruits is controlled by copper oxychloride (Dry Bordox) in the clean-up spray. A further spray in autumn at leaf fall is advisable. Brown rot is a destructive disease of stone fruits. The clean-up spray in winter helps to control brown rot, but this should be followed by spraying with Benlate at full bloom, three weeks before harvest and again just before harvest for complete control. Collect and destroy any fruit affected with brown rot.

PESTS AND DISEASES OF CITRUS TREES

The important pests of citrus trees are scale insects (white wax, red, brown and white louse scale), bronze orange bug, aphids, citrus mite and fruit fly. Control measures for these pests are given in previous pages. One other serious pest is the citrus gall wasp. This small wasp, about 6 mm long, lays eggs in soft stems in spring. The larvae tunnel through the stems forming swellings or galls. There is no chemical control. Remove any stems or twigs with galls, cutting well behind them. Burn them before the end of August because a new generation of wasps emerges from the galls about this time.

A serious disease of citrus trees in coastal districts is lemon scab. Spray with Dry Bordox and white oil when trees commence flowering in spring, followed by a second spray of Zineb and white oil in summer after flowering has finished. Melanose is also a common citrus disease causing rotting of the fruit at the stem end and die-back of branches and twigs. The sprays to control lemon scab should also control Melanose.

Sooty mould is caused by a black fungus which covers the leaves of citrus. It does very little damage as the fungus survives on the sugary secretions from aphids or scale insects. If these pests are controlled, the sooty mould will disappear.

Collar rot attacks the trunks of citrus trees just above soil level. Yellowing leaves and cracking bark are usual symptoms. Cut the bark back to healthy growth and paint the area with a solution of Zineb or Dry Bordox. Yellowing and dwarfing of leaves may also be due to lack of nitrogen and perhaps trace element deficiencies. See Chapter 16 for fertilizer recommendations. If the condition persists, seek advice from the local office of your Department of Agriculture.

LAWN PROBLEMS

Black beetles, grass grubs, ants and mole crickets are the main pests of lawns. For control, see previous section under the heading Beetles, Weevils, Grasshoppers, Crickets.

The main diseases of lawns are brown patch and dollar spot. Both are caused by a fungus and usually occur in late spring, summer and autumn when the weather is warm and humid. Brown patch starts with small discoloured patches of grass which later spread to form irregular dead patches a metre or more in diameter. Bentgrass is very susceptible. Dollar spot is very similar in appearance but the spots usually remain circular. Both bentgrass and Queensland blue couch are susceptible. A number of turf fungicides are available to control both diseases. Zineb, Captan and Mancozeb will give good control if applied when the diseases are first noticed. Benlate is also effective against dollar spot.

Fairy ring forms a ring of green grass surrounded by an outer ring of dead grass with mushroom-like growths appearing occasionally in the affected area. The fungus responsible is very deep in the soil and the only complete cure is to remove the turf and soil to a depth of 20 cm from the affected area.

Algae (a green or black scum) and moss are usually problems in shaded, over-wet or badly drained sections of the lawn. If these conditions are alleviated the grass will again cover the area. Slime moulds form small, steel-grey or black mounds on leaves and stems of grass in warm, moist weather. These are not parasites and do not injure the grass although they may look unsightly. Slime moulds can be brushed off with a stiff broom but they usually disappear in a few days.

WEEDS

A weed is often defined as a plant growing out of place. Apart from their untidy appearance in a garden, weeds compete with useful plants for space, light, moisture and nutrients. They may also harbour pests and diseases.

WEEDS IN GARDEN BEDS

Weeds in the garden, especially annual weeds, can be eliminated to a great degree if they are prevented from flowering and forming seed. They can be controlled by hand weeding or hoeing, preferably when they are quite small. Pre-emergence weedicides such as Dacthal can be sprayed on the soil and watered in, and will control a wide range of weeds. Dowpon (Dalapon) is a systemic weed killer for the control of grasses only. It does not harm established trees or shrubs if used as directed. It is recommended for use on gardens if grass weeds are a problem.

WEEDS NEAR FENCES, TREES AND BORDERS

Weeds, both grasses and others, are a nuisance if they trail over fences, around trunks of woody shrubs and trees or across borders and edges of gardens. Desiccant weedicides such as Tryquat or Weedex are extremely useful in these situations. They can be watered, or preferably sprayed, onto all weeds which then shrivel and die in a day or two. These chemicals kill all green tissue (containing chlorophyll) but do not affect the woody parts of plants. The desiccant weedicides have no residual effect on the soil and are quickly broken down into harmless compounds. For this reason, perennial weeds may re-grow and new annual weeds will germinate from seed but two or three sprays each year will keep the area relatively weed-free. For small areas, pressure-pack spot weeders are available.

Another extremely effective weedicide is Zero (also known as Glyphosate). This is a non-selective chemical which is ideal to control persistent weeds like paspalum, couch grass, nutgrass and sorrel. The chemical is translocated from the above-ground parts of the plant to the persistent underground parts. There is no residual weedicide action when Zero is sprayed on to the soil. The chemical is safe to use but take care to prevent spray or drift reaching useful plants.

WEEDS IN PATHS AND DRIVEWAYS

On areas such as paths, driveways, courtyards or tennis hardcourts, total weed killers can be used. There are many brands, such as Weedazol Total, K.O. Weedkiller, New Camellia Weedkiller (Arzeena Weedkiller), Erase and Vorox. It is important to stress that all these have a residual effect on the soil—usually for a period of at least three months—so that during heavy rain there may be some movement of the chemicals into adjacent areas.

PERENNIAL WOODY WEEDS

Large woody weeds like blackberry, briars, scrub and trees can be killed with hormone weedicides containing esters of 2,4-D and 2,4,5-T. Overall sprays control blackberries and other woody weeds. For trees and larger scrub growth, the chemical can be poured into holes bored in the trunk or into a frill ring cut around the trunk. Hormone sprays may drift in the wind and damage nearby plants so take care to spray on a calm day.

WEEDS IN LAWNS

The best way to prevent weeds in lawns is to have healthy vigorous turf which resists weed invasion. This is achieved by correct application of fertilizers, adequate watering and other maintenance practices. (See Chapter 7.) However, weeds may still be a problem in lawns and a wide range of selective weedicides are available to control them. Lawn weeds can be divided into four groups:

Group 1—Grassy Weeds Any grass which differs from the grass composing the lawn can be considered a weed. Common grassy weeds are paspalum, summer grass, winter grass and Mullumbimby couch. In fact couch grass in a bent lawn or carpet grass in a couch lawn could be regarded as weeds. Regular spraying with Passtox or Methar will control grasses such as paspalum, summer grass, crab grass, Mullumbimby couch and sedges in lawns of couch, bentgrass and fescue, or mixtures containing these species. These weedicides cannot be used on lawns composed of Queensland blue couch, buffalo, kikuyu or carpet grass because they will kill these grasses.

Passtox or Methar should be sprayed on the lawn when the weeds are growing vigorously. After 7–10 days mow the lawn and repeat the spray when the weeds' reserves are low. Several applications may be needed for complete eradication of grassy weeds. Both chemicals can be spot-sprayed on grass weeds—for example carpet grass or kikuyu in a lawn of Queensland blue couch. The blue couch will be slightly affected on the edges of the sprayed area but it will recover.

Sprays of Endothal will control actively growing winter grass and pre-emergence sprays of Dacthal will kill the germinating seeds of winter grass if sprayed in early autumn. Seeds of summer grass are also killed by Dacthal when the lawn is sprayed in spring.

Group 2—Broad-leaf Weeds Marshmallow, dandelion, cat's-ear, lamb's tongue, cudweed and chickweed are some of the common broad-leaf weeds found in lawns. Weedicides such as Clovotox, Nocweed A10, Lane's Lawn Weeder and Weedoben-M are effective in eradicating them.

Group 3—Clover-like Weeds Clovers, medics and creeping oxalis are the main weeds in this group. Clovotox and Lane's Lawn Weeder will control clovers and Lane's Creeping Oxalis and Clover Killer will kill both clovers and oxalis.

Group 4—Fine-leaf Weeds Carrot weed and bindii are the main fine-leaf weeds found in lawns. Control both with Lane's Lawn Weeder or Weedoben-M.

The weedicides listed in Groups 2, 3 and 4 can be used on all lawn grasses. They should not be used on lawns composed of mixtures including strawberry clover or white clover.

The use of Sulphate of Ammonia, or lawn foods which contain it, will discourage weeds, especially broad-leaf weeds and clovers. The weedicide effect is increased if the fertilizer is applied dry to a lawn which is damp with dew. Lawn sand—a mixture of equal parts of sulphate of ammonia, sulphate of iron and sand—is still a recommended method of lawn weed control. The mixture is applied dry at 4 kg per 100 square metres and the lawn should not be watered for a day or two. The grass may suffer a temporary burn but will recover rapidly.

The pesticides mentioned in this chapter are generally available throughout Australia. For advice on specific treatment of pests in your district consult your State agricultural department. Before buying or using any pesticide read the label carefully.

Plantain treated with Lane Lawn Weeder

RECOMMENDED HOME GARDEN INSECTICIDES

Common Name or Reg. Trade Name	Active Ingredient Spray or Dust	Witholding Period (days)	Used to Control	Remarks
Carbaryl, Sevin®	Carbaryl spray (wettable powder) or dust	3	Beetles, bugs, borers, caterpillars, cutworms	Safe insecticide, component of many all-purpose sprays and tomato or cabbage dusts
Chlordane	Chlordane spray (liquid)	30	Ants, borers, black beetle, lawn grubs, mole crickets	Persistent chemical but very effective for all lawn pests
Derris	Rotenone dust	1	Beetles, caterpillars	Very safe insecticide for leaf-eating insects but frequent dusting is necessary
Dipel® HG	Bio-insecticide containing bacteria spray (wettable powder)	0	Caterpillars	Very safe insecticide, specific control of caterpillars
House Plant Spray	Pyrethrum spray (liquid) or aerosol	1	Aphids, caterpillars, jassids, thrips	Very safe insecticide, good for balcony gardens and indoor plants
Kelthane®	Dicofol spray (liquid)	7	Red spider, mites	Specific miticide, component of many all-purpose sprays
Lebaycid®	Fenthion spray (liquid)	7	Aphids, bugs, jassids, fruit fly, red spider, mites, leaf miners, leaf eelworm, thrips	Effective systemic insecticide for all sap-sucking pests
Malathion®	Maldison spray (liquid)	3	Aphids, beetles, bugs, borers, caterpillars, cutworms, jassids, scale insects, thrips, white fly	Safe contact insecticide, component of many all-purpose sprays

Common Name or Reg. Trade Name	Active Ingredient Spray or Dust	Witholding Period (days)	Used to Control	Remarks
Metasystox®	Demiton-S-methyl spray (liquid)	21	Aphids, bugs, jassids, red spider, mites, leaf miner, leaf eelworm, thrips	Effective but persistent systemic insecticide for all sap-sucking pests. Not recommended for vegetables close to harvest
Rogor®	Dimethoate spray (liquid)	7	Aphids, bean fly, bugs, fruit fly, jassids, red spider, mites, leaf miner, thrips, white fly	Safe systemic insecticide for all sap-sucking pests
Thiodan®	Endosulphan spray (liquid)	7	Aphids, beetles, bugs, borers, caterpillars, cutworms, jassids, thrips	Effective all-purpose but persistent insecticide. Not recommended for vegetables close to harvest
White Oil	Paraffin oil spray (liquid)	1	Scale insects	Addition of Malathion or Rogor increases effectiveness

® Registered Trade Mark.

RECOMMENDED HOME GARDEN FUNGICIDES AND NEMATICIDES

Common Name or Reg. Trade Name	Active Ingredient Spray or Dust	Witholding Period (days)	Used to Control	Remarks
Benlate® (Bayleton)	Benomyl spray (wettable powder)	5	Leaf spots, leaf blights, powdery mildew, stem and root rots, dollar spot (lawns)	Very effective systemic fungicide but does not control downy mildew or rust
Captan	Captan spray (wettable powder)	1	Leaf spots, leaf blights, downy mildew, damping-off of seeds (dust) or seedlings (spray), brown patch and dollar spot (lawns)	Component of many all-purpose sprays
Dry Bordox	Copper oxychloride spray (wettable powder)	1	Leaf spots, leaf blights, downy mildew, rusts, fruit tree diseases, damping-off of seeds (dust) or seedlings (spray)	All-purpose fungicide, usually a component of tomato dusts
Harola	Lime sulphur spray (liquid)	3	Leaf spots, leaf blights, powdery mildew, rusts, also controls red spider, mites and some scale insects	Useful fungicide for diseases of fruit trees. May cause leaf scorch in vine crops
Karathane®	Dinocap spray (wettable powder)	21	Powdery mildew	Component of many all-purpose sprays
Mancozeb Dithane M45®	Mancozeb spray (wettable powder)	14	Leaf spots, leaf blights, downy mildew, rust, stem and root rots, damping-off of seedlings, brown patch and dollar spot (lawns)	Effective all-purpose fungicide
Nemacur	Fenamiphos granules	—	Root eelworms	Soil application. Use before planting

Common Name or Reg. Trade Name	Active Ingredient Spray or Dust	Witholding Period (days)	Used to Control	Remarks
Sulphur	Colloidal sulphur spray (wettable powder) or dust	0	Powdery mildew, rusts. Also controls red spider, mites	Usually component of tomato dusts. May cause leaf scorch of vine crops
Zineb	Zineb spray (wettable powder)	7	Leaf spots, leaf blights, downy mildew, rust, stem and root rots, damping-off of seeds (dust) and seedlings (spray), brown patch and dollar spot (lawns)	Effective all-purpose fungicide for control of most diseases except powdery mildew. Component of many all-purpose sprays

® Registered Trade Mark.

PEST AND DISEASE CONTROL CHART FOR VEGETABLES

Plant	Pest or Disease	Symptoms	Control
Broad beans	Black aphids	Small, black insects on underside of leaves and shoots	Malathion, Rogor
French and runner beans	Red spider	Bronzing and unthriftiness of foliage	Kelthane, Rogor
	Bean fly	Larvae burrow into stalks and stems	Rogor
	Green vegetable bug	Large green shield bugs	Carbaryl, Malathion, Rogor, Lebaycid
	Rust	Red-brown blisters on leaves and pods	Zineb, Mancozeb
Broccoli, Brussels sprouts, cabbage, cauliflower	Larvae of cabbage moth and white butterfly	Holes in leaves	Carbaryl, Malathion, Derris, Dipel, Thiodan
	Grey aphis	Small grey insects on underside of leaves	Malathion, Rogor
	Cutworms	Seedlings eaten at ground level	Carbaryl drench
Capsicum	Fruit fly	Larvae burrow through fruit	Rogor, Lebaycid
Carrot	Aphids	Small insects usually underneath leaves	Malathion, Rogor
Celery	Leaf spot	Brown spots over leaves	Benlate, Zineb, Mancozeb
Cucumber	Powdery mildew	Powdery white film on leaves	Benlate, Karathane
	Downy mildew	Leaf spots and downy tufts underneath	Zineb, Mancozeb
	28-spotted ladybird and pumpkin beetle	Leaf damage	Carbaryl, Malathion
Eggplant	Fruit fly	*See Capsicum*	
Lettuce	Aphids	*See Carrots*	
	Downy mildew	Leaf spots and downy tufts underneath	Zineb, Mancozeb
Marrows (zucchinis), melon, squash	*See Cucumber*		
Onion	Thrips	White flecks on foliage	Malathion, Rogor, Lebaycid
	Downy mildew	Leaves die from tips and downy tufts on leaves	Zineb, Mancozeb

PEST AND DISEASE CONTROL CHART FOR VEGETABLES

Plant	Pest or Disease	Symptoms	Control
Parsnip	Powdery mildew	*See Cucumber*	
Peas	Red spider, mites	*See Beans*	
Potato	Potato moth	Leaves and tubers infested	Carbaryl, Derris
	28-spotted ladybird	*See Cucumber*	
	Late blight	Large black areas on leaves	Zineb, Mancozeb, Dry Bordox
Pumpkin	*See Cucumber*		
Seeds and seedlings	Damping-off	Seeds fail to germinate or seedlings fall over at soil level	Captan, Zineb, Dry Bordox, dust seeds before sowing or spray seedlings
Silver beet, spinach	Leaf miner	White tunnel streaks on leaves	Rogor, Lebaycid
	Leaf spot	Small brown spots on foliage	Zineb, Mancozeb
Sweet corn	Corn ear worm	Caterpillar in top of cob	Dust or spray with Carbaryl
Tomato	Tomato cater-pillar	Caterpillar attacks fruit at stalk end	Carbaryl, Malathion, Derris
	Spotted wilt	Browning of young foliage	Remove infected plants and control thrips—*see Onions*
	Fruit fly	*See Capsicum*	
	Root eelworm	Swellings on roots	Nemacur granules before planting
	Leaf spot	Brown target-like spots on foliage and fruit	Zineb, Mancozeb

PEST AND DISEASE CONTROL CHART FOR FLOWERS AND ORNAMENTALS

Plant	Pest or Disease	Symptoms	Control
Aster	Aphids	Small insects under foliage and on flower buds	Malathion, Thiodan, Rogor, Metasystox, Lebaycid
	Jassid	Small leaf-hopping insects which spread virus	*see above*
	Caterpillars	Caterpillar in leaves which come together	Carbaryl, Malathion, Thiodan, Derris

PEST AND DISEASE CONTROL CHART FOR FLOWERS AND ORNAMENTALS

Plant	Pest or Disease	Symptoms	Control
	Red spider	Foliage turns bronze and is unthrifty	Kelthane, Rogor, Lebaycid, Metasystox
	Virus	Flowers become green	Control jassids
Azaleas and rhododendrons	Lace bug	Foliage becomes silver-bronze	Malathion, Rogor, Lebaycid, Metasystox
	Leaf miner	*See Cineraria*	
	Red spider	*See Aster*	
	Petal blight	Flecking and shrivelling of flowers	Bayleton
	Leaf gall	Pale green swellings develop on foliage. Foliage becomes distorted	Remove and burn infected foliage, spray with Zineb, Mancozeb or Dry Bordox
Begonia	Powdery mildew	Powdery growths on leaves	Benlate, Karathane
Bellis (English daisy)	Rust	Bronze blisters on foliage	Zineb, Mancozeb
	Red spider	*See Aster*	
Calendula	Rust	*See Bellis*	
	Red spider	*See Bellis*	
	Powdery mildew	*See Begonia*	
Camellias	Dieback	Large, dark, dead areas on branches	Remove and burn infected twigs, spray with Benlate, Zineb or Mancozeb
	Balling	Flower buds do not open and bud scales are damaged due to bud mite	Spray with Kelthane, Rogor, Lebaycid, Metasystox
Carnation	Thrips	White flecks on petals	Malathion, Thiodan, Rogor, Lebaycid, Metasystox
	Rust	*See Bellis*	
Chrysanthemum	Black aphids	Small, black insects underneath leaves and on flower buds	*See Aster*
	Leaf spot and rust	Various spots on leaves	Benlate, Zineb, Mancozeb
	Leaf eelworm	Triangular black areas on leaves	Lebaycid, Metasystox Rogor
Cineraria	Leaf miner	Irregular yellow tracks on leaves	Rogor, Lebaycid, Metasystox
	Leaf spot	Brown target spots on leaves	Zineb, Mancozeb
Cornflower	Aphids	*See Aster*	
	Powdery mildew	*See Begonia*	
	Collar rot	Plant rots at ground level	Remove and burn

PEST AND DISEASE CONTROL CHART FOR FLOWERS AND ORNAMENTALS

Plant	Pest or Disease	Symptoms	Control
Dahlia	Jassids	*See Aster*	
	Red spider	*See Aster*	
	Powdery mildew	Powdery film over foliage	*See Begonia*
Delphinium	Powdery mildew	*See Begonia*	
Ferns	Mealy bug	White insects with filaments under fronds	Malathion, Rogor, Lebaycid, Metasystox
	Scale insects	Swellings along stem	*As above*
	Aphids	Soft insects attack young growth	*See Aster*
	Red spider	*See Aster*	
	Staghorn beetle	Holes in fronds and tips die	Carbaryl, Malathion, Thiodan
	Leaf eelworm	Dark streaks on foliage	Lebaycid, Metasystox Rogor
Foxglove	Aphids	Insects on flower spikes and leaves	*See Aster*
Geranium	Rust	*See Bellis*	
Gerbera	Thrips	Flowers deformed and flecked	*See Carnation*
	Powdery mildew	*See Begonia*	
Gladioli	Thrips	Flecks on leaves and flowers	*See Carnation*
Hibiscus	Caterpillars	Leaves stick together, holes in leaves	*See Aster*
	Aphids	Young foliage distorted	*See Aster*
	Beetles, weevils	Holes in foliage	Carbaryl, Malathion, Thiodan
	Thrips	Flowers damaged	*See Carnation*
	Dieback	Plants become unproductive	Thorough watering is best control
Hollyhock	Red spider	Flowers become bronzed and dry	*See Aster*
	Rust	*See Bellis*	
Larkspur	Powdery mildew	*See Begonia*	
Marigold	Jassids	*See Aster*	
	Red spider	Foliage becomes red-brown and unthrifty	*See Aster*
Pansy	*See Violas*		
Penstemon	Red spider	*See Aster*	

PEST AND DISEASE CONTROL CHART FOR FLOWERS AND ORNAMENTALS

Plant	Pest or Disease	Symptoms	Control
Polyanthus	Red spider	*See Aster*	
	Mealy bug	White insects under leaves and on roots	Malathion sprays and soil drenches
	Caterpillars	Holes in leaves	*See Aster*
Roses	Aphids	Young foliage distorted	*See Aster*
	Thrips	Flowers damaged by flecking	*See Carnation*
	Caterpillars	Leaf roll and holes in leaves	*See Aster*
	Red spider	*See Aster*	
	Black spot	Large black spots on leaves	Benlate
	Powdery mildew	*See Begonia*	
	Downy mildew	*See Stock*	
	Rust	*See Bellis*	
	Flower and bud blight	Grey fluffy growths on flowers and buds	Benlate
	Rose wilt virus	Leaves droop and bush is unthrifty	Control aphids and remove and burn infected bushes
	Root eelworm	Large swellings on roots	Nemacur granules forked into soil around roots
Snapdragon	Rust	*See Bellis*	
	Caterpillars	Green caterpillars on flower buds	*See Aster*
Stock	Aphids	Young plants deformed	*See Aster*
	Caterpillars	On leaves and buds	*See Aster*
	Downy mildew	Seedlings very unthrifty and white downy growths under leaves	Zineb, Mancozeb
Sweet pea	Broad bean wilt virus	Plants wither and die, foliage becomes puckered	Control aphids and plant later in season to avoid attack
	Thrips	*See Carnation*	
Violas	Aphids	Distorted foliage and shoots	*See Aster*
Zinnia	Caterpillars	Holes in leaves and flower buds	*See Aster*
	Powdery mildew	White powdery growth on older leaves	*See Begonia*
	Leaf spot	Brown target spots on leaves and stems	Zineb, Mancozeb

Chapter 19

KEEPING CUT FLOWERS FRESH

Most flowers last better if you pick them in the early morning or the cool of the evening when they are sappy and contain a lot of moisture. Once flowers are picked, don't leave them out of water a minute longer than necessary. It is a good idea to have a bucket of water handy to put them in immediately. Most flowers benefit from a good drink by standing them undisturbed in fairly deep water for an hour or two before arranging.

Always use clear, fresh water which is not too cold. In cooler weather, a little warm water may be added. Flower stems absorb tepid water more easily and it is quickly distributed to the flowers. Stems cut on the slant will absorb water better than those with a straight cut. Always use a sharp knife or secateurs to make a clean cut without ragged edges.

Vases and bowls should be thoroughly clean before use because accumulated dirt and vegetable matter will quickly cause the water to become stale. Needle or pin holders can be cleaned up with a stiff brush. Water will quickly become stale from decayed submerged foliage, so strip off the lower leaves to prevent this. Some people suggest that the water should be changed every day or two but this is rarely necessary except, perhaps, with flowers such as chrysanthemums, marigolds or daisies which may produce an unpleasant smell. Never pack the container too tightly because the stems and flowers must breathe. It is a good idea to remove any wilted flowers and replace them with fresh ones. Flower arrangements should be kept out of hot sun and away from radiators or cold draughts.

Many additives to the water have been suggested in the past, including copper coins, sugar, aspirin and gin. The benefit of these materials in prolonging the life of cut flowers is doubtful, but there are a few chemical preparations available which do help. One is Lane Formula 20, a hormone preparation, and another is a cut flower food called Chrysal. There are some very practical ways to make your flowers last longer. These are based on the type of stem—soft, hard, woody, hollow or milky. A summary of ways to keep cut flowers fresh is given below.

Anemone: Scrape ends of stems and place in water for three-quarters of their length for half an hour.

Aster: Remove most of the leaves, scrape or slit ends of stems and place in deep water for an hour.

Azalea: Scrape or slit stems, handle blooms as little as possible.

Calendula: Pick when flowers are fully open, scrape stems.

Camellia: Scrape or slit stems. Spray flowers once or twice a day with water.

Canterbury Bells: Place ends of stems in boiling water for 30 seconds. Then place in cold water.

Carnation: Stand stems in deep water for an hour, then dip flowers in water for a few minutes.

Chrysanthemum: Pick flowers when fully open. Crush the ends of the woody stems and then stand in boiling water for one minute.

Cornflower: Cut stems at an angle and place in water immediately.

Dahlia: Scald end of stems for 30 seconds then stand in plenty of water for two hours.

Daffodil and Jonquil: Cut stems at an angle and stand in shallow water for one hour.

Daisy (Marguerite and Shasta): Scald ends of stems or sear over a flame.

Elegant arrangement emphasizes form and line

Delphinium: Stand small spikes in boiling water for 30 seconds. Cut large spikes to required size, hold upside down and fill hollow stem with water and plug with cotton wool.

Dianthus: Stand stems in deep water for an hour, then dip flowers in water for a few minutes.

Foxglove: Cut stems to required length, then dip ends in hot water for one minute and then into cold water.

Gaillardia: Cut stems at an angle and

This attractive arrangement would be a focal point in any room

place in water immediately.

Gardenia: Lasts well without water but flowers mark easily if touched.

Geranium and Pelargonium: Crush ends of stems and stand in water for one hour.

Gladiolus: Cut flower spikes when the first florets open. Cut spike to required length and stand in deep water. Snap off top buds.

Hibiscus: Usually only last one day, possibly better out of water.

Hydrangea: Crush ends of stems and scorch over flame, then stand in water for two hours.

Iris: Pick when first bud opens fully. Scrape stems.

Irish Green Bell: Crush ends of stems and stand in water for one hour.

Larkspur: Cut when flowers are three-quarters open. Scrape or slit ends of stems. Snap off top buds.

Lilies: Cut stems at an angle and stand in deep water for one hour.

Lupin: Place stems in boiling water for 15–20 seconds, then into cold water. Fill the hollow stems of Russel Lupin with water and plug with cotton wool.

Marigold: Scrape stems and place immediately in water for one hour. Re-cut stems before arranging. Change water frequently.

Phlox: Place stems in shallow hot water for 30 seconds, then into cold water for half an hour.

Polyanthus and Primula: Place stems in tepid water immediately after picking.

Pansy and Viola: Dip flowers completely in water before arranging.

Poppy (Iceland): Pick in full bud stage. Scorch cut ends of stems over a flame or dip in boiling water for 30 seconds. Then stand in deep water for half an hour.

Rose: Cut in bud as colour shows. Scrape ends of stems for 2–3 cm and split them. Stand ends in shallow hot water for 30 seconds.

Snapdragon (Antirrhinum): Crush or split ends of stems and place in deep water for half an hour.

Stock: Split ends of stems and crush lightly. Place in boiling water for 30 seconds, then in deep water for half an hour. Snap off top buds.

Sweet Pea: Stand stems in deep water for 2–3 hours before arranging.

Violet: Absorbs water through leaves and flowers as well as stems. Immerse bunch completely in water, then wrap in newspaper for an hour or two.

Wallflower: Split ends of stems and crush lightly. Place in boiling water for 30 seconds, then in deep water for half hour.

Zinnia: Pick when flowers are fully open. Scorch cut ends in flame or dip in boiling water for 30 seconds. Plunge into cold water quickly.

Chapter 20

QUICK FREEZING VEGETABLES

Even the most experienced of home vegetable gardeners will have surplus vegetables at one time or another. If you have a deep-freeze unit you can conserve this surplus for later use. If you or your family are fond of a particular vegetable such as peas or sweet corn you can plant a larger crop so you can have your favourite 'out of season'. Many vegetables can be deep frozen successfully by following a few simple rules.

1. Always select vegetables at their peak of quality for best texture, flavour and colour.
2. Pick them and freeze them as soon as possible after harvest.
3. Wash vegetables in cold water to remove dirt and insects.
4. Prepare them as you would for cooking.
5. Before quick freezing, vegetables need to be 'blanched' to fix the colour and inactivate plant enzymes which cause 'off' flavours. Place the prepared vegetables in a wire basket or piece of cloth and plunge them into rapidly boiling water in a large saucepan. Do not blanch large quantities of vegetables at a time because it will take too long for the water to return to the boil and the vegetable will overcook. If it takes longer than two minutes for the water to boil again then reduce the quantity of vegetables in the basket.
6. At the end of the blanching time, lift out the wire basket and plunge it into ice-cold water. It is important to chill the vegetables thoroughly, otherwise they will continue to cook in their own heat. Break a piece of vegetable open and check that it has cooled to the centre.
7. Lift the chilled vegetables from the cold water and transfer them to another wire basket or colander to drain.
8. After all the water has drained away, pack the vegetables quickly into suitable containers. Special non-stick plastic freezer bags are the most commonly used. Place sufficient vegetables for one meal in each bag. When frozen vegetables have thawed out, they cannot be re-frozen as this damages the vegetables and makes them soft.
9. When packing the vegetables, expel as much air as possible from the bags; twist the top tightly, fold over and tie with a twist-tie or string. Rubber bands are not satisfactory as they tend to perish in the freezer.
10. Place the bags in the freezer as soon as packaged. Most frozen vegetables will keep in perfect condition for a period of 12–18 months. The vegetables should be placed in the freezer in such a manner as to allow maximum air circulation. This produces the quickest possible freezing and thus the highest quality product. If all packages are tightly stacked it takes a far longer time for the heat to be removed from the centre packages and they are slowly frozen. This results in a product of softer texture and less satisfactory flavour.

GUIDE TO QUICK FREEZING VEGETABLES

Vegetable	Blanching Time (minutes)	Preparation for Freezing
Asparagus	2–4	Asparagus becomes tough and loses flavour quickly after picking, so freeze as soon as possible. Trim stalks using only the tender portion. Wash well, blanch, chill, drain, pack and freeze.
Beans	3	Select young tender beans only. Wash and prepare as for cooking. Blanch, chill, drain, pack and freeze.
Broccoli	3–4	Wash thoroughly and trim off leaves. Cut centre heads into serve-size pieces or any side shoots larger than 2 cm thick. Blanch, chill, drain, pack and freeze.
Brussels sprouts	4–5	Trim off outer leaves and wash thoroughly to remove insects. Then treat as for broccoli but note longer blanching time.
Cabbage	2–3	Discard coarse outer leaves and wash well. Shred or slice hearts, blanch, chill, drain, pack and freeze.
Capsicum (halves)	3–4	Wash well, cut fruit in half and remove the stalk and seeds. Rinse to remove loose seeds. Blanch, chill, drain, pack and freeze.
Capsicum (slices)	2–3	As above but cut into slices for use in stews, casseroles or for frying.
Carrots	4–6	Wash and scrape well. Prepare as for cooking by slicing or dicing. Small Baby carrots can be used whole. Blanch, chill, drain, pack and freeze.
Cauliflower	4–5	Wash thoroughly and break florets into serve-size pieces. Then treat as for Brussels sprouts.
Celery	3	Frozen celery can only be used in cooked dishes as it tends to collapse on thawing. Only use crisp tender celery. Wash, trim and string stalks and cut into 2–3 cm lengths. Blanch, chill, drain, pack and freeze.
Kohl Rabi	2	Select small, tender bulbs 5–7 cm in diameter. Cut off tops, wash, peel and dice into 1–2 cm pieces. Blanch, chill, drain, pack and freeze.
Mixed Soup Vegetables	—	Prepare each vegetable separately, using the correct blanching time. Chill and drain before mixing together for packaging and freezing.
Parsnip	2–3	Remove tops, peel and wash well. Cut into 1 cm slices or dice into 1 cm cubes. Blanch, chill, drain, pack and freeze.

Vegetable	**Blanching Time** (minutes)	**Preparation for Freezing**
Peas	½–1	Shell the peas and discard old, discoloured or split peas. Wash and drain, then blanch, chill, drain, pack and freeze.
Potatoes	—	Must be cooked before freezing. French fried potatoes are best. Peel the potatoes then cut as desired. Scald for 1–2 minutes then drain. Fry in pan to a light brown. Allow fat to drain and when cold, pack and freeze.
Sweet Corn (on the cob)	6–9	Select young tender cobs and prepare as soon after picking as possible. Husk and de-silk cobs, then blanch, chill and drain. Wrap each cob separately and freeze.
Sweet Corn	5	Husk and de-silk the cobs, blanch, chill and drain. Cut the corn from the cob, package and freeze.
Tomato	—	Must be cooked before freezing. Wash tomatoes, scald and remove skins. Place in saucepan with a small amount of water and simmer gently until soft. Allow to stand until quite cold then pack in rigid containers (leave 2–3 cm head space for expansion) and freeze. For tomato juice, press the cooked tomatoes through a fine sieve. Allow juice to become quite cold then pour into rigid containers as before with 2–3 cm head space. To serve, allow juice to thaw slowly in the refrigerator. When liquid add sugar, salt, lemon or mint to taste.

GLOSSARY

Some gardening words may be unfamiliar to the average reader. These words are used in gardening books or by specialist garden writers in magazines or newspapers. The glossary should help the non-expert reader to understand the gardening terms used in this guide.

Acid: Soil deficient in lime.
Air-layering: Method of propagation to stimulate roots from an above-ground part of the plant.
Alkaline: Soil rich in lime.
Annual: A plant which grows from seed, flowers, forms seed and dies in one season; more loosely, a plant which can be treated this way.
Aquatic: A plant which grows in water, sometimes floating (duckweed), sometimes with roots in mud (water lily).
Axil: The angle between a stem and a leaf, the normal position for a lateral bud.

Bacteria: Microscopic, one-celled organisms mainly responsible for decomposition of plant and animal remains. Some are the cause of plant disease.
Banding (fertilizer): Scattering fertilizer in a band alongside plants or the line where seeds are to be sown.
Bare root: The term applied to trees (usually deciduous) and roses when sold by nurseries in winter.
Bedding plant: Flowering plant (usually annual) used for planting in masses or in a bed of its own.
Bicoloured: A flower having two colours on the same petals.
Biennial: A plant which grows from seed one year, over-winters and flowers the next year, then dies.
Bolt (*verb*): A plant, usually a vegetable, which runs to seed prematurely.
Border plant: Plant which lends itself to planting in borders or clumps in the foreground of garden beds.
Bract: A modified leaf at the base of a flower, e.g. bougainvillea.
Broadcast (*verb*): To scatter seeds or fertilizers in a random fashion.
Bud: Compact, undeveloped shoot (leaves or flowers).
Budding: Method of grafting in which the grafted part is a bud; **bud-union** is the position of grafted bud.
Bulb: A bud or shoot, usually underground, enclosed by modified leaves or fleshy scales for food storage, e.g. onion, daffodil.
Bulbil: A small immature bulb.

Callus: Tissue developed in woody plants in response to wounding (and grafting).
Calyx: Outer covering of a flower consisting of sepals. Usually green but may be decorative e.g. Christmas bush.
Cambium: A thin, unbroken layer of growing tissue which connects all parts of a plant to allow for growth in thickness of roots, stems and leaves.
Catkin: A spike of pendulous flowers on some trees and shrubs, e.g. willow.
Cell: The individual building block of plant and animal life. Each cell contains a nucleus and cell sap surrounded by a cell wall. Cells in growing tissue can divide to form new cells.
Chlorophyll: The green, light-absorbing pigment of plants.
Clove: A small bulb of some plants, e.g. garlic.
Clump: (1) A few seeds sown in a group or station at a specified distance apart.
(2) Several plants of the same kind grown in a group, e.g. clump planting.
Colloid: Very small particles of soil or humus which can hold nutrients and build soil crumbs.
Compost: Decomposed organic material (usually plant remains), or more loosely, a mixture of soil and other materials used to grow plants in containers.
Container plant: A plant which is grown and sold by nurseries in a pot or other container, or more loosely, any plant which can be grown in a container.

Conifer: A tree or shrub bearing cones, e.g. pine, cypress, spruce.
Corm: Swollen base of plant stem serving the same storage purpose as a bulb, e.g. gladiolus.
Cormel: A small immature corm.
Cotyledon: A leaf forming part of embryo of seeds. Some seeds have one seed-leaf, others two seed-leaves. Cotyledons may store food for the seed, e.g. peas, beans.
Cutting: A piece of plant used to grow a new plant.

Deciduous: A plant which sheds its leaves at the end of the growing season.
Dibble: Small stick for making holes for planting seeds or seedlings.
Disbudding: Removing side flower buds to concentrate growth in a single flower.
Division: Method of dividing perennial plants which form a crown.
Dormant: A plant which has stopped growing, usually in winter.
Drip-line: The line underneath the outer foliage of trees where the feeding roots are usually located.
Drill: A shallow furrow in the soil for sowing seed or placing fertilizer.

Edging plant: A dwarf or creeping plant for dividing beds or borders from lawns, paths or paved areas.
Embryo: Young plant enclosed in a seed.
Epiphyte: A plant supporting itself without soil, usually growing on trees or shrubs but not a parasite, e.g. orchids.
Espalier: Shrub, tree or vine trained in two directions, usually against a wall.
Evergreen: A plant that retains its leaves all year round.
Everlastings: Plants with dry straw-like flowers which can be preserved for long periods, e.g. acroclinium.

Family: A broad grouping of plants which have similarities of flower structure and other botanical characteristics.
Fertilizer: Material to supply plant nutrients, whether of organic or inorganic origin. This term, however, does not cover animal manure or compost.
Field capacity: The quantity of water a soil can hold after drainage water is removed.
Foliage plant: A plant grown for the decorative value of its leaves.
Friable (soil): Easily crumbled.
Frond: The leaf and leaf stem of a fern.
Fungus: A form of plant life which includes mushrooms and toadstools but also those which are parasites on plants causing disease.
Fungicide: A chemical used to control fungus diseases.

Genus: A name given to plants (belonging to the same family) which have similar field characteristics. It has been described as a botanical surname. The plural is *genera*.
Germination: The commencement of growth of an embryo within a seed.
Grafting: Joining a detached bud or stem of one plant on to the stem of another.
Gravitational water: Water lost from a soil by drainage.
Green manure: A crop grown to be incorporated in soil to increase organic matter.
Growing tissue: Parts of plants where cells are dividing, e.g. root tips, shoots, buds.

Habit: The general growth form of a plant, e.g. upright, trailing, bushy.
Hardening off: Exposing young plants (especially seedlings) gradually to harder conditions in the open garden.
Hardy: An English term used for plants which can be grown outdoors in all seasons, i.e. they tolerate frost and low temperatures. Also half-hardy, meaning they are less tolerant. In Australia, the word hardy has been used by garden writers to mean tolerance to tough conditions generally—heat, cold, drought, poor soil or whatever.
Herb: A non-woody plant or a plant grown for flavouring purposes.
Herbaceous perennial: A non-woody plant which dies back to the crown or roots in the dormant (winter) period, to grow again the following season.
Hill: A term to describe a clump or station of one or more plants spaced widely apart, e.g. cucumbers or pumpkins. Also used as a verb, to draw up soil around plants (especially when sown in rows), e.g. sweet corn, potatoes.
Hilum: Scar on seed coat marking former point of attachment to parent plant, e.g. beans.
Humus: The remains of decomposed organic matter, a vital component for garden soil with good structure.

Hybrid: A seed or plant which is the result of breeding from two selected but unlike parents. Hybrid plants are very uniform, vigorous and tolerate adverse conditions. *F1 Hybrid* refers to the first generation of this breeding programme. *F2* is the second generation, and so on.

Hydroponics: A system for growing plants without soil.

Hygroscopic water: Water in the soil which is unavailable to plants.

Insecticide: Chemical used to control insect pests.

Larva: Pre-adult form of insect, e.g. caterpillar.

Lateral: A secondary shoot or stem.

Layering: Pinning a branch or stem to the soil to produce roots.

Leaching (of soils): Loss of nutrients in drainage water.

Legume: A plant with pea-like flowers and seeds in a pod or the seeds of such a plant.

Major nutrient: A plant nutrient required in large quantity. Nitrogen, phosphorus and potassium are major nutrients.

Manure: Animal excreta which may also contain straw and other litter. Manure does not include animal or plant by-products, e.g. blood and bone. These are more correctly fertilizers.

Microclimate: A local combination of climatic conditions.

Micro-organism: Forms of life too small to be seen with the naked eye, e.g. bacteria.

Minor element: A plant nutrient required in very small quantities. Also called *trace element*.

Mites: Small spider-like animals which are plant pests.

Miticide: Chemical used to combat mites.

Mulch: A natural or artificial soil covering which protects plant roots and conserves soil moisture.

N.P.K. ratio: The percentage of major nutrients in mixed fertilizers, e.g. Gro-Plus Complete (N.P.K. 5:7:4).

Nematode: Small worm-like animals (eel-worms), some of which attack plant roots causing swellings or galls.

Nematicide: Chemical used to control nematodes.

Node: Joint in plant stem from which leaves or roots may arise.

Nodule: Swellings (containing bacteria) on the roots of legumes. The bacteria are able to supply the legume plant with nitrogen.

Nucleus: Of a plant or animal cell containing the chromosomes.

Offset: A small plant which grows from its parent and can be detached and grown separately.

Organic: Composed of live or formerly living tissue.

Osmosis: The ability of dissolved substances to diffuse through a membrane.

Ovule: A structure in seed plants containing the female or egg-cells which are fertilized by the male pollen cells to form a seed.

pH: A unit for measuring the acidity or alkalinity of substances such as soil.

Pathogen: An organism causing disease, e.g. fungus, bacterium or virus.

Perennial: A plant which lives for a number of years.

Petiole: A leaf-stalk.

Pinch back: To prune soft, leading shoots to encourage branching.

Plant food: Commonly used to describe plant nutrients applied as fertilizers.

Photosynthesis: The conversion of water and carbon dioxide into carbohydrate (sugar) by plants in the presence of light and green chlorophyll.

Pollen: Small grains containing a male cell.

Pollination: Transfer of pollen, usually by insects, to the female part of the flower.

Pore space: Space between soil particles.

Respiration: Breathing; all parts of plants take in oxygen and give off carbon dioxide. This releases energy for growth.

Rhizome: Root-like, underground stem by which many plants increase in size or can be propagated.

Root hair: Tubular outgrowth from outside root cells which absorb nutrients from the soil.

Runner: A trailing stem which forms roots (and leaves) at intervals, e.g. couch grass.

Run-off: The surface flow of water from an area.

Scale leaves: Fleshy segments of bulbs.

Scion: Bud or piece of stem for grafting on to a growing plant.

Self-coloured: A flower having a single pure colour.
Shoot: A combination of young leaves and stem.
Side dressing: Application of fertilizer adjacent to growing plants.
Skeletonize: Severe pruning of fruit trees, e.g. citrus.
Soil texture: The proportion of different sizes of particles in soils, e.g. sandy soils, clay soils, loam.
Soil structure: The formation of aggregates or crumbs of soil.
Species: A name given to plants (belonging to the same genus) which differ in characteristics such as size, plant form or shape of flowers. It has been described as a botanical Christian name. Pollination between different species is relatively uncommon in nature.
Spore: The equivalent of a seed in lower plants, e.g. ferns, mosses, fungi.
Stamen: The male or pollen-bearing part of a flower.
Standard: (1) The upright petal of some flowers, e.g. sweet pea.
(2) A shrubby plant grafted on to a single tall stem, e.g. standard rose.
Stigma: The female part of a flower which receives the male pollen grains.
Stock: Stem or branch of growing plant on which a piece of stem or bud (the scion) is grafted.
Stolon: Horizontal, above-ground stem which develops roots and leaves at each node. Commonly called *runner*.
Stoma (stomate): Breathing pore on the surface of leaves and sometimes young stems.
Stopping: Breaking off the growing tip of a plant to promote the development of lateral buds.
Strain: A specially selected group of plants from a seed-grown variety.
Sucker: A shoot which develops from an underground root.
Succulent: Plants with fleshy leaves or stems acting as reservoirs against drought and evaporation, e.g. cactus.
Subsoil: Soil 15–20 cm (spade depth) below the surface.

Tap root: Prominent main root bearing smaller lateral roots. May be used for food storage, e.g. carrot.
Tender: A plant which is susceptible to low temperatures and frost. Usually grown indoors or under glass in cold climates.
Tendril: A small curling stem which clings to a support and helps plants to climb, e.g. sweet pea, passionfruit.
Tetraploid: A plant in which the normal number of chromosomes has been doubled.
Thatch: A spongy mat of roots and stems in a lawn.
Transpiration: Loss of moisture from leaves and other above-ground parts.
Tuber: A swollen, underground stem which stores food and can be used for propagation, e.g. potato, dahlia.
Turf: Generally a grass sward, but particularly planting material for a lawn which is purchased in squares or rolls.

Variegated: Leaves which are patterned, blotched or spotted with contrasting colours, generally cream or white but sometimes other colours.
Variety: A distinctly different member of a plant species.
Vermiculite: Granules of the mineral called mica which have been expanded under heat to form a moisture-holding material for soil improvement.
Virus: Extremely small particles of living matter which cause disease in plants, usually with symptoms of striping, blotching or dwarfing. Virus diseases are usually transmitted by sap-sucking insects.

Water-soluble fertilizer: A mixture of completely soluble chemicals which can be watered onto or around plants.
Water shoot: A vigorous, sappy shoot which develops from the crown of a plant, e.g. roses.
Wilting point: A soil which contains some moisture but this is unavailable to plant roots.

INDEX